AF252035

CCN PROTEINS
A New Family of Cell Growth and Differentiation Regulators

CCN PROTEINS

A New Family of Cell Growth and Differentiation Regulators

Editors

Bernard Perbal
Université Paris, France

Masaharu Takigawa
Okayama University, Japan

Imperial College Press

Published by

Imperial College Press
57 Shelton Street
Covent Garden
London WC2H 9HE

Distributed by

World Scientific Publishing Co. Pte. Ltd.

5 Toh Tuck Link, Singapore 596224

USA office: 27 Warren Street, Suite 401-402, Hackensack, NJ 07601

UK office: 57 Shelton Street, Covent Garden, London WC2H 9HE

British Library Cataloguing-in-Publication Data
A catalogue record for this book is available from the British Library.

ISBN 1-86094-552-X

Typeset by Stallion Press
Email: enquiries@stallionpress.com

Printed by FuIsland Offset Printing (S) Pte Ltd, Singapore

Dedicated to Annick
and Chikako

Contents

Preface

CCN is an acronym that is derived from $\underline{C}$yr61, $\underline{C}$TGF, and $\underline{N}$ov, the names originally given to the first three proteins to be identified in what is now a six-member family. These proteins comprise a fascinating and important group of homologous proteins that have not received the attention that they deserve from cell and molecular biologists. The publication of this book should do much to rectify this oversight.

In their excellent overview (Chapter 1) Perbal and Takigawa describe the early history of these proteins and the various names that were assigned to them, most of which were based on their perceived functions. As might be expected for proteins with multiple, diverse functions that were discovered almost simultaneously by different investigators, several names were often assigned to the same protein. In some cases, as with CCN2 for which the term connective tissue growth factor (CTGF) did not accurately describe its function, these names were a potential source of confusion. It is therefore a tribute to the field, and of substantial assistance to investigators in related areas of research with an interest in CCN proteins, that agreement on a systematic nomenclature was reached.

The practice of categorization in biology is a time-honored one that can be traced to Carolus Linnaeus, the father of taxonomy. In the case of CCN proteins, a grouping on the basis of homologous genetic structure is sensible, and is in keeping with the principles of protein evolution. There are, in addition, many other similarities among the six CCN proteins, which are summarized in Chapter 1, and are considered in greater detail in other Chapters throughout the book.

A common feature of CCN proteins is that they are all secreted proteins and are therefore present in the extracellular milieu. Since there is currently no evidence that any of the CCN proteins functions as an integral component of a structural entity such as a fibril or a basement membrane, it is presumed that these proteins function predominantly by interaction with cell-surface receptors. Indeed, as discussed by Lau and Lam (Chapter 3), there is good evidence for functional interactions of CCN1, 2, and 3 with many integrins, and this property is likely to extend to the other CCNs. Several members of the CCN family have been shown to influence the Wnt signaling pathway, which includes the receptor LRP6 (Latinkic, Chapter 8), and interactions with LRP1 and Notch have also been demonstrated. In addition to cell-surface receptors, CCN proteins can interact with many different cytokines, growth factors, and extracellular matrix proteins.

The similarity of CCN proteins to matricellular proteins has not escaped the notice of several authors in this book. The term 'matricellular' was coined to call attention to the unusual properties of a group of proteins that, while resident in the extracellular space, do not appear to play a structural role in the matrix, at least postnatally in vertebrates, but rather serve to modulate cell-matrix interactions and cell function (Bornstein, 1995). The realization that the extracellular matrix contains regulatory proteins that are distinct from growth factors and cytokines grew from an initial consideration of the properties of thrombospondin (TSP)-1, SPARC, and tenascin-C (Sage and Bornstein, 1991). This notion was later supported by studies of TSP-2, osteopontin, and tenascin-X (Bornstein, 2001; Bornstein and Sage, 2002). There are many similarities between the functions of members of the CCN and matricellular protein families. These include the regulation of angiogenesis, cell adhesion, migration, and proliferation, and control of collagen fibrillogenesis. Their fundamental mode of action is also similar in that it results from interactions with bioeffector molecules (cytokines, growth factors, and in some cases proteases), and with a multitude of cell-surface receptors.

However, matricellular and CCN proteins differ in that the latter are grouped on the basis of protein homology, whereas matricellular proteins include structurally dissimilar proteins that are functionally analogous. Within the matricellular family, TSPs 1 and 2 are also closely related structurally, and lessons learned from the studies of these two proteins may be particularly relevant to the CCN family. Despite their very similar structures and almost identical properties when the purified proteins are tested *in vitro*, the phenotypes of the TSP1- and TSP2-null mice are very different. This finding very likely reflects the different spatial and temporal patterns of expression of the two proteins, and the fact that their functions, as well as that of other matricellular proteins, are contextual and depend on the cell-surface receptors and bioeffector molecules with which they interact. Both the phenotypes of the TSP1- and TSP2-null mice, and the fact that there is no evidence in these mice for compensation of one TSP by the other, are consistent with the very different sequences of the promoters in the two genes. Thus far, only the phenotypes of

CCN1- and CCN2-null mice have been published. It will be of interest to determine the degree of overlap, or lack thereof, among the six mouse knockouts and to correlate this information with studies of the promoters of the relevant genes. Nevertheless, with the data that are currently at hand, it is probably safe to say that the information that is gained from experiments *in vitro* with purified CCN proteins can only serve as a rough guide to the physiological functions of these proteins.

On balance, it would seem advantageous for scientists in both fields to consider CCN proteins as members of the matricellular family. However, it should be remembered that categorization of proteins on the basis of function, and the resulting assumptions that are made, are human tendencies that Nature does not necessarily share.

REFERENCES

Bornstein P, Diversity of function is inherent in matricellular proteins: an appraisal of thrombospondin 1, *J Cell Biol* **130**:503–506, 1995.

Bornstein P, Thrombospondins as matricellular modulators of cell function, *J Clin Invest* **107**:929–934, 2001.

Bornstein P, Sage EH, Matricellular proteins: extracellular modulators of cell function, *Curr Opin Cell Biol* **14**:608–616, 2002.

Sage EH, Bornstein P, Extracellular proteins that modulate cell-matrix interactions: SPARC, tenascin, and thrombospondin, *J Biol Chem* **266**:14831–14834, 1991.

Paul Bornstein

Professor, Departments of Biochemistry and Medicine,
University of Washington,
Seattle, WA 98195, USA.

CHAPTER 1

THE CCN FAMILY OF PROTEINS: AN OVERVIEW

Bernard Perbal* and Masaharu Takigawa[†]

*Laboratoire d'Oncologie Virale et Moléculaire
Tour 54, Case 7048, Université Paris 7-D.Diderot
2 Place Jussieu, 75005 Paris, France
perbal@ccr.jussieu.fr

[†]Department of Biochemistry and Molecular Dentistry
Okayama University Graduate School of Medicine and Dentistry
2-5-1, Shikata-cho, Okayama, 700-8525, Japan
takigawa@md.okayama-u.ac.jp

In this chapter, we introduce the general structures and functions of CCN proteins and briefly review major questions regarding the biological properties of this new family of signaling regulators.

1. INTRODUCTION

The CCN family of proteins originally consisted of 3 members, namely CTGF (connective tissue growth factor, in human), CYR61 (cysteine rich 61, in mouse) and NOV (nephroblastoma overexpressed, in chicken). Hence the "CCN" family acronym, which uses the first letter of each member's name. However, when the same proteins were discovered independently by other groups working with different biological systems, they were given several different names. Later, other proteins such as ELM-1/WISP-1, rCop-1/WISP-2/CTGF-L and WISP-3 were reported to share structural identity with CTGF, CYR61 and NOV and therefore belong to the CCN family of proteins (Brigstock, 1999; Lau and /Lam, 1999; Perbal, 2001; Takigawa, 2003). In addition to the confusion that resulted from the variety of names given to the same proteins, this nomenclature sometimes turned out to be misleading. For example, several lines of evidence

Table 1.   Nomenclature for CCN proteins

CCN nomenclature	Current names
CCN1	CYR61/CEF10/βIG-M1/IGFBP9/IGFBP-rP4
CCN2	CTGF/FISP12/Hcs24/βIG-M2/HBGF-0.8/ecogenin/ IGFBP8/IGFBP-rP2
CCN3	NOV/IGFBP9/IGFBP-rP3
CCN4	ELM-1/WISP-1
CCN5	rCOP-1/WISP-2/CTGF-L/HICP
CCN6	WISP-3

CCN names were given according to the order in which they were described in the literature.

indicated that CTGF is not a genuine "growth factor" and elevated expression of nov is not a feature of human nephroblastomas. A consensus was, therefore, reached to propose a unifying nomenclature for the CCN family, numbering the proteins CCN1-CCN6 in the order in which they were first described in the literature (Brigstock *et al.*, 2003). The various names that were used previously and the correspondence of old names to the new CCN nomenclature are summarized in Table 1.

2. STRUCTURES OF CCN PROTEINS AND GENES

The prototypic CCN protein is encoded by five exons and is constituted by the assembly of 4 modules, i.e. the IGFBP, von Willebrand factor type C repeat (VWC), thrombospondin type 1 repeat (TSP), and carboxy-terminal (CT) (Fig. 1). All CCN proteins but CCN6 contain 38 conserved cysteine residues and share this tetra-modular structure with the exception of CCN5, which lacks the C-terminal (CT) module (Fig. 2). (Brigstock, 1999; Lau and Lam, 1999; Perbal, 2001; Takigawa, 2003). The mRNAs encoding the prototypic CCN protein is characterized by a long 3'-UTR (Kubota *et al.*, 1999).

However, the CCN family is now known to be complex, with biologically active CCN isoforms generated by either post-translational processing or alternative splicing (Perbal, 2004). For example, CCN5 lacks the CT module (Fig. 1); and the VWC module of CCN6 contains only 6 cysteine residues (Brigstock, 1999; Lau and Lam, 1999; Perbal, 2001; Takigawa, 2003). Moreover, two CCN2 variants consisting of TSP1 and CT modules only, and of module CT alone, are believed to result from post-translational processing (Brigstock, 1999). A similar post-translationally processed variant of CCN3 comprising only modules TSP1 and CT was also reported (Perbal, 2004). An amino-truncated CCN3 isoform was first found in tumors and nuclear CCN3

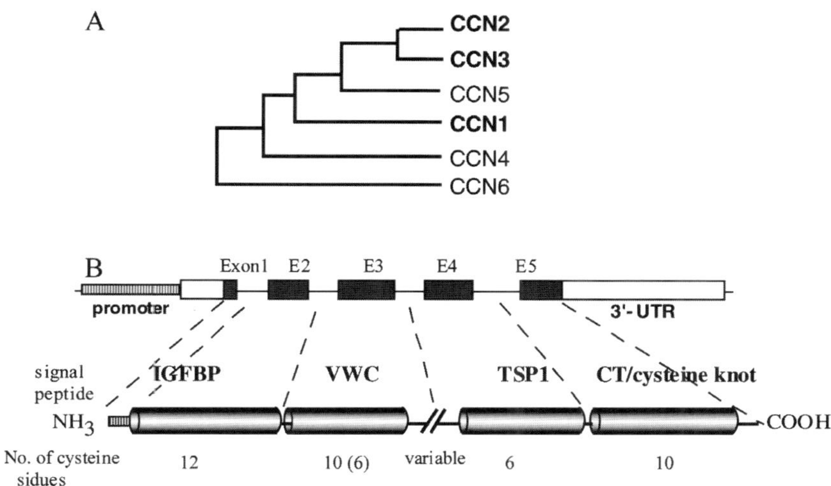

F:g. 1.   Dendrogram of CCN family members (A) and schematic organization of CCN genes and proteins (B). 3′-UTR: 3′-untranslated region. The protoypic CCN protein is encoded by 5 exons and is characterized 4 modules. IGFBP: insulin-like growth factor binding protein-like, VWC: von Willebrand factor type C, TSP: thrombospondin type 1 repeat, CT: C-terminal. The schematic structure of each protein is represented in Fig. 2.

isoforms deprived of module 1 were also described in several tumor cell lines (Perbal, 2001). In addition, a CCN4 variant lacking the VWC module and a CCN6 variant lacking both TSP and CT modules have recently been reported (Perbal, 2004; Tanaka *et al.*, 2001; Tanaka *et al.*, 2002).

The existence of four potentially functional domains in these proteins raised fundamental questions about their contribution to the various biological properties of the CCN proteins. It is now currently accepted that each of the four modules act both independently and interdependently, and that the multimodular structure of the CCN proteins provide the basis for a wide range of interactions with different partners.

3. FUNCTIONS OF CCN PROTEINS

The CCN genes are expressed in a variety of embryonic and adult tissues. Major sites of expression include the nervous system, the musculoskeletal system, the blood vessels and the adrenal. The expression of the CCN genes is tightly controlled, both temporally and spatially, suggesting that CCN proteins play important biological functions.

The various functions attributed to the CCN proteins are summarized in Table 2. They can be similar, complementary or antagonistic.

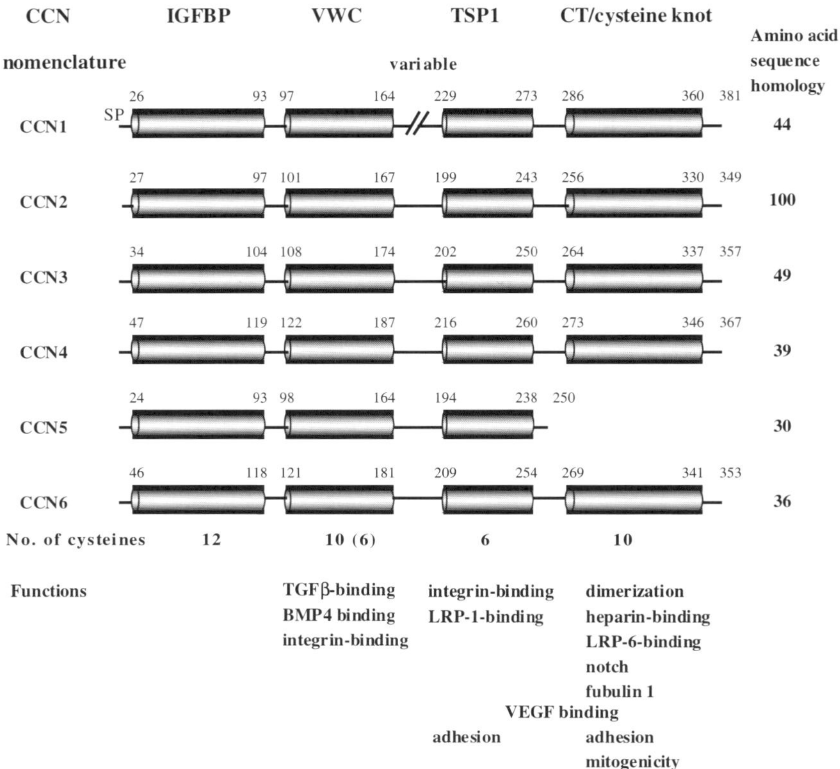

Fig. 2.   Modular structure of each CCN protein. SP: signal peptide, BMP: bone morphogenetic protein, LRP: low density lipoprotein receptor-related protein, VEGF: vascular endothelial cell growth factor.

The high degree of homology exhibited by the CCN proteins suggested that they might have similar or redundant functions. However, this view has been challenged by many recent results. The different CCN proteins share some biological functions; the functional overlap probably resulting from the presence of common structural motifs among the various members. Thus, the CCN proteins are now being defined as a novel family of growth and differentiation regulators acting on a large variety of cell types.

On the other hand, despite their similar structure, the CCN proteins also show significant functional divergence. They exhibit a many biological functions that are highly dependent upon the type of cells and the cellular context and that depend upon multiple interactions with a variety of regulatory molecules.

Table 2.   Functions of the CCN proteins

---

1) Biological processes in which the CCN proteins are involved
    a) Growth, development, differentiation
        CCN2, CCN1: angiogenesis, chondrogenesis, endochondral ossification
        CCN3: kidney development
    b) Wound healing, tissue repair, and tissue regeneration
        CCN2: skin wound healing, fracture healing,
    c) Uterine function
    d) Fibrosis (skin, kidney, lung, liver, blood vessels)
    e) Inflammation
    f) Tumor growth

2) Regulation of cellular functions by the CCN proteins
    a) Control of cell cycle
        CCN2, CCN1: immediate early genes
        CCN3, 4, 5: growth arrest/tumor suppressor
    b) Cell adhesion and migration (bindings to integrins, expression of integrins)
    c) Production of extracellular matrix (collagen, proteoglycans, laminin)
    d) Degradation of extracellular matrix (gene expression of matrix
        metalloproteinase and plasmin etc.)
    e) Tissue remodeling resulting from c) and d)

3) Molecular interaction of the CCN proteins and other proteins
    a) Extracellular matrix components: collagens, proteoglycans etc.
    b) Growth factors
    c) Integrins, LRPs, 240 kDa cell surface protein

These proteins are known to be involved in various biological processes such as normal growth and development, wound healing, tissue regeneration and uterine function, fibrosis in various tissues/organs, inflammation such as arthritis and tumor growth (Table 2) (Brigstock, 1999; Brigstock, 2003; Lau and Lam, 1999; Perbal, 2001; Perbal, 2004; Takigawa, 2003; Takigawa *et al.*, 2003). Generally speaking, the regulatory functions of CCN1 and 2 are those of immediate early genes, whereas CCN3-5 are putative growth arrest/suppression genes (Table 2). CCN proteins are also involved in cell adhesion and migration, cell differentiation, production of extracellular matrix (ECM) components and their degrading enzymes (Hashimoto *et al.*, 2002), suggesting their involvement in tissue remodeling. Their abilities to bind ECM components, various growth factors, and receptors in a way that can result in modifying their actions (Abreu *et al.*, 2002; Inoki *et al.*, 2002) suggest that the CCN proteins form multifunctional regulatory complexes as matricellular proteins. The multiple biological functions of the CCN proteins are exerted through various signaling pathways involving cell-surface receptors such as integrins (Gao and Brigstock, 2003; Lau and Lam, 1999), 240-kDa protein (Nishida *et al.*, 1998; Takigawa, 2003), LRPs (Gao and Brigstock, 2003; Mercurio *et al.*, 2004; Segarini *et al.*, 2001), Notch (Sakamoto *et al.*, 2002), as well as connexins and calcium channels (Li *et al.*, 2002; Gelhaus *et al.*, 2004; Fu *et al.*, 2004). The regulation of intracellular calcium concentration by CCN3 and CCN2 has established these proteins as genuine signaling factors. CCN proteins have been proposed to coordinate signaling pathways governing intercellular and intracellular communication needed for efficient control of cell growth, growth arrest and differentiation (Perbal, 2004; Lombet *et al.*, 2003).

4. THE CCN FAMILY OF PROTEINS

4.1. CCN1

CCN1 was first identified as *Cyr61* and *cef-10* immediate-early genes in mouse fibroblasts and chicken embryo fibroblasts, respectively (Simmons *et al.*, 1989; O'Brien *et al.*, 1990). It is composed of 355 amino acids, contains 38 cysteines, and has a long variable region between modules II (VWC) and III (TSP). This stretch has been proposed to represent a hinge between the two pairs of modules constituting the CCN proteins. However, conformation prediction analyses (Perbal B, unpublished) revealed that this region is unique to CCN1 and is unlikely to adopt similar 3D structure in other CCN family members.

CCN1 is expressed in multiple adult and embryonic tissues and is induced by epidermal growth factor (EGF), platelet-derived growth factor (PDGF), transforming growth factor-β (TGF-β), interleukin(IL)-1, -2 -6, 1,25-dihydoroxyvitamin D_3, and dexamethasone. Serum responsive elements, but not TGF-β responsive element, are found in its promoter region (Brigstock, 1999; Lau and Lam, 1999). CCN1 stimulates cell proliferation, chemotaxis, adhesion, ECM production, chondrogenesis and angiogenesis (Lau and Lam, 1999). These functions are also attributed to CCN2. However, higher doses of CCN1 are generally required to produce these effects (Nakanishi *et al.*, 2000; Wong *et al.*, 1997). Moreover, knockout mutants of CCN1 are lethal at the embryonic stage, due to defective placental angiogenesis (Mo *et al.*, 2002), whereas those of CCN2 are neonatal lethal due to impaired endochondral ossification (Ivkovic *et al.*, 2003). The above differences suggest similar but different functions of these proteins. Among the different CCN proteins, CCN1 has the highest affinity for heparin, followed by CCN2 and CCN3 (Brigstock, 1999). Interactions of CCN1 with integrins and ECM have been investigated in great detail (Lau and Lam, 1999) and are reviewed in Chapter 3.

The implication of CCN1 in tumorigenesis has been extensively studied, in a variety of human tumors, where its expression appears to be either increased or decreased (Perbal, 2001).

Increased expression of CCN1 has been documented in the case of breast tumors and brain tumors (Tsai *et al.*, 2000; Xie *et al.*, 2001). Forced expression of CCN1 in breast tumor cells can overcome the requirement for estrogen, and promotes cell migration and invasion. However, an antiproliferative activity of CCN1 was reported in the case of non-small cell lung carcinoma (NSCLC) in which expression of CCN1 is largely decreased. The pro- and anti-proliferative activities of CCN1 are discussed in Chapter 15 of this book.

4.2. CCN2

The human CCN2 was originally discovered as a single polypeptide with a molecular weight of 38-kDa secreted by cultured vein endothelial cells, and its cDNA was shown to encode a 349-amino acid protein (Bradham *et al.*, 1991). Because the partially purified protein was mitogenic and chemotactic for fibroblasts, it was named connective tissue growth factor (CTGF). Before the discovery of CTGF, the cDNA of the mouse CTGF ortholog had been isolated as an immediate early gene from serum-stimulated NIH3T3 cells and designated "fibroblast-inducible secreted protein-12" (*fisp12*) (Ryseck *et al.*, 1991). The mouse ortholog was also isolated as a TGF-β inducible gene from mouse ARK-2B cells and named βIG-M2. Fisp12/βIG-M2 encodes 348 amino acids, contains 39 cysteine residues, and is 91% homologous to CTGF (Brunner *et al.*, 1991). Independently of these studies, a gene, named *hcs24* (a hypertrophic chondrocyte-specific gene 24), was isolated from a human chondrocytic cell line, HCS-2/8, and shown to be identical to CTGF (Nakanishi *et al.*, 1997). All these genes are now recognized as being the same and their proteins are thus called CCN2.

Early studies had suggested the involvement of CCN2 in fibrosis, because its gene expression is induced by TGF-β and is found in various fibrotic disorders and because CCN2 promotes the proliferation, migration, and adhesion of fibroblasts (Brigstock, 1999; Moussad and Brigstock, 2000). However, once CCN2 was isolated from chondrocytes and shown to be highly expressed in hypertrophic chondrocytes, various physiological functions were uncovered (Brigstock, 1999; Moussad and Brigstock, 2000; Takigawa, 2003; Takigawa *et al.*, 2003). For example, it promotes the proliferation and differentiation of chondrocytes and osteoblasts *in vitro* (Nakanishi *et al.*, 2000; Nishida *et al.*, 2002; Nishida *et al.*, 2000). It also promotes adhesion, migration, and proliferation of endothelial cells *in vitro* and angiogenesis *in vivo* (Babic *et al.*, 1999; Shimo *et al.*, 1998; Shimo *et al.*, 1999). From these findings, it was suggested that, because CCN2 promotes endochondral ossification in various stages, it should be called "ecogenin" (Takigawa *et al.*, 2003). The validity of this hypothesis was proven by examining knockout mutant mice of CCN2, which were shown to be neonatal lethal due to defective chest development as a consequence of impaired endochondral ossification (Ivkovic *et al.*, 2003). Among the biochemical functions of CCN2, the most important one may be the promotion of extracellular matrix (ECM) production, which is closely related to physiology in ECM-rich tissues/organs such as cartilage and to pathology in tissues/organs, where ECM production is undesirable under physiological conditions (Takigawa, 2003; Takigawa *et al.*, 2003).

The expression of CCN2 is the highest in hypertrophic chondrocytes in the physiological state and it is also moderately expressed in developing vascular endothelial cells and osteoblasts (Takigawa *et al.*, 2003). Additionally, it is detected in neurons in the brain and in kidney at lower levels (Takigawa, 2003; Takigawa *et al.*, 2003). Genetic elements such as promoter and silencer elements and the intracellular counterparts that control CCN gene expression have been much investigated especially for *ccn2* (Eguchi *et al.*, 2002; Grotendorst *et al.*, 1996; Kubota *et al.*, 1999). Detailed reviews about this aspect of CCN2 biology can be found in Chapter 2 of this book.

CCN2 gene expression is induced in various pathological states such as fibrosis, inflammation, and arthritis and during wound healing (Moussad and Brigstock, 2000; Nishida *et al.*, 2004; Omoto *et al.*, 2004; Takigawa, 2003; Takigawa *et al.*, 2003) and is stimulated by various other growth factors such as TGF-β, bone morphogenetic protein (BMP), PDGF, EGF, bFGF, and CCN2 itself (Takigawa, 2003; Takigawa *et al.*, 2003). Moreover, CCN2 can bind some of these growth factors and modify their actions (Abreu *et al.*, 2002). These interactions may contribute to at least some of the multiple functions of CCN2. Several extracellular counterparts (Gao and Brigstock, 2003; Mercurio *et al.*, 2004; Nishida *et al.*, 2003), cell-surface receptors (Gao and Brigstock, 2003; Mercurio *et al.*, 2004; Nishida *et al.*, 1998; Segarini *et al.*, 2001), and intracellular signal transduction pathways (Yosimichi *et al.*, 2004; Yosimichi *et al.*, 2001; Li *et al.*, 2003) have been reported to interact with CCN2, suggesting that they may also be involved in its various functions.

In addition to the classical roles of CCN2 as an extracellular messenger, the intracellular distribution and possible functions of CCN2 have been recently indicated (Kubota *et al.*, 2000; Wahab *et al.*, 2001). In light of these findings, the direct binding of cytoskeletal actin to CCN2 (Yosimichi *et al.*, 2002) is particular of interest. Moreover, overexpressed CCN2 was found to negatively modulate the cell cycle (Kubota *et al.*, 2000), which is consistent with the findings that CCN2 is accumulated in terminally-differentiated hypertrophic chondrocytes *in vivo* (Takigawa, 2003) and overexpressed CCN2 induced apoptosis (Hishikawa *et al.*, 1999) and attenuated growth of cancer-derived cell lines (Moritani *et al.*, 2003), as was also observed in the case of CCN3 (Gupta *et al.*, 2001; Benini *et al.*, unpublished).

Detailed and updated reviews about the physiological and pathological significances and molecular mechanism of actions of CCN2 are presented in other chapters of this book.

4.3. CCN3

CCN3 was first isolated as an integration site of Myeloblastosis Associated Virus type 1(N) (MAV-1(N)) in a chicken nephroblastoma induced by this virus (Joliot *et al.*, 1992). The gene in which MAV was inserted in this tumor was highly expressed and was found to encode a truncated version of a new secretory protein showing partial identity with CTGF and CYR61. Because the expression of this gene was high in all MAV-induced nephroblastomas, it was designated *nov* (nephroblastoma overexpressed).

The discovery of *nov* was pivotal in several respects: i) the high degree of identity shared by NOV, CTGF and CYR61 suggested the existence of a broader family of structurally related proteins, ii) contrary to the situation reported for *ctgf* and *cyr61*, the expression of *nov* was repressed upon serum induction of starved fibroblasts, and upon virus-induced proliferation (Scholtz *et al.*, 1996); and iii) the amino-truncated nov protein that was produced following MAV integration in the tumor DNA, showed oncogenic properties whereas the full length CCN3 protein was found to inhibit cell growth when tested in chicken embryo fibroblasts.

These observations indicated that in spite of their similar structural organization and striking identity, these proteins might have quite distinct functions.

Furthermore these results provided the first clue for a role of CCN proteins in tumorigenesis and showed for the first time that a full length CCN protein was acting as a negative regulator of cell growth, whereas a truncated version was acting as an oncogene.

We now know that the "nov" name is unfortunately misleading. In none of the other tumors (a total of 150 were analyzed since then) MAV found to be integrated within or in the vicinity of the ccn3 gene (Li *et al.*, submitted for publication), and elevated expression of ccn3 seems to be a feature of the

MAV target cells that expand during the tumorigenic process (Cherel *et al.*, manuscript in preparation). Cloning of the human and the mouse ccn3 gene revealed that it was well conserved during evolution. However, over-expression of CCN3 was not observed in Wilms tumor, the human equivalent of MAV-induced nephroblastomas. On the contrary, increased expression of CCN3 was associated to heterotypic differentiation in these tumors, an observation which is in full agreement with the implication of CCN3 in chondrocytic and muscular differentiation in normal conditions (Perbal, 2001).

For historical reasons, CCN3 expression has been studied in pathological conditions associated to cancer development. The great deal of information that has been obtained along the analysis of several different types of human tumors, has permitted to consider that CCN3 dosage and manipulation can represent a new tool in molecular medicine, especially for cancer diagnosis and therapy (Perbal, 2003). A detailed presentation of CCN3 expression in human tumors, and its potential usefulness in pathological studies can be found in Chapters 13 and 14.

These studies pointed out that an elevated expression of CCN3 could be associated to either a good or a bad prognosis. Analysis of CCN3 expression in tumors of the musculo skeletal system and of the nervous system provided such opposite results. In the case of Ewing's sarcoma, assessment of CCN3 expression in primary tumors and corresponding metastases permitted to establish that CCN3 expression was associated to a higher risk of developing metastases (Manara *et al.*, 2002). Conversely, in the case of glioblastomas, it has been shown that the expression of CCN3 is inversely correlated to the metastatic potential of the tumor cells which were freshly explanted from tumors of various grades (Li *et al.*, 1996).

Recent results indicated that, in spite of this apparent discrepancy, CCN3 might play common roles in these two systems. Indeed, in all the cases that have been studied thus far, including Ewing and glioblastomas, the CCN3 protein was found to inhibit growth of tumor cells when assayed *ex vivo*. This observation is in agreement with previous observations, and with the association of CCN3 with differentiation. Furthermore, tumor cells stably transfected with CCN3 expression vectors showed a decreased tumorigenicity when injected in nude mice. These observations confirmed that the full length CCN3 protein exhibited an antiproliferative activity.

The first clues for mechanisms of action were obtained by means of the two-hybrid and co-immunoprecipitation strategies which identified several key regulatory proteins as CCN3 partners. The striking feature was that many different proteins appeared to interact with CCN3. Among them, the finding of

ECM proteins including CCN2, receptors and channels was in accordance with the secreted nature of CCN3. More surprising was the finding that cytoplasmic and nuclear proteins also physically interact with CCN3. The biological relevance and significance of these interactions is discussed in Chapter 13. It is worth noting that the CCN2 protein was also found at the nucleus of cells (Wahab *et al.*, 2001) and to interact with calcium channels (Lombet *et al.*, 2003).

These studies attributed for the first time a biological function to CCN3 and designated CCN proteins as genuine signaling factors.

4.4. CCN4

CCN4 was first isolated as a gene expressed in a low-metastatic type of murine melanoma and called ELM1 (expressed in low-metastatic type 1 cells) (Hashimoto *et al.*, 1998). Its human orthologue was identified later as a Wnt-induced secreted protein, named WISP-1 (Pennica *et al.*, 1998). Consisting of 367 amino acids, it was induced within 3 h of serum stimulation and overexpressed in the stroma of tumors (Hashimoto *et al.*, 1998; Pennica *et al.*, 1998). Interaction with decorin and biglycan may be related to its localization there (Desnoyers *et al.*, 2001). CCN5 was found to inhibit tumor growth and metastasis *ex vivo* and *in vivo*. A variant lacking VWC module was reported (Tanaka *et al.*, 2001) and suggested to be involved in the aggressive progression of scirrhous gastric carcinoma and in the invasion phenotype of cholangiocarcinoma (Tanaka *et al.*, 2003; Tanaka *et al.*, 2001) (see Chapter 16 of this book).

4.5. CCN5

ccn5 was first identified as a gene of normal rat embryo fibroblasts (rCop1) whose expression was undectable after transformation (Zhang *et al.*, 1998). Its human ortholog was identified later as encoding a Wnt-induced secreted protein (Wisp-2) (Pennica *et al.*, 1998). The human orthologue was also isolated as a gene encoding a CTGF-like protein (CTGF-L) from human osteoblasts (Kumar *et al.*, 1999). Heparin-induced CTGF-like protein isolated from heparin-treated vascular smooth muscle cells is also identical to CCN5 (Brigstock, 1999). CCN5 lacks the C-terminal module and has an overall amino acid homology of ~ 30% with CCN2 (Brigstock, 1999). Gene expression is high in quiescent cells and low during proliferation. Its functions include growth arrest, tumor suppression, osteoblast adhesion and inhibition of osteocalcin production by osteoblasts (Brigstock, 1999; Delmolino *et al.*, 2001; Kumar *et al.*, 1999; Zhang *et al.*, 1998). Recent studies have established the relevance of CCN5 in human breast disease. Undetectable in normal breast epithelial cells, the expression of CCN5 is high in tumor samples and breast tumor derived cell lines (Zoubine *et al.*, 2001; Sanexa *et al.*, 2001). Silencing of CCN5 functions interferes with serum-induced proliferation of breast tumor cells (Banerjee *et al.*, 2003).

Other aspects of CCN5 biology are presented in Chapter 12.

4.6. CCN6

ccn6 was first identified by screening expressed sequence tag databases. The corresponding protein was designated WISP-3 (Pennica *et al.*, 1998). Unlike the prototypic CCN protein, it lacks 4 cysteines in the VWC module. Various mutations of this gene were reported in cases of pseudorheumatid dysplasia (Hurvitz *et al.*, 1999). Transfection of *ccn6* into chondrocytic cells increased the expression of genes encoding cartilage-specific matrix molecules such as type II collagen and aggrecan. A mutation in the IGFBP module of *ccn6* impaired the effects of *ccn6* on these molecules, suggesting a role for CCN6 in cartilage integrity (Sen *et al.*, 2004). A variant, lacking modules TSP and CT was identified in gastrointestinal carcinomas (Perbal, 2004; Tanaka *et al.*, 2002) (see Chapter 16).

5. CONCLUDING COMMENTS

In this chapter, we have briefly introduced the general characteristics of CCN proteins. Other chapters of this book, written by acknowledged experts in this field, are up-to-date, comprehensive and authoritative, affording insights into the biological roles of CCN proteins. As already stressed in this introduction, the CCN family members appear as multifunctional proteins which are major actors in the control of cell proliferation and differentiation during normal development, wound repair, and regeneration of various types of tissues including those of cartilage, bone, and the vascular system. The involvement of CCN proteins in pathological processes such as fibrosis, renal diseases and cancer development opens new promising avenues in various fields of human molecular medicine.

This volume will be valuable for all those interested in CCN proteins, growth factors, extracellular matrix proteins, signaling molecules, and mechanism of gene expression, whether from biochemical, molecular, biological, biomedical, physiological, pathological or clinical points-of-view.

ACKNOWLEDGMENTS

The work performed in Professor Perbal's laboratory was supported in part by the Ministère de L'Education Nationale, de la Recherche et de la Technologie and by grants of, La Ligue Nationale Contre le Cancer (Comités de Paris, du Cher et de l'Indre), and Association pour la Recherche contre le Cancer to B.P. The work in Professor Takigawa's laboratory was supported in part by Grants-in-Aid for Scientific Research(s) from the Ministry of Education, Culture, Sports, Science and Technology of Japan (to M.T.).

REFERENCES

Abreu JG, Ketpura NI, Reversade B and De Robertis EM (2002) Connective-tissue growth factor (CTGF) modulates cell signalling by BMP and TGF-beta. *Nat Cell Biol* **4**:599–604.

Babic AM, Chen CC and Lau LF (1999) Fisp12/mouse connective tissue growth factor mediates endothelial cell adhesion and migration through integrin alphavbeta3, promotes endothelial cell survival, and induces angiogenesis *in vivo*. *Mol Cell Biol* **19**:2958–2966.

Banerjee S, Saxena N, Sengupta K, Tawfik O, Mayo MS and Banerjee SK (2003) WISP-2 gene in human breast cancer: estrogen and progesterone inducible expression and regulation of tumor cell proliferation. *Neoplasia* **5**:63–73.

Bradham DM, Igarashi A, Potter RL and Grotendorst GR (1991) Connective tissue growth factor: a cysteine-rich mitogen secreted by human vascular endothelial cells is related to the SRC-induced immediate early gene product CEF-10. *J Cell Biol* **114**:1285–1294.

Brigstock DR (1999) The connective tissue growth factor/cysteine-rich 61/nephroblastoma overexpressed (CCN) family. *Endocr Rev* **20**:189–206.

Brigstock DR (2003) The CCN family: a new stimulus package. *J Endocrinol* **178**:169–175.

Brigstock DR, Goldschmeding R, Katsube KI, Lam SC, Lau LF, Lyons K, Naus C, Perbal B, Riser B, Takigawa M and Yeger H (2003) Proposal for a unified CCN nomenclature. *Mol Pathol* **56**:127–128.

Brunner A, Chinn J, Neubauer M and Purchio AF (1991) Identification of a gene family regulated by transforming growth factor β. *DNA Cell Biol* **10**:293–300.

Delmolino LM, Stearns NA and Castellot JJ Jr (2001) COP-1, a member of the CCN family, is a heparin-induced growth arrest specific gene in vascular smooth muscle cells. *J Cell Physiol* **188**:45–55.

Desnoyers L, Arnott D and Pennica D (2001) WISP-1 binds to decorin and biglycan. *J Biol Chem* **276**:47599–47607.

Eguchi T, Kubota S, Kondo S, Kuboki T, Yatani H and Takigawa M (2002) A novel cis-element that enhances connective tissue growth factor gene expression in chondrocytic cells. *Biochem Biophys Res Commun* **295**:445–451.

Fu CT, Bechberger JF, Ozog MA, Perbal B and Naus CC (2004) CCN3 (NOV) interacts with connexin43 in C6 glioma cells: possible mechanism of connexin-mediated growth suppression. *J Biol Chem* **279**:36943–36950.

Gao R and Brigstock DR (2003) Low density lipoprotein receptor-related protein (LRP) is a heparin-dependent adhesion receptor for connective tissue growth factor (CTGF) in rat activated hepatic stellate cells. *Hepatol Res* **27**:214–220.

Gellhaus A, Dong X, Propson S, Maass K, Klein–Hitpass L, Kibschull M, Traub O, Willecke K, Perbal B, Lye SJ and Winterhager E (2004) Connexin43 interacts with NOV: a possible mechanism for negative regulation of cell growth in choriocarcinoma cells. *J Biol Chem* **279**:36931–36942.

Grotendorst GR, Okochi H and Hayashi N (1996) A novel transforming growth factor beta response element controls the expression of the connective tissue growth factor gene. *Cell Growth Differ* **7**:469–480.

Gupta N, Wang H, McLeod TL, Naus CC, Kyurkchiev S, Advani S, Yu J, Perbal B and Weichselbaum RR (2001) Inhibition of glioma cell growth and tumorigenic potential by CCN3 (NOV). *Mol Pathol* **54**:293–299.

Hashimoto G, Inoki I, Fujii Y, Aoki T, Ikeda E and Okada Y (2002) Matrix metalloproteinases cleave connective tissue growth factor and reactivate angiogenic activity of vascular endothelial growth factor 165. *J Biol Chem* **277**:36288–36295.

Hashimoto Y, Shindo–Okada N, Tani M, Nagamachi Y, Takeuchi K, Shiroishi T, Toma H and Yokota J (1998) Expression of the Elm1 gene, a novel gene of the CCN (connective tissue growth factor, Cyr61/Cef10, and neuroblastoma overexpressed gene) family, suppresses In vivo tumor growth and metastasis of K-1735 murine melanoma cells. *J Exp Med* **187**:289–296.

Hishikawa K, Oemar BS, Tanner FC, Nakaki T, Fujii T and Luscher TF (1999) Overexpression of connective tissue growth factor gene induces apoptosis in human aortic smooth muscle cells. *Circulation* **100**:2108–2112.

Hurvitz JR, Suwairi WM, Van Hul, El-Shanti H, Superti–Furga A, Roudier J, Holderbaum D, Pauli RM, Herd JK, Van Hul EV, Rezai-Delui H, Legius, Le Merrer M, Al-Alami J, Bahabri SA and Warman ML (1999) Mutations in the CCN gene family member WISP3 cause progressive pseudorheumatoid dysplasia. *Nat Genet* **23**:94–98.

Inoki I, Shiomi T, Hashimoto G, Enomoto H, Nakamura H, Makino K, Ikeda E, Takata S, Kobayashi K and Okada Y (2002) Connective tissue growth factor binds vascular endothelial growth factor (VEGF) and inhibits VEGF-induced angiogenesis. *Faseb J* **16**:219–221.

Ivkovic S, Yoon BS, Popoff SN, Safadi FF, Libuda DE, Stephenson RC, Daluiski A and Lyons KM (2003) Connective tissue growth factor coordinates chondrogenesis and angiogenesis during skeletal development. *Development* **130**: 2779–2791.

Joliot V, Martinerie C, Dambrine G, Plassiart G, Brisac M, Crochet J and Perbal B (1992) Proviral rearrangements and overexpression of a new cellular gene (nov) in myeloblastosis-associated virus type 1-induced nephroblastomas. *Mol Cell Biol* **12**:10–21.

Kubota S, Hattori T, Nakanishi T and Takigawa M (1999) Involvement of cis-acting repressive element(s) in the 3′-untranslated region of human connective tissue growth factor gene. *FEBS Lett* **450**:84–88.

Kubota S, Hattori T, Shimo T, Nakanishi T and Takigawa M (2000) Novel intracellular effects of human connective tissue growth factor expressed in Cos-7 cells. *FEBS Lett* **474**:58–62.

Kumar S, Hand AT, Connor JR, Dodds RA, Ryan PJ, Trill JJ, Fisher SM, Nuttall ME, Lipshutz DB, Zou C, Hwang SM, Votta BJ, James IE, Rieman DJ, Gowen M and Lee JC (1999) Identification and cloning of a connective tissue growth factor-like cDNA from human osteoblasts encoding a novel regulator of osteoblast functions. *J Biol Chem* **274**:17123–17131.

Lau LF and Lam SC (1999) The CCN family of angiogenic regulators: the integrin connection. *Exp Cell Res* **248**:44–57.

Li WC, Martinerie C, Zumkeller W, Westphal M and Perbal B (1996) Differential expression of novH and CTGF in human glioma cell lines. *Mol Pathol* **49**:91–97

Li CL, Martinez V, He B, Lombet A and Perbal B (2002) A role for CCN3 (NOV) in calcium signalling. *Mol Pathol* **55**:250–261.

Lombet A, Planque N, Bleau AM, Li CL and Perbal B (2003) CCN3 and calcium signaling. *Cell Commun Signal* **1**:1.

Manara MC, Perbal B, Benini S, Strammiello R, Cerisano V, Perdichizzi S, Serra M, Astolfi A, Bertoni F, Alami J, Yeger H, Picci P and Scotlandi K (2002) The expression of ccn3(nov) gene in musculoskeletal tumors. *Am J Pathol* **160**:849–859.

Mercurio S, Latinkic B, Itasaki N, Krumlauf R and Smith JC (2004) Connective-tissue growth factor modulates WNT signalling and interacts with the WNT receptor complex. *Development* **131**:2137–2147.

Mo FE, Muntean AG, Chen CC, Stolz DB, Watkins SC and Lau LF (2002) CYR61 (CCN1) is essential for placental development and vascular integrity. *Mol Cell Biol* **22**:8709–8720.

Moritani NH, Kubota S, Nishida T, Kawaki H, Kondo S, Sugahara T and Takigawa M (2003) Suppressive effect of overexpressed connective tissue growth factor on tumor cell growth in a human oral squamous cell carcinoma-derived cell line. *Cancer Lett* **192**:205–214.

Moussad EE and Brigstock DR (2000) Connective tissue growth factor: what's in a name? *Mol Genet Metab* **71**:276–292.

Nakanishi T, Kimura Y, Tamura T, Ichikawa H, Yamaai Y, Sugimoto T and Takigawa M (1997) Cloning of a mRNA preferentially expressed in chondrocytes by differential display-PCR from a human chondrocytic cell line that is identical with connective tissue growth factor (CTGF) mRNA. *Biochem Biophys Res Commun* **234**:206–210.

Nakanishi T, Nishida T, Shimo T, Kobayashi K, Kubo T, Tamatani T, Tezuka K and Takigawa M (2000) Effects of CTGF/Hcs24, a product of a hypertrophic chondrocyte-specific gene, on the proliferation and differentiation of chondrocytes in culture. *Endocrinology* **141**:264–273.

Nishida T, Kubota S, Fukunaga T, Kondo S, Yosimichi G, Nakanishi T, Takano-Yamamoto T and Takigawa M (2003) CTGF/Hcs24, hypertrophic chondrocyte-specific gene product, interacts with perlecan in regulating the proliferation and differentiation of chondrocytes. *J Cell Physiol* **196**:265–275.

Nishida T, Kubota S, Kojima S, Kuboki T, Nakao K, Kushibiki T, Tabata Y and Takigawa M (2004) Regeneration of Defects in Articular Cartilage in Rat Knee Joints by CCN2 (Connective Tissue Growth Factor). *J Bone Miner Res* **19**:1308–1319.

Nishida T, Kubota S, Nakanishi T, Kuboki T, Yosimichi G, Kondo S and Takigawa M (2002) CTGF/Hcs24, a hypertrophic chondrocyte-specific gene product, stimulates proliferation and differentiation, but not hypertrophy of cultured articular chondrocytes. *J Cell Physiol* **192**:55–63.

Nishida T, Nakanishi T, Asano M, Shimo T and Takigawa M (2000) Effects of CTGF/Hcs24, a hypertrophic chondrocyte-specific gene product, on the proliferation and differentiation of osteoblastic cells in vitro. *J Cell Physiol* **184**: 197–206.

Nishida T, Nakanishi T, Shimo T, Asano M, Hattori T, Tamatani T, Tezuka K and Takigawa M (1998) Demonstration of receptors specific for connective tissue growth factor on a human chondrocytic cell line (HCS-2/8). *Biochem Biophys Res Commun* **247**:905–909.

O'Brien TP, Yang GP, Sanders L and Lau LF (1990) Expression of cyr61, a growth factor-inducible immediate-early gene. *Mol Cell Biol* **7**:3569–3577.

Omoto S, Nishida K, Yamaai Y, Shibahara M, Nishida T, Doi T, Asahara H, Nakanishi T, Inoue H and Takigawa M (2004) Expression and localization of connective tissue growth factor (CTGF/Hcs24/CCN2) in osteoarthritic cartilage. *Osteoarthritis Cartilage*, in press.

Pennica D, Swanson TA, Welsh JW, Roy MA, Lawrence DA, Lee J, Brush J, Taneyhill LA, Deuel B, Lew M, Watanabe C, Cohen RL, Melhem MF, Finley GG, Quirke P, Goddard AD, Hillan KJ, Gurney AL, Botstein D and Levine AJ (1998) WISP genes are members of the connective tissue growth factor family that are up-regulated in wnt-1-transformed cells and aberrantly expressed in human colon tumors. *Proc Natl Acad Sci USA* **95**:14717–14722.

Perbal B (2001) NOV (nephroblastoma overexpressed) and the CCN family of genes: structural and functional issues. *Mol Pathol* **54**:57–79.

Perbal B (2003) The CCN3 (NOV) cell growth regulator: a new tool for molecular medicine. *Expert Rev Mol Diagn* **3**:597–604.

Perbal B (2004) CCN proteins: multifunctional signalling regulators. *Lancet* **363**: 62–64.

Ryseck RP, Macdonald–Bravo H, Mattei MG and Bravo R (1991) Structure, mapping, and expression of fisp-12, a growth factor-inducible gene encoding a secreted cysteine-rich protein. *Cell Growth Differ* **2**:225–233.

Sakamoto K, Yamaguchi S, Ando R, Miyawaki A, Kabasawa Y, Takagi M, Li CL, Perbal B and Katsube K (2002) The nephroblastoma overexpressed gene (NOV/ccn3) protein associates with Notch1 extracellular domain and inhibits myoblast differentiation via Notch signaling pathway. *J Biol Chem* **277**:29399–29405.

Saxena N, Banerjee S, Sengupta K, Zoubine MN and Banerjee SK. (2001) Differential expression of WISP-1 and WISP-2 genes in normal and transformed human breast cell lines. *Mol Cell Biochem* **228**:99–104.

Segarini PR, Nesbitt JE, Li D, Hays LG, Yates JR, 3rd and Carmichael DF (2001) The low density lipoprotein receptor-related protein/alpha2-macroglobulin receptor is a receptor for connective tissue growth factor. *J Biol Chem* **276**:40659–40667.

Sen M, Cheng YH, Goldring MB, Lotz MK and Carson DA (2004) WISP3-dependent regulation of type II collagen and aggrecan production in chondrocytes. *Arthritis Rheum* **50**:488–497.

Scholz G, Martinerie C, Perbal B and Hanafusa H (1996) Transcriptional down regulation of the nov proto-oncogene in fibroblasts transformed by p60v-src. *Mol Cell Biol.* **16**: 481–486.

Shimo T, Nakanishi T, Kimura Y, Nishida T, Ishizeki K, Matsumura T and Takigawa M (1998) Inhibition of endogenous expression of connective tissue growth factor by its antisense oligonucleotide and antisense RNA suppresses proliferation and migration of vascular endothelial cells. *J Biochem (Tokyo)* **124**:130–140.

Shimo T, Nakanishi T, Nishida T, Asano M, Kanyama M, Kuboki T, Tamatani T, Tezuka K, Takemura M, Matsumura T and Takigawa M (1999) Connective tissue growth factor induces the proliferation, migration, and tube formation of vascular endothelial cells in vitro, and angiogenesis in vivo. *J Biochem (Tokyo)* **126**: 137–145

Simmons DL, Levy DB, Yannoni Y and Erikson RL (1989) Identification of a phorbol ester-repressible v-src-inducible gene. *Proc Natl Acad Sci USA* **86**:1178–1182.

Takigawa M (2003) CTGF/Hcs24 as a multifunctional growth factor for fibroblasts, chondrocytes and vascular endothelial cells. *Drug News Perspect* **16**:11–21.

Takigawa M, Nakanishi T, Kubota S and Nishida T (2003) Role of CTGF/HCS24/ecogenin in skeletal growth control. *J Cell Physiol* **194**:256–266.

Tanaka S, Sugimachi K, Kameyama T, Maehara S, Shirabe K, Shimada M, Wands JR and Maehara Y (2003) Human WISP1v, a member of the CCN family, is associated with invasive cholangiocarcinoma. *Hepatology* **37**:1122–1129.

Tanaka S, Sugimachi K, Saeki H, Kinoshita J, Ohga T, Shimada M and Maehara Y (2001) A novel variant of WISP1 lacking a Von Willebrand type C module overexpressed in scirrhous gastric carcinoma. *Oncogene* **20**:5525–5532.

Tanaka S, Sugimachi K, Shimada M, Maehara Y and Sugimachi K (2002) Variant WISPs as targets for gatrointestinal carcinomas. *Gastroenterology* **123**:392–393

Tsai MS, Hornby AE, Lakins J and Lupu R (2000) Expression and function of CYR61, an angiogenic factor, in breast cancer cell lines and tumor biopsies. *Cancer Res* **60**:5603–5607.

Wahab NA, Brinkman H and Mason RM (2001) Uptake and intracellular transport of the connective tissue growth factor: a potential mode of action. *Biochem J* **359**: 89–97.

Xie D, Nakachi K, Wang H, Elashoff R and Koeffler HP (2001) Elevated levels of connective tissue growth factor, WISP-1, and CYR61 in primary breast cancers associated with more advanced features. *Cancer Res* **61**:8917–8923.

Wong M, Kireeva ML, Kolesnikova TV and Lau LF (1997) Cyr61, product of a growth factor-inducible immediate-early gene, regulates chondrogenesis in mouse limb bud mesenchymal cells. *Dev Biol* **192**:492–508.

Yosimichi G, Kubota S, Hattori T, Nishida T, Nawachi K, Nakanishi T, Kamada M, Takano-Yamamoto T and Takigawa M (2002) CTGF/Hcs24 interacts with the cytoskeletal protein actin in chondrocytes. *Biochem Biophys Res Commun* **299**:755–761.

Yosimichi G, Kubota S, Nishida T, Kondo S, Nakao K, Takano-Yamamoto T and Takigawa M (2004) CTGF/CCN2 promotes proliferation and differentiation of chondrocytes through multiple signal transduction pathways: Involvement of PKC-mediated p38 MAPK, MEK1/ERK, and JNK and of PI3K-mediated PKB. *J Bone Miner Res* submitted.

Yosimichi G, Nakanishi T, Nishida T, Hattori T, Takano–Yamamoto T and Takigawa M (2001) CTGF/Hcs24 induces chondrocyte differentiation through a p38 mitogen-activated protein kinase (p38MAPK), and proliferation through a p44/42 MAPK/extracellular-signal regulated kinase (ERK). *Eur J Biochem* **268**:6058–6065.

Zhang R, Averboukh L, Zhu W, Zhang H, Jo H, Dempsey PJ, Coffey RJ, Pardee AB and Liang P (1998) Identification of rCop-1, a new member of the CCN protein family, as a negative regulator for cell transformation. *Mol Cell Biol* **18**:6131–6141.

Zoubine MN, Banerjee S, Saxena NK, Campbell DR and Banerjee SK (2001) WISP-2: a serum-inducible gene differentially expressed in human normal breast epithelial cells and in MCF-7 breast tumor cells. *Biochem Biophys Res Commun* **282**:421–425.

CHAPTER 2

ROLES OF CCN2/CTGF IN THE CONTROL OF GROWTH AND REGENERATION

Masaharu Takigawa*, Takashi Nishida and Satoshi Kubota

Department of Biochemistry and Molecular Dentistry
Okayama University Graduate School of Medicine and Dentistry
Okayama 700-8525, Japan
**takigawa@md.okayama-u.ac.jp*

CCN2/CTGF is a multifunctional growth factor for fibroblasts, chondrocytes, osteoblasts, and vascular endothelial cells. Depending on the type of cell with which it interacts, this factor promotes chemotaxis, migration, adhesion, proliferation, differentiation and/or extracellular matrix formation. Because of the very high level of gene expression of CCN2 in hypertrophic chondrocytes in the physiological state, a major physiological role for this factor has been suggested to be the promotion of endochondral ossification. However, recent observations of its gene expression during embryonic development and wound healing suggest the involvement of this factor in growth and regeneration of various types of tissues including those of the skeletal and vascular systems. Also, the results of administration of recombinant CCN2 *in vivo* suggest the factor to be therapeutically useful for the regeneration and reconstruction of skeletal and vascular tissues. Moreover, CCN2 promotes at least chondrogenic differentiation of mesenchymal stem cells, suggesting that this factor is applicable to cell therapy. CCN2 is also known to be involved in fibrotic disorders, angiogenic diseases, and malignancies; but these pathological states may be due to uncontrolled overexpression of this growth factor. This chapter focuses on the role of CCN2 in the control of growth and regeneration, including that at the molecular level.

1. INTRODUCTION

Connective tissue growth factor/hypertrophic chondrocyte specific gene product 24 (CTGF/Hcs24) is the second member of the CCN protein family (CCN2), which is characterized by 4 homologous modules, i.e. the IGFBP (insulin-like growth factor binding protein)-like, von Willebrand factor type C repeat (VWC), thrombospondin type 1 repeat (TSP), and C-terminal (CT) modules and conservation of all 38 cysteine residues among its members (Fig. 1) (Bork, 1993; Brigstock, 1999; Lau and Lam, 1999; Perbal, 2004; Takigawa, 2003; Takigawa *et al.*, 2003). Human CCN2 was originally discovered due to the cross-reactivity of a platelet-derived growth factor (PDGF) antiserum with a single polypeptide having a molecular weight of 38-kDa and secreted by cultured human vein endothelial cells (HUVEC), and its cDNA was isolated from a HUVEC cDNA expression library with anti-PDGF and shown to encode a 349-amino acid protein (Bradham *et al.*, 1991). In that study, the partially purified protein was found to be both mitogenic and chemotactic for fibroblast-like cells *in vitro*; and so this protein was named "CTGF" (Bradham *et al.*, 1991). Before the discovery of CTGF, the cDNA of the mouse CCN2 ortholog had been isolated as an immediate early gene by differential

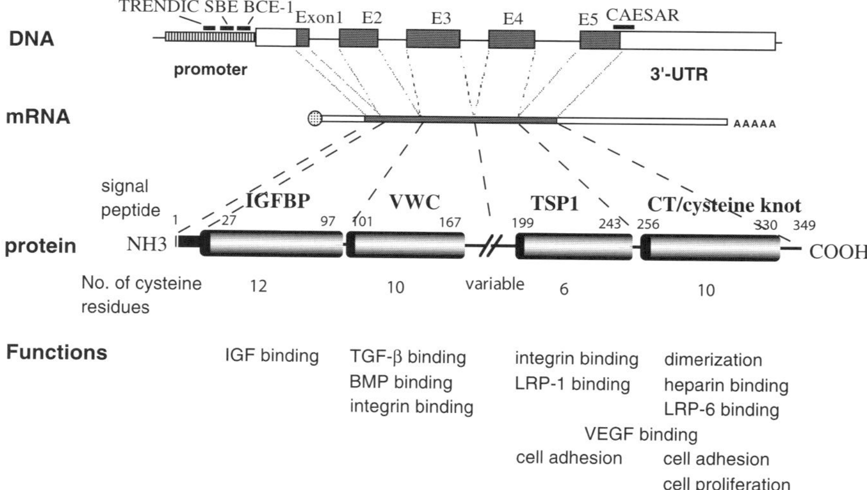

Fig. 1.   Modular structure of CCN2 protein and structure of its gene. The upper part of the figure illustrates the structure of the human *ccn2* gene. Exons are indicated by shading. Abbreviations: TRENDIC, transcriptional enhancer dominant in chondrocytes; SBE, Smad binding element; BCE-1/TbRE, basal control element 1/transforming growth factor-β responsive element; CAESAR, *cis*-acting element of structure-anchored repression; 3'-UTR, 3'-untranslated region. The mature *ccn2* mRNA is also shown with a 5'-capped structure and polyadenyl tail. The *ccn2* mRNAs are characterized by their long 3'-UTR. The middle part of the figure illustrates the structure of the human CCN2 protein. IGFBP: IGF binding protein-like module, VWC: von Willebrand factor type C module, TSP1: thrombospondin type I repeat, CT: C-terminal module. The lower part of the figure shows the function of each domain. BMP, bone morphogenetic protein; LRP, low-density lipoprotein receptor-related protein; VEGF, vascular endothelial cell growth factor.

screening of a cDNA library from serum-stimulated NIH3T3 cells and named "fibroblast-inducible secreted protein-12" (*fisp12*) (Ryseck *et al.*, 1991). The mouse ortholog was also isolated from TGF-β-stimulated mouse ARK-2B cells and named βIG-M2 (Brunner *et al.*, 1991). Fisp12/βIG-M2 encodes 348 amino acids, contains 39 cysteine residues, and is 91% homologous to human CCN2 (CTGF). Independently of these studies, we cloned an mRNA highly expressed in hypertrophic chondrocytes from a human chondrocytic cell line, HCS-2/8 (Enomoto and Takigawa, 1992; Takigawa *et al.*, 1991; Takigawa *et al.*, 1989) and found that its gene product, which we named Hcs24, was identical with that of human CCN2 (Nakanishi *et al.*, 1997; Takigawa, 2003; Takigawa *et al.*, 2003). We also found it to be a regulatory molecule involved in many stages of the process of skeletal and vascular growth and development. In addition, evidence showing the role of CCN2 in the regeneration of these tissues has been recently accumulated.

This chapter describes the role of CCN2 in the control of growth and regeneration at tissue, cellular, and molecular/gene levels.

2. ROLES OF CCN2 IN GROWTH CONTROL

It is no exaggeration to say that growth control of vertebrates is highly dependent on the growth control of the skeletal tissues. During skeletal growth, coordinate blood vessel formation also occurs. Therefore, the roles of CCN2 in skeletal growth and angiogenesis are firstly and secondly, respectively, described in this section. The role of CCN2 in other cells and tissues is mentioned lastly.

2.1. Roles of CCN2 in Skeletal Growth

Bone is formed by 2 pathways (Fig. 2). One is intramembranous bone formation; and the other, endochondral ossification. In the process of intramembranous bone formation, mesenchymal stem cells differentiate into osteoblasts; and then these cells directly deposit bone matrix. Mesenchymal stem cells can also differentiate into chondrocytes. These chondrocytes follow 2 distinct developmental pathways towards 2 distinct fates and functions (Fig. 2). Chondrocytes that give rise to articular cartilage acquire their permanent phenotype to maintain normal joint function throughout life (Buckwalter, 1998). In contrast, other chondrocytes, much more numerous, become organized into growth plates and undergo maturation, hypertrophic differentiation, calcification, and apoptosis, and are replaced by bone cells during the final stages of endochondral ossification (Cancedda, 1995; Hunziker, 1994; Takigawa *et al.*, 2003).

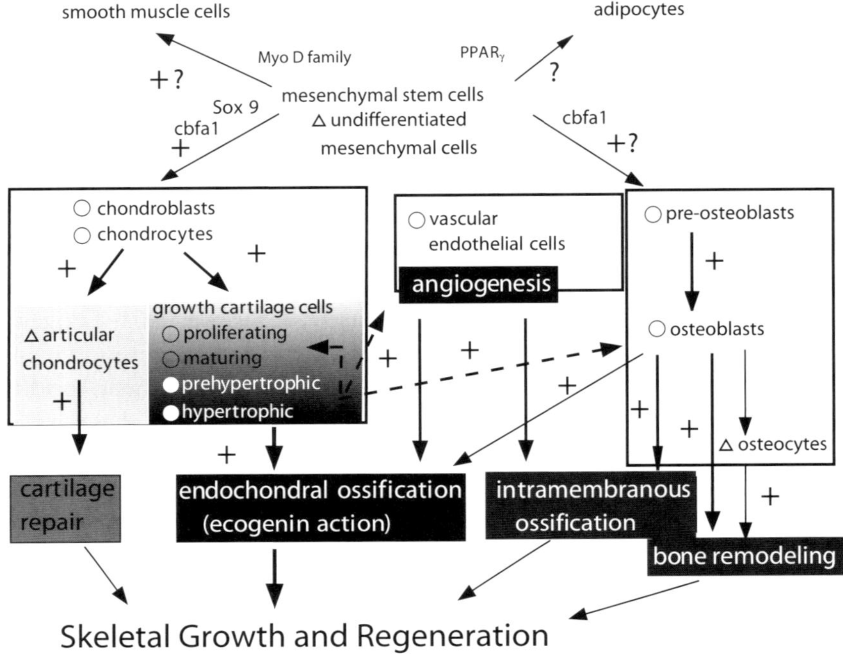

Fig. 2. Possible roles of CCN2 in skeletal development and regeneration. The level of CCN2 expression is shown with marks. Spontaneously high (closed circle); spontaneously low but high when stimulated (open circle); expressed only by stimulation (open triangle). + indicates the presence of direct evidence. Broken arrows indicate ecogenin (endochondral ossification genetic factor) action: The arrows point to the target cells, and the root of arrows is next to CCN2/CTGF-producing cells. CCN2 released by the cells marked with "open circle and triangle" may act in an autocrine/paracrine manner.

2.1.1. *The role of CCN2 in chondrocytes*

As described above, CCN2 was found to be highly expressed in hypertrophic chondrocytes (Fig. 2 and Table 1). *In situ* hybridization using both whole-mount neonates (P0) and their longitudinal sections revealed the expression of CCN2 in hypertrophic chondrocytes and in cells in the zone of calcifying cartilage (Nakanishi *et al.*, 1997; Nakanishi *et al.*, 2000; Takigawa, 2003; Takigawa *et al.*, 2003). Whole-mount *in situ* hybridization of embryonic day 17 (E17) mouse embryos showed that CCN2 was selectively expressed in the phalanges of the forelimbs (Takigawa *et al.*, 2003; Yamaai *et al.*, 2005). Moreover, *in situ* hybridization using tissue sections of E17 and neonatal mice revealed gene expression of CCN2 in hypertrophic chondrocytes in the long bones, ribs, and

Table 1.　Gene expression of CCN2

In vivo

Physiological state: +++

Hypertrophic chondrocytes	(Ivkovic *et al.*, 2003; Kadota *et al.*, 2003; Mukudai *et al.*, 2003; Nakanishi *et al.*, 1997; Nakata *et al.*, 2002)

Physiological state: ++

Proliferating and maturing chondrocytes	(Ivkovic *et al.*, 2003; Kadota *et al.*, 2003; Mukudai *et al.*, 2003; Nakanishi *et al.*, 1997; Nakata *et al.*, 2002)
Developing vascular endothelial cells	(Ivkovic *et al.*, 2003; Takigawa, 2003)
Endothelial cells in injured tissues	(Igarashi *et al.*, 1993; Kanyama *et al.*, 2003; Nakata *et al.*, 2002)
Osteoblasts & pre-osteoblasts in embryo	(Kadota *et al.*, 2003; Kanyama *et al.*, 2003; Nakata *et al.*, 2002)
Periosteal cells proliferating during wound healing	(Nakata *et al.*, 2002)
Osteocytes stimulated by mechanical stress	(Yamashiro *et al.*, 2001)
Stimulated undifferentiated mesenchymal cells	(Kadota *et al.*, 2003; Kanyama *et al.*, 2003; Nakata *et al.*, 2002)
Periodontal cells stimulated by mechanical stress	(Kanyama *et al.*, 2003; Yamashiro *et al.*, 2001)

Physiological state: +

Tooth germ	(Shimo *et al.*, 2002; Yamaai *et al.*, 2005)
Neural tissues	(Heuer *et al.*, 2003; Ivkovic *et al.*, 2003; Kondo *et al.*, 1999; Yamaai *et al.*, 1999)
Uterus	(Moussad *et al.*, 2002)
Kidney	(Ito *et al.*, 1998)

Pathological states in which CCN2 is induced:

Fibrosis [experimental skin fibrosis (Mori *et al.*, 1999), scleroderma
　(Igarashi *et al.*, 1996), systemic sclerosis (Igarashi *et al.*, 1995), renal fibrosis
　(Ito *et al.*, 1998), pulmonary fibrosis (Sato *et al.*, 2000), liver fibrosis (Abou–Shady
　et al., 2000; Paradis *et al.*, 1999)]

Inflammatory bowel disease (Dammeier *et al.*, 1998b)

Atherosclerosis (smooth muscle cells; Oemar *et al.*, 1997)

Osteoarthritis (Omoto *et al.*, 2004)

Tumors [endothelial cells in tumors (Shimo *et al.*, 2001b), fibrous stromas in tumors
　(Frazier and Grotendorst, 1997), chondrosarcomas (Shakunaga *et al.*, 2000),
　vascular tumors (Igarashi *et al.*, 1998), desmoplastic malignant melanoma
　(Kubo *et al.*, 1998), glioblastoma (Pan *et al.*, 2002), pancreatic cancer
　(Wenger *et al.*, 1999), breast cancer (Kang *et al.*, 2003; Shimo *et al.*, 2001b;
　Xie *et al.*, 2001)]

Table 1. *(Continued)*

In culture
In general: growing phase > quiescent phase
              (Safadi *et al.*, 2003; Shimo *et al.*, 1998; Shimo *et al.*, 2001b)
Chondrocytes: bi-phasic, becoming high again in the highly differentiated
              state (Nakanishi *et al.*, 1997)
Osteoblasts: bi-phasic, becoming high again in the highly differentiated state
              (Safadi *et al.*, 2003)

Typical cells, tissues, and organs that express CCN2 are listed.
+ + +: strong, ++: moderate, +: weak.

vertebral columns, which cells were destined to form calcified tissues (Takigawa *et al.*, 2003, Yamamai *et al.*, 2005). Even in 3-week-old mice, hypertrophic chondrocytes in the growth plate of long bones expressed CCN2 (Oka M and Takigawa M, unpublished). Also in an *in vitro* study, CCN2 was highly expressed in rabbit growth cartilage cells in culture but was little expressed in cultured resting cartilage cells or osteoblastic cells (Nakanishi *et al.*, 1997; Takigawa *et al.*, 2003) (Table 1). The expression of CCN2 increased during the continuous culture after confluence of rabbit growth cartilage cells and peaked in the hypertrophic phase (Nakanishi *et al.*, 1997; Takigawa *et al.*, 2003). These findings agree with the results showing that its *in vivo* expression was predominant in hypertrophic chondrocytes in growth plate cartilage, suggesting an autocrine/paracrine action of CCN2.

In fact, endogenous overexpression of CCN2 in human chondrocytic HCS-2/8 cells, achieved by using recombinant adenoviruses that generated CCN2 sense RNA (mRNA), resulted in enhanced cellular proliferation and expression of aggrecan and type X collagen (Nakanishi *et al.*, 2000). Similarly, recombinant CCN2 (rCCN2) protein (20–50 ng/mL) promoted the proliferation of HCS-2/8 cells and rabbit growth cartilage (RGC) cells in sparse and growing cultures (Nakanishi *et al.*, 2000). It also increased the proteoglycan synthesis and gene expressions of aggrecan and collagen type II, typical markers of chondrocyte maturation, in both types of cells in confluent cultures in which they were maturing. Furthermore, the rCCN2 effectively stimulated the gene expression of collagen type X, a marker of chondrocyte hypertrophy, in RGC cells in overconfluent cultures in which the cells were in the prehypertrophic stage (Nakanishi *et al.*, 2000). Moreover, the recombinant protein stimulated alkaline phosphatase (ALPase) activity, a marker of calcification, and indeed induced matrix calcification of RGC cells in culture (Nakanishi *et al.*, 2000). Considering the high expression of CCN2 in hypertrophic chondrocytes, these results indicate that CCN2 produced by hypertrophic chondrocytes promotes the proliferation and differentiation of growth cartilage cells toward endochondral ossification (Takigawa *et al.*, 2003) (Fig. 2).

Like rabbit growth cartilage (RGC) cells, rabbit articular cartilage (RAC) cells transfected with recombinant adenoviruses generating mRNA for CCN2 synthesized more proteoglycan than the control cells (Nishida *et al.*, 2002). Also, RAC cells treated with rCCN2 (20–50 ng/mL) showed increased synthesis of DNA and proteoglycans. Although the rCCN2 stimulated the gene expression of type II collagen and aggrecan core protein, which are markers of chondrocyte maturation, in both RGC and RAC cells, the gene expression of type X collagen, a marker of hypertrophic chondrocytes, was stimulated by rCCN2 in RGC cells, but not in RAC cells (Nishida *et al.*, 2002). Oppositely, gene expression of tenascin-C, a marker of articular chondrocytes, was stimulated by rCCN2 in RAC cells, but not in RGC cells (Nishida *et al.*, 2002). Moreover, rCCN2 effectively increased both ALPase activity and matrix calcification of RGC cells, but not those of RAC cells (Nishida *et al.*, 2002). These results indicate that CCN2 promotes the proliferation and differentiation of articular chondrocytes, but does not promote their hypertrophy or calcification (Nishida *et al.*, 2002) (Fig. 2). The difference between the effects of CCN2 on RGC and RAC would be dependent on differences in the characteristics of these 2 types of chondrocytes.

Another molecule of the CCN family, Cyr61, was also shown to act on chondrocytes but seems to be related to early chondrogenesis (Wong *et al.*, 1997). The effective concentration ranges of its recombinant protein was 0.3–5 µg/mL, one much higher than that of CCN2 (20–50 ng/mL) (Nakanishi *et al.*, 2000).

2.1.2. *The role of CCN2 in osteoblasts*

Gene expression of CCN2 in osteoblasts in adult mice was minimal; but it was also observed in osteoblasts in the primary spongiosa in embryonic mice, although its level was lower than that in the hypertrophic chondrocytes (Table 1 and Fig. 2). In addition, the expression of the factor was detected in both ameloblasts and odontoblasts in the developing tooth germ (Table 1) (Shimo *et al.*, 2002; Takigawa *et al.*, 2003, Yamaai *et al.*, 2005). Moreover, rCCN2 promoted the proliferation of osteoblastic cell lines Saos-2 and MC3T3E1 (Nishida *et al.*, 2000), and increased the mRNA expression of type I collagen, alkaline phosphatase, osteopontin, and osteocalcin, as well as the activity of alkaline phosphatase, in these cells (Nishida *et al.*, 2000). It also stimulated collagen synthesis in and matrix mineralization by MC3T3-E1 cells (Nishida *et al.*, 2000). These findings indicate that osteoblasts are at least target cells for CCN2 and that in some cases CCN2 acts on osteoblasts in an autocrine manner in addition to a paracrine manner (Fig. 2).

2.1.3. *Involvement of CCN2 in endochondral ossification*

At the final stage of endochondral ossification, calcifying cartilage is invaded by blood vessels, which leads to the recruitment of perivascular osteoblast-progenitor cells. Therefore, it is feasible that CCN2 would act on endothelial cells in addition to chondrocytes and osteoblasts. In fact, as described in the Section 2.2, CCN2 promoted migration, adhesion, and proliferation of vascular endothelial cells and their tube formation *in vitro* as well as angiogenesis *in vivo* (Lau and Lam, 1999; Shimo *et al.*, 2001a, b; Shimo *et al.*, 1998; Shimo *et al.*, 1999). These findings, taken together with the high expression level of CCN2 in hypertrophic chondrocytes *in vivo*, indicate that CCN2 is a novel, paracrine regulator that promotes the entire process of endochondral ossification by stimulating the proliferation and differentiation of chondrocytes, osteoblasts, and endothelial cells. Thus it may also be called "ecogenin" (Fig. 2).

The embryonic and neonatal growth of transgenic mice that overproduced CCN2 under the control of the mouse type XI collagen promoter occurred normally, but such animals showed dwarfism within a few months of birth. X-ray analysis revealed that their bone density was decreased compared with that of normal mice and that the femora in the hindlimbs in particular showed an apparent low density (Nakanishi *et al.*, 2001). These results indicate that overexpression of CCN2 adversely affects certain steps of endochondral ossification, further supporting our hypothesis described above. In addition, the stimulatory role of CCN2 in endochondral ossification shown by our *in vitro* and *in vivo* studies was confirmed by the results of gene knockout experiments in mice, which showed *ccn2*-deficient mice to display skeletal dysmorphisms due to impaired endochondral ossification (Ivkovic *et al.*, 2003).

2.1.4. *Possible role of CCN2 in intramembranous ossification and bone remodeling*

As mentioned above, CCN2 mRNA is minimally expressed in osteoblasts when compared with its expression level in hypertrophic chondrocytes. However, even in normal rats, osteoblasts and osteocytes in the alveolar bone adjacent to the periodontal ligament express a moderate level of CCN2 (Yamashiro *et al.*, 2001). This may be due to a high level of bone-forming activity of these cells in this area because of physiological tooth movement (Fig. 2; Table 1). In fact, little or no expression was observed in osteocytes embedded deeper in the alveolar bone (Yamashiro *et al.*, 2001). CCN2 was also shown to be expressed in normal long bones during the period of growth or modeling (Safadi *et al.*, 2003). Moreover, CCN2 was over-expressed in osteopetrotic (op) rats, animals which exhibit defects in osteoclast function, severely high bone density, and up-regulation of bone matrix and mineralization-related genes (Xu *et al.*, 2000). This finding is consistent with the *in vitro* data mentioned above and suggests some role for this factor in normal skeletal modeling/remodeling. The dramatic over-expression in the op mutant skeleton may be secondary to the uncoupling of bone resorption and bone formation, resulting in dysregulation of osteoblast gene expression and function (Xu *et al.*, 2000).

Interestingly, experimental tooth movement increased the expression of CCN2 mRNA in osteocytes and osteoblasts around the periodontal ligament, and intense expression of CCN2 extended to osteocytes situated deep in the alveolar bone matrix apart from the periodontal ligament (Yamashiro *et al.*, 2001) (Fig. 2; Table 1). These findings suggest that osteoblasts activated by certain stimuli also express CCN2 and that the expressed CCN2 may play some roles in bone remodeling as well as in bone formation.

2.2. Role of CCN2 in Angiogenesis

Angiogenesis is an important biological process for development; tissue growth, regeneration, and remodeling; and organ formation; as well as for the growth of solid tumors. The central player in this process is the vascular endothelial cell. Therefore, the role of CCN2 in vascular endothelial cells is first described and then followed by that of physiological and pathological angiogenesis.

2.2.1. *Role of CCN2 in vascular endothelial cells*

Although CCN2 had been originally identified in the conditioned medium from HUVEC (Bradham *et al.*, 1991), no detectable *in vivo* expression was found in normal adult human arteries (Igarashi *et al.*, 1996; Oemar and Luscher, 1997; Oemar *et al.*, 1997). However, the presence of CCN2 protein and/or mRNA was detected in endothelial cells during human embryonic development (Surveyor *et al.*, 1998; Wandji *et al.*, 2000). This difference may be due to the difference between rapidly growing embryos and adults; because actively migrating and proliferating endothelial cells in culture, but not quiescent ones, were shown to express CCN2 (Shimo *et al.*, 1998; Takigawa, 2003; Yamaai *et al.*, 1999). However, the expression of CCN2 mRNA in endothelial cells, even in mouse embryos, was lower than that in hypertrophic chondrocytes (Takigawa, 2003; Yamaai *et al.*, 1999) (Fig. 2).

Since suppression of endogenous expression of CCN2 in cultured endothelial cells by the addition of antisense oligomer or by transfection with expression vectors that generated its antisense RNA markedly inhibited the abilities of the cells to proliferate and migrate, CCN2 should be involved in the proliferation and migration of the cells *via* an autocrine pathway (Shimo *et al.*, 1998; Takigawa, 2003). In fact, rCCN2 promoted the adhesion, proliferation, and migration of vascular endothelial cells and induced tube formation by them (Shimo *et al.*, 1999; Takigawa, 2003) and recombinant mouse CTGF (Fisp12) also stimulated endothelial cell adhesion and migration through integrin $\alpha v\beta 3$ and promoted endothelial cell survival (Babic *et al.*, 1999). These findings indicate that endothelial cells are a target for CCN2.

The expression of CCN2 in endothelial cells was induced by wounding of monolayer cultures (Shimo *et al.*, 1998). Moreover, the expression was induced by other angiogenic factors such as vascular endothelial cell growth factor (VEGF), basic fibroblast growth factor (bFGF), and CCN2 itself (Shimo *et al.*, 2001b). These findings suggest that upregulation of CCN2 by angiogenic stimuli is a common phenomenon in endothelial cells. In other words, CCN2 might be a common autocrine mediator of all angiogenesis factors. If this hypothesis proves to be correct, since CCN2 is involved in physiological and pathological angiogenesis (described in the Sections 2.2.2 and 2.2.3), CCN2 would be a very important therapeutic target for the regeneration of vascular systems and the treatment of angiogenic diseases.

2.2.2. *Role of CCN2 in physiological angiogenesis*

In ovo application of rCCN2 to the chicken chorioallantoic membrane resulted in a gross angiogenic response, and rCCN2 injected along with collagen gel into the back of mice induced strong angiogenesis (Shimo *et al.*, 1999; Takigawa, 2003). It also induced neovascularization in rat corneal micropocket implants (Babic *et al.*, 1999). Moreover, impaired angiogenesis at the final stage of endochondral ossification was also observed in CCN2-deficient mice. These findings, in conjunction with the *in vitro* data described above, indicate that CCN2 is a potent angiogenesis factor.

In contrast, Inoki *et al.* (2002) reported recently that recombinant CCN2 and the thrombospondin type-1 repeat domain of CCN2 inhibited VEGF165-induced angiogenesis by forming a complex with VEGF165 (Fig. 1), although they also showed that their recombinant CCN2 itself had a weak angiogenic activity. However, since the angiogenic activity of CCN2 has been confirmed by several groups, the relation between CCN2 and VEGF should be investigated in more detail.

Similar angiogenic activity was reported for CCN1 (Lau and Lam, 1999). However, although CCN1-deficient mice were shown to be embryonic lethal due to failure of angiogenesis during placental development (Mo *et al.*, 2002), CCN2-deficient mice were reported to show only diminished angiogenesis in growth plates (Ivkovic *et al.*, 2003). The mechanism underlying the effect on angiogenesis between CCN1 and CCN2 may be different and should be investigated in greater detail.

2.2.3. *Role of CCN2 in pathological angiogenesis*

With respect to pathogenic processes, endothelial cells of atherosclerotic blood vessels expressed CCN2 (Oemar and Luscher, 1997; Oemar *et al.*, 1997). Moreover, human breast cancer (MDA231) cells, which express a high level of CCN2 mRNA and secrete much CCN2 protein, showed a high level of neovascularization when xenografted in nude mice; whereas human squamous cell carcinoma cells (A431), which express a low level of CCN2 mRNA and secrete little CCN2 protein, showed a low level of neovascularization (Shimo *et al.*, 2001b). Furthermore, CCN2 also increased the expression by MDA231 cells of a number of metalloproteinases, which are involved in endothelial cell invasion (Kondo *et al.*, 2002). These findings indicate that CCN2 is a novel, potent autocrine/paracrine factor that functions in multi-stages in tumor angiogenesis. Hypoxic induction of angiogenesis by human breast cancer cells (MDA-231) can be ascribed at least in part to CCN2 (Shimo *et al.*, 2001a; Shimo *et al.*, 2001b). A role for CCN2 in tumor angiogenesis is also suggested in glioblastomas (Pan *et al.*, 2002).

2.3. Possible Roles of CCN2 in the Growth of Other Tissues

After midgestation, major producers of CCN2 are fibroblasts in addition to chondrocytes, endothelial cells, and osteoblasts. In addition, fibroblasts are a major interstitial component of various organs. Therefore, the role of CCN2 in fibroblasts are mentioned first followed by that of embryonic development. The role of CCN2 in tumor growth is also discussed.

2.3.1. *The role of CCN2 in fibroblasts*

In *in vitro* studies, CCN2 was reported to be both mitogenic and chemotactic for fibroblast-like cells (Brigstock *et al.*, 1997; Frazier *et al.*, 1996) and its mitogenic action to be synergized by other growth factors and heparin (Brigstock *et al.*, 1997; Frazier *et al.*, 1996; Jin *et al.*, 2004; Kireeva *et al.*, 1997). It was also reported that CCN2 directly stimulated adhesion of fibroblasts, and this adhesion-promoting effect was proposed to be mediated by integrins (Chen *et al.*, 2001; Kireeva *et al.*, 1997) (Fig. 1). With respect to biochemical phenomena, CCN2 also stimulated the production of extracellular matrix components such as collagens and fibronectin, and of integrin $\alpha 5$ (Dammeier *et al.*, 1998b; Frazier *et al.*, 1996; Murphy *et al.*, 1999; Riser and Cortes, 2001; Riser *et al.*, 2000; Twigg *et al.*, 2002). These findings indicate that fibroblasts are target cells for CCN2.

Concerning gene expression and localization of CCN2, there have been many reports on overexpression of CCN2 in fibrotic skin disorders and in fibrotic lesions of various organs and tissues including the lung, kidney, liver, cardiovascular system, pancreas, lens, bowel, and gingiva, as well as in the stroma of various tumors (Takigawa, 2003); but few reports have appeared on its expression in fibroblasts or fibroblast-like cells in the physiological state. These findings strongly suggest the pathological significance of CCN2 in these fibrotic disorders (Leask *et al.*, 2002), which is described in other chapters in detail. The involvement of CCN2 in fibrosis is in line with the fact that CCN2 is highly expressed during the wound healing process (Igarashi *et al.*, 1993), which is described in Section 3.5.

2.3.2. *Possible roles of CCN2 in embryonic development and growth*

During normal development of the mouse embryo (E17), gene expression was also observed in the brain and kidney, although the level was much lower than that in the hypertrophic zone of various cartilages (Takigawa *et al.*, 2003; Yamaai, 2005) (Table 1). In addition, CCN2 gene expression was detected during tooth development (Yamaai *et al.*, 2005); and neutralizing antibody against CCN2 inhibited the proliferation and differentiation of epithelium and mesenchymal cells and delayed the cytodifferentiation of ameloblasts and odontoblasts in tooth germ explants (Shimo *et al.*, 2002). These findings suggest the role of CCN2 in the growth and development of these organs.

In earlier mouse embryos, CCN2 protein was detectable in various organ systems and tissues including in mesenchymal and epithelial cells in them (Kireeva *et al.*, 1997; Surveyor and Brigstock, 1999; Surveyor *et al.*, 1998). Moreover, in pregnancy, CCN2 protein was detected in both the uterus and embryo (Surveyor and Brigstock, 1999; Surveyor *et al.*, 1998). Therefore, these findings suggest that during prenatal life, especially before midgestation, various types of cells produce CCN2 in a specific temporospacial pattern and support a role for CCN2 in cellular differentiation and development.

2.3.3. *CCN2 and tumor growth*

CCN2 has been shown to be overexpressed in various types of malignant tumors such as chondrosarcoma (Shakunaga *et al.*, 2000), vascular tumors (Igarashi *et al.*, 1998), desmoplastic malignant melanoma (Kubo *et al.*, 1998), pancreatic cancer (Wenger *et al.*, 1999), glioblastoma (Pan *et al.*, 2002), breast cancer (Kang *et al.*, 2003; Shimo *et al.*, 2001b; Xie *et al.*, 2001), and to be significantly correlated with the prognosis in particular types of tumors (Kubo *et al.*, 1998; Shakunaga *et al.*, 2000; Wenger *et al.*, 1999; Xie *et al.*, 2001). However, the role of CCN2 in malignancy is quite controversial. For example, expression of CCN2 gene in chondroma is higher than that in chondrosarcoma, although there is a correlation between the expression level of CCN2 and tumor grade of chondrosarcomas (Shakunaga *et al.*, 2000). It is noteworthy that the tumors listed in Table 1 have something to do with the expression and function of CCN2 that we have mentioned in this chapter. CCN2 is expressed in chondrocytes, vascular endothelial cells and neural tissues and overexpressed in chondrosarcoma (Shakunaga *et al.*, 2000), vascular tumors (Igarashi *et al.*, 1998) and glioblastoma (Kubo *et al.*, 1998). The melanoma (Kubo *et al.*, 1998) and pancreatic cancer (Wenger *et al.*, 1999) reported are also fibrous types. The breast cancer is known to be metastatic to bone (Kang *et al.*, 2003). Therefore, the overexpression of CCN2 in these tumors might be only a secondary outcome after tumorigenesis. Moreover, forced overexpression of CCN2 causes apoptosis (Hishikawa *et al.*, 1999a, b), growth suppression (Moritani *et al.*, 2003b), and benign conversion (Moritani *et al.*, 2003b) in several types of malignant tumors. Because hypertrophic chondrocytes, which show the highest expression of CCN2 in the physiological state, are fated for death, forced overexpression may cause growth arrest or apoptosis. Although further investigation is required to determine the exact role of CCN2 in tumorigenesis, it is unlikely that CCN2 is an oncogenic protein.

3. ROLES OF CCN2 IN TISSUE REGENERATION

Wound healing is defined as a kind of "tissue regeneration" and is believed to occur by reproducing normal developmental processes. In an early study,

CCN2 was shown to be expressed in dermal fibroblasts and endothelial cells during wound healing of the skin (Igarashi *et al.*, 1993). However, concerning the role of CCN2 in fibroblasts, much attention has been paid to the pathological state such as fibrosis, which is believed to be due to over- and persist expression of CCN2. In contrast, skeletal tissues such as bone and cartilage, which are rich in matrix, have been targets for investigating the role of CCN2 in tissue regeneration.

Expression and production of CCN2 by cultured cells are generally higher in the sparse, growing phase than in the confluent, quiescent phase (Shimo *et al.*, 2001b) (Table 1). For example, as described above, actively migrating and proliferating endothelial cells, but not quiescent ones, express CCN2 (Shimo *et al.*, 1998). Even chondrocytic cells express a higher level of CCN2 in sparse cultures than in confluent ones (Takigawa *et al.*, 2003), but gene expression of CCN2 goes up again after the cells reach confluence, at which time they differentiate to the hypertrophic state (Nakanishi *et al.*, 1997) (Table 1). A similar bi-phasic change was also found in osteoblasts in culture (Safadi *et al.*, 2003) (Table 1).

As such, the high-level expression of CCN2 in rapidly growing and remodeling tissues *in vivo* and in activated cells in culture strongly suggests that CCN2 is an autocrine factor for tissue regeneration.

3.1. Possible Role of CCN2 in Bone Regeneration

During fracture healing, both endochondral and intramembranous ossification processes occur. During fracture healing in a mouse rib model, the expression of CCN2 mRNA revealed by Northern blot analysis significantly increased on day 2 of fracture healing, reaching its peak on day 8, and then declined. CCN2 mRNA and protein were remarkably detected especially in hypertrophic chondrocytes and in proliferating chondrocytes in the regions of regenerating cartilage on days 8 and 14 after the fracture (Nakata *et al.*, 2002), suggesting that proliferating chondrocytes can also express CCN2 in rapidly regenerating cartilage. CCN2 mRNA was also expressed in proliferating periosteal cells in the vicinity of the fracture sites on days 2 and 8, and in cells in the fibrous tissue around the callus on day 8 (Nakata *et al.*, 2002) (Table 1). CCN2 was also detected in active osteoblasts in the regions of intramembranous ossification (Nakata *et al.*, 2002) and in cells in fibrous tissue, in vascular endothelial cells in the callus, and in periosteal cells around the fracture sites, thus suggesting the involvement of CCN2 in the regeneration of chondrocytes, osteoblasts, and endothelial cells from precursor cells or mesenchymal stem cells.

CCN2 is also involved in alveolar bone regeneration after tooth extraction (Kanyama *et al.*, 2003). At the early healing stage after tooth extraction, CCN2 was expressed strongly in the residual periodontal ligament and in the endothelial cells and undifferentiated fibroblast-like cells, possibly mesenchymal stem cells, migrating into the granulation tissue at the bottom of the sockets (Table 1). At a later stage, osteoblast-like cells proliferated in the sockets, which showed low CCN2 expression. These findings suggest that CCN2 produced by residual periodontal cells, endothelial cells and undifferentiated mesenchymal cells, acts as an autocrine/paracrine factor in the sockets, causing the mesenchymal cells to differentiate into osteoblasts or the recruitment of osteoblasts from bone, which then replace the granulation tissue with bone.

To clarify the role of CCN2 in tissue regeneration of bone, we also investigated the localization and expression of CCN2 during distraction osteogenesis, which is a bone regeneration therapy to expand bone length (Kadota *et al.*, 2003). We performed osteotomy in the midshaft of the right femur and then started the distraction. After a 7-day lag phase, the distraction was done for 21 days (distraction phase) by using a small external fixator; and it was followed by a 7-day consolidation phase. Immunostaining showed the localization of CCN2 in various cells located in the bone-forming area around the osteotomy site. During the distraction phase, *in situ* hybridization showed that CCN2 mRNA was expressed not only in hypertrophic chondrocytes and osteoblasts but also in fibroblast-like cells and mesenchymal cells in sites of endochondral ossification, and not only in osteoblasts but also in pre-osteoblasts and fibroblasts-like cells in sites of intramembranous ossification. RT-PCR revealed a higher level of CCN2 mRNA expression in the distracted group than in the non-distracted group. These results suggest that CCN2 is involved in not only endochondral but also intramembranous ossification and plays important roles in regeneration of bone during distraction osteogenesis (Fig. 2).

The role of CCN2 in bone regeneration via intramembranous ossification was also reported, in which *in vivo* delivery of rCCN2 into the femoral marrow cavity induced osteogenesis in an animal model (Safadi *et al.*, 2003).

In addition to that in mammalians, gene expression of CCN2 was also found during regeneration of the newt limb (Cash *et al.*, 1998). The gene was expressed in hypertrophic chondrocytes in the restored cartilaginous skeleton.

3.2. Role of CCN2 in Regeneration of Articular Cartilage

CCN2 is normally not expressed in articular cartilage cells of adult rodents, although its weak expression is observed in these cells in growing, young animals (Nawachi *et al.*, 2002). However, even in adult rodents, quantitative real-time reverse-transcription polymerase chain reaction (RT-PCR) assays showed a significant increase in the level of CCN2 mRNA in the monoiodoacetic acid (MIA)-induced OA model (Nishida *et al.*, 2004). Immunohistochemical analysis and *in situ* hybridization revealed that the clustered chondrocytes, in which case clustering indicates an attempt to repair the damaged cartilage, produced CCN2 (Nishida *et al.*, 2004). Similarily, gene expression and localization of CCN2 was observed in human osteoarthritic cartilage (Omoto *et al.*, 2004). Therefore, CCN2 was suspected to play critical roles in cartilage regeneration. In fact, rCCN2 incorporated in gelatin hydrogel and injected as a single dose into the joint cavity of MIA-induced OA model rats repaired their articular cartilage to the extent that it became histologically similar to normal articular cartilage (Nishida *et al.*, 2004). When rCCN2-hydrogel-collagen complex was implanted into the defects on the surface of articular cartilage *in situ*, new cartilage filled the defect 4 weeks postoperatively, whereas only soft tissue repair occurred when the PBS-hydrogel-collagen was implanted (Nishida *et al.*, 2004) (Fig. 3). These findings suggest the utility of CCN2 in the regeneration of articular cartilage.

3.3. Effect of CCN2 on Bone Marrow-Derived Mesenchymal Stem Cells

To investigate whether rCCN2 could stimulate chondrogenic cell differentiation or not, we cultured bone marrow-derived stem cells (MSCs) for 14 days in α-MEM with or without 30 or 50 ng/mL of rCCN2 (Nishida *et al.*, 2004). The cultures were thereafter processed for histochemical detection of cartilage matrix by using toludine blue staining. Some cartilage matrix was present in control cultures and in 30 ng/mL rCCN2-treated cultures of MSCs, but a much larger amount of it was detected in 50 ng/mL rCCN2-treated ones. In addition, to clarify the role of CCN2 in the chondrocytic differentiation of MSCs, we used Northern blot analysis to investigate the effect of rCCN2 on the mRNA expression of type II collagen and aggrecan in cultured MSCs. Chondrogenesis, as represented by the expression of type II collagen and aggrecan mRNAs, was clearly induced in MSCs treated with 50 ng/mL rCCN2. These findings suggest that CCN2 can induce chondrogenic differentiation of mouse bone marrow-MSCs (Nishida *et al.*, 2004) (Fig. 2).

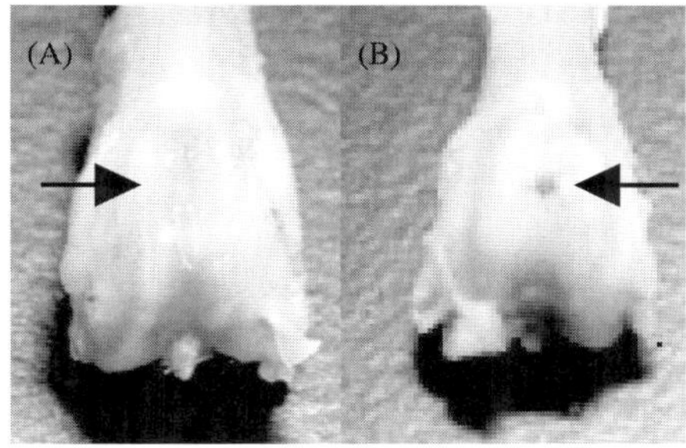

Fig. 3.   Gross appearance of the CCN2- (A) and PBS- (B) hydrogel-collagen-treated articular cartilage 4 weeks after the operation. A defect, 2 mm in diameter and penetrating the subchondral bone plate, was prepared on the patellar groove of the femur with a microdrill. One microgram of rCCN2 that had been incorporated into gelatin hydrogel contained in a collagen sponge was lyophilized and implanted into the defect. Then, the patella was repositioned, and the medial aspect of the capsule was closed with a nylon suture. By 4 weeks after the operation, the rCCN2-treated defect (A) was filled with cartilage, the surface of which was smooth; and the reparative tissue consisted of hyaline cartilage-like cartilage (Nishida *et al.*, 2004). The PBS-treated defect remained filled with soft tissue which appears in dark (B) (Nishida *et al.*, 2004). Allows indicate the positions of defects made with a microdrill.

It is a well-known fact that MSCs can differentiate into osteoblasts, myoblasts, and adipocytes, as well as into chondrocytes (Fig. 2). The findings described in the Section 3.1 on bone regeneration also indicate a possible role of CCN2 in the differentiation of MSCs into osteoblasts and chondrocytes. The *cbfa1* gene encodes a well-known transcription factor that directs mesenchymal stem cells to differentiate into osteoblasts (Komori *et al.*, 1997) and is also important in endochondral ossification (Enomoto *et al.*, 2000; Inada, 1999). The expression of *cbfa1* was detected from the zone of proliferating chondrocytes to the calcification zone (Enomoto *et al.*, 2000; Inada, 1999). Whole-mount *in situ* hybridization using E17 *cbfa1*-null mice revealed that the expression of CCN2 in the phalanges of the embryonic forepaws was completely undetectable (Takigawa *et al.*, 2003; Yamaai *et al.*, 2005). Also by *in situ* hybridization using tissue sections, the expression of CCN2 in ribs and cervical vertebrae found in normal embryos was undetectable in *cbfa1*-null embryos (Takigawa *et al.*, 2003; Yamaai *et al.*, 2005). These findings suggest that CCN2 may be a downstream regulator of *cbfa1* in chondrocytic differentiation and osteoblastic differentiation of mesenchymal stem cells during embryonic development (Fig. 2).

However, so far there is no report on the role of CCN2 in differentiation of MSCs into adipocytes or myoblasts. Concerning myogenic differentiation, inhibition of CCN2 expression resulted in decreased differentiation of human rhabdomyosarcoma cells (Croci *et al.*, 2004). Therefore, it would be interesting to investigate whether CCN2 determines the direction of differentiation of MSCs or promotes the differentiation of MSCs who have already been commited to a given direction.

3.4. Possible Role of CCN2 in Regeneration of Vascular Tissues

As described in Section 2.2, CCN2 is not only a paracrine but also an autocrine angiogenesis factor that is expressed in activated, but not quiescent, endothelial cells *in vivo* (Table 1; Fig. 2). In fact, the expression of CCN2 in endothelial cells was induced by wounding of monolayer cultures of endothelial cells (Shimo *et al.*, 1998) and by other angiogenic factors such as VEGF, bFGF, and CCN2 itself (Shimo *et al.*, 2001b), suggesting that upregulation of CCN2 by angiogenic stimuli is a common phenomenon in endothelial cells. Therefore, as mentioned before CCN2 might be a common autocrine mediator of all angiogenesis factors. If this hypothesis proves to be correct, CCN2 would be a very important therapeutic target for the regeneration of vascular systems in addition to being useful for the treatment of angiogenic diseases.

Since CCN2 can bind to an artificial slow-releaser gelatin hydrogel, which can be made to assume any shape, its strong angiogenic activity may be applied to tissue regeneration, especially to tissue engineering of the vascular system.

3.5. Possible Role of CCN2 in Regeneration of Other Tissues

CCN2 is expressed in fibroblasts or fibroblast-like cells during wound healing of the skin (Igarashi *et al.*, 1993) and cornea (Blalock *et al.*, 2003). Therefore, considering the biological actions of CCN2 described in Section 2.3.1, the roles of CCN2 in wound healing may be summarized as follow: First of all, fibroblast-derived CCN2 stimulates the migration of fibroblasts to the injured tissue. Then, it mediates the adhesion of these cells to settle them at the defect site and initiates tissue regeneration by increasing the proliferation of the cells to provide an adequate number of them. Finally, CCN2 promotes the deposition of extracellular matrix (ECM) components to repair the tissue defect. So, all of the activities of CCN2 just described contribute to the regeneration of the injured tissues. In addition, as mentioned in Section 2.2, CCN2 is capable of inducing angiogenesis, which is essential for most tissues. Therefore, it is clear that CCN2 is a key player involved in the wound-healing cascade in connective tissues.

However, control of its gene expression is very important for proper regeneration. If up-regulated expression of CCN2 is appropriately terminated, normal wound healing would occur. In contrast, if the overexpression continues, it would cause fibrosis and scaring. No disease in skeletal tissues with continuous overexpression of CCN2 has been reported, although there is a report that *ccn2* transgenic mice in which *ccn2* was forced to be overexpressed in cartilage have skeletal disorders (Nakanishi *et al.*, 2001). Presumably, unless this gene is dramatically overexpressed, nothing would happen in these tissues in which the high level of ECM synthesis occurs in the normal state.

In addition, CCN2 is also suggested to be involved in repair of the intestinal mucosa (Dammeier *et al.*, 1998b) and in *in vitro* wound healing of renal epithelium cells (Pawar *et al.*, 1995). Therefore, CCN2 may be important for regeneration of not only connective tissues but also other tissues as well, although the latter have been little investigated.

3.6. Platelet-Derived CCN2 May Be a Trigger for Tissue Regeneration

In many tissues/organs, blood clotting is the first step of wound healing after an injury. Since CCN2 is induced in and produced by undifferentiated mesenchymal cells, probably including mesenchymal stem cells, which migrate into the defect, it may play the role as an autocrine/paracrine regeneration factor throughout the process of wound healing. However, at the initial step, CCN2 may be supplied by blood clots, because CCN2 was recently found to be abundant in platelets and to be released from activated platelets (Cicha *et al.*, 2004; Kubota *et al.*, 2004). The CCN2 released from platelets may be important for the initiation of proliferation, migration, and adhesion of undifferentiated mesenchymal cells, maybe including stem cells, to regenerate wounded tissues.

4. MOLECULAR REGULATION OF CCN2 FOR DRIVING THE PROPER GROWTH AND REGENERATION OF TISSUES

Almost all organs except avascular tissues such as cartilage comprise a mixture of cell types, and regeneration thus requires the coordinated growth of various tissue components. As described above, CCN2 is an important regeneration factor at least for the connective tissues. Therefore, if, due to elevated expression of CCN2, the regeneration rate of connective tissue framework becomes higher than that of the parenchymal cells in organs such as liver, kidney and lung, disturbance of the regeneration of parenchymal cells may occur, as seen in liver cirrhosis. CCN2 is a multi-functional growth factor that potentiates either growth or differentiation of several types of mesenchymal cells, but its expression *in vivo* is normally low. However, the expression in certain tissues is upregulated during development and cytodifferentiation and is induced in certain pathological states. Therefore, it is very important to understand the molecular regulatory mechanism of action and gene expression of CCN2, and then to control them for establishment of a successful strategy for tissue regeneration.

4.1. Multiple Mechanisms for Multiple Actions

4.1.1. *Cell-surface receptors*

Since CCN2 is considered to be a growth factor, a number of efforts have been made to identify its specific receptor(s) that transduces the signal from CCN2 to intracellular messengers. In general, extracellular signals from typical growth factors, such as epidermal growth factor (EGF) and PDGF, are transmitted through kinase-type receptors. In fact, a 280-kDa receptor (CSP240)-ligand complex, which was first found on chondrocytes (Nishida *et al.*, 1998) (Fig. 4), was phosphorylated upon CCN2 stimulation of osteoblastic cells (Nishida *et al.*, 2000). Therefore, it is likely that major signals by CCN2 are mediated by this type of receptor. However, albeit a few candidates have been nominated (Nawachi *et al.*, 2002), so far no distinct kinase-type receptor specific for CCN2 has been identified yet.

Because of the multiple functionality of CCN2, it is also reasonable for this factor to use multiple types of receptors, probably through differential interactions with its 4 modules. Apart from kinase-type receptors, several other cell-surface molecules have been found to interact with CCN2 (Fig. 4). The low-density lipoprotein receptor-related protein (LRP-1), which is a quite large member of an ancient family of endocytic receptors, was revealed to be one such molecule (Segarini *et al.*, 2001). LRP-1 is known to bind to a variety of biologically diverse ligands to exert multiple functions. In particular, phosphatidylinositol-3 kinase-mediated phosphorylation of a tyrosine residue in LRP-1 was reported to be induced by PDGF (Herz and Strickland, 2001). Moreover, the finding that CCN2 may be internalized by endocytosis further suggests an active role for LRP-1 in CCN2 function (Wahab *et al.*, 2001a). In addition to LRP-1, another family member LRP-6, which is a coreceptor for Wnt, was also found to interact with CCN2, which resulted in modulation of Wnt signaling in *Xenopus* (Mercurio *et al.*, 2004).

Finally, the best-characterized cell-surface CCN2 counterparts are integrins (Fig. 4). Integrins $\alpha_M\beta_2$, $\alpha_6\beta_1$ and $\alpha_{IIb}\beta_3$ were reported to mediate the cell adhesion of monocytes, fibroblasts, and platelets, respectively, caused

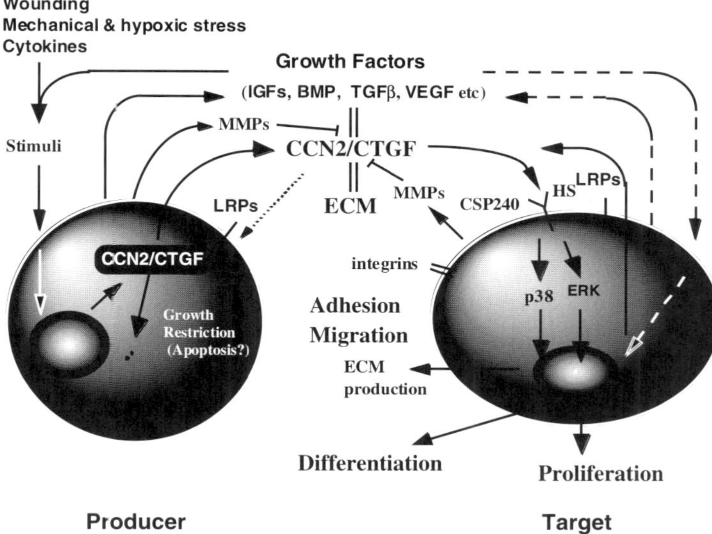

Fig. 4. Molecular mechanism of CCN2 action. CCN2 is induced by various stimuli such as growth factors, wounding, mechanical stress and hypoxia. Produced CCN2 is released from cells and binds to ECM, and thus may act in a matricrine manner. The CCN may also bind to other growth factors and modify their activity. Various types of receptor candidates and receptor-like molecules such as 240-kDa cell surface protein (CSP240), LRPs, and integrins are reported. Integrins are believed to be involved in the cell adhesion activity of CCN2. Some intracellular signal transduction pathways have been uncovered; i.e. P38 and ERK mediate differentiation and proliferation of chondrocytes, respectively. Intracrine action of CCN2 is also suggested in some cases such as hypertrophic chondrocytes and certain types of tumor cells. ECM: extracellular matrix, MMPs: matrix metalloproteinases, HS: heparan sulfate proteoglycans.

by CCN2 (Jedsadayanmata *et al.*, 1999; Schober *et al.*, 2002); whereas $\alpha_v\beta_3$ was shown to be involved in CCN2-induced angiogenesis (Babic *et al.*, 1999; Chen *et al.*, 2001) and adhesion of hepatic stellete cells (Gao and Brigstock, 2004). Therefore, integrins may be important partners of CCN2 in the processes of tissue growth and regeneration. Furthermore, the interaction between CCN2 and integrin $\alpha_6\beta_1$ actually activates intracellular phosphorylation signaling cascades, which obviously supports significant portions of the diverse CCN2 functions.

4.1.2. *Signal transduction cascades*

Downstream of the multiple cell-surface ligand-receptor interactions follows the activation of multiple signal transduction pathways. To date, analysis of CCN2-induced phosphorylation signaling cascades has been mainly carried out by using chondrocytes. As described in Section 2.1.1, CCN2 promotes both proliferation and differentiation of chondrocytes (Nakanishi *et al.*, 2000; Nishida *et al.*, 2002). According to recent reports, these apparently contradictory effects of CCN2 are enabled by the differential use of intracellular second messengers. The major mediators of the promotion of chondrocytic differentiation were found to be protein kinase C (PKC) and p38 mitogen-activated kinase (p38MAPK), whereas extracellular regulated kinases (ERK) promote cell proliferation without the positive contribution of PKC (Yosimichi *et al.*, 2001; Yoshimichi *et al.*, 2005) (Fig. 4). Also, the results of a quite recent investigation suggested that other secondary messengers may not be disregarded in accounting for the entire intracellular signaling process provoked by CCN2 (Yosimichi *et al.*, 2005). It should be noted that the situation indicated above has been proven in chondrocytes only. What is happening in CCN2-stimulated fibroblasts, vascular endothelial cells and other CCN2 target cells still remain to be clarified.

4.1.3. *Extracellular counterparts*

All of the 4 modules that comprise CCN2 are highly interactive with a variety of other molecules, which provides a firm basis for CCN2's ability to have multiple functions. Among the extracellular counterparts known today, growth factors are particularly of note, since the biological outcome yielded by CCN2 should be considered to be the net effects together with the functional modulation of such counterparts through direct interactions. Indeed, direct interaction of CCN2 with TGF-β was found to enhance the ligand-receptor interaction, in which the VWC module acted as the interface (Abreu *et al.*, 2002) (Figs. 1 and 4). Contrarily, it was also reported that CCN2 inhibited the strong angiogenic effects of VEGF through direct binding, which was mediated by the VWC and TSP modules of CCN2 (Inoki *et al.*, 2002) (Figs. 1 and 4); although CCN2 itself possesses angiogenic activities (Babic *et al.*, 1999; Shimo *et al.*, 1999; Lau and Lam, 1999; Takigawa, 2003). Negative regulation by the CCN2 direct binding was described in the case of BMP-4 as well (Abreu *et al.*, 2002), in which case the interaction was also mediated by the VWC module (Figs. 1 and 4). As clearly represented in its terminology, the IGFBP module is believed to play a central role in the direct interaction between CCN2 and IGFs, although the affinity between the two was shown to be quite low (Brigstock, 2003) (Figs. 1 and 4).

CCN2 is a matrix-associated extracellular molecule. In this regard, the association of CCN2 with ECM was indicated to depend on the CT module (Kubota *et al.*, 2001). Since the CT module directly binds to heparin, the involvement of heparan sulfate proteoglycans as CCN2-associated molecules in CCN2 actions has been suggested. Recently, we found that heparan sulfate proteoglycans were critically required for CCN2 to exert its function in growth plate chondrocytes (Nishida *et al.*, 2003) (Figs. 1 and 4). Enzymatic removal of heparan sulfate proteoglycans from chondrocytes resulted in the loss of their responsiveness to stimulation by CCN2 to promote total proteoglycan synthesis and aggrecan gene expression. Moreover, among a number of distinct proteoglycans therein, we could specify perlecan as a key molecule that mediates CCN2 function as a co-receptor. Perlecan directly interacted with CCN2, and the distribution of perlecan in differentiating chondrocytic populations was quite similar to that of CCN2 binding sites. Considered together with the malformation of the growth plate observed in prelecan gene KO mice (Arikawa–Hirasawa *et al.*, 1999), these findings indicate the critical roles of perlecan in endochondral ossification as an extracellular functional counterpart of CCN2.

4.1.4. *Intracellular CCN2 and its molecular counterparts*

In addition to the classical roles of CCN2 as an extracellular messenger, the intracellular distribution and possible functions of CCN2 have been recently indicated (Fig. 4). Generally 2 pathways are available for proteins to be inside the plasma membrane. Nascent proteins may be sequestered by certain molecules after translation, while extracellular proteins can be taken up by endocytosis. In the case of CCN2, both pathways are suggested to be used. CCN2 overexpressed in monkey kidney Cos-7 cells did not undergo efficient secretion (Kubota *et al.*, 2000a), whereas extracellular CCN2 was taken up by mesangial fibroblasts (Wahab *et al.*, 2001a). In light of these findings, the direct binding of cytoskeletal actin to CCN2 (Yosimichi *et al.*, 2002) and the functional property of LRP-1 as a endocytotic key molecule are particular of interest.

The cell biological effects caused by intracellular CCN2 were also described previously. Overexpressed CCN2 was found to negatively modulate the cell cycle of Cos-7 cells (Kubota *et al.*, 2000a), which is consistent with the findings that CCN2 is accumulated in terminally-differentiated hypertrophic chondrocytes *in vivo* (Moritani *et al.*, 2003a); and overexpressed CCN2 induced apoptosis (Hishikawa *et al.*, 1999a) and attenuated growth of cancer-derived cell lines (Moritani *et al.*, 2003b), as was also observed in the case of CCN3 (Gupta *et al.*, 2001) (Fig. 4).

4.1.5. *Processing of CCN2 and the role of its subfragments*

Post-translational processing is one of the common regulatory systems for proteins with critical functions. It is widely known that a number of proteases and hormones are produced primarily as inactive pro-forms that require specific proteolysis in order for them to become active. Interestingly, involvement of such restricted proteolysis is also suggested in the case of CCN2. In early studies, the presence and functional significance of C-terminal fragments of CCN2 were already indicated. A CT module-containing processed fragment of 10–13 kDa was reported to be mitogenic for fibroblasts (Steffen *et al.*, 1998), and ECM-associated in chondrocytes (Kubota *et al.*, 2001). In contrast, little is known with regard to N-terminal processed fragments, except for the formation of such a fragment in CCN2-overexpressing mesangial cells (Wahab *et al.*, 2001b). However, the accumulation of such an N-terminal CCN2 fragment in the vitreous of patients of proliferative diabetic retinopathy was recently uncovered (Hinton *et al.*, 2004). This discovery not only suggests the pathological roles of the N-terminal CCN2 fragment in a particular disease, but also represents the utility of the fragment as a diagnostic tool (Kawaki *et al.*, 2003). Although precise functions of the processed CCN2 fragments are currently unclear, a significant portion of the multiple functions of CCN2 can be specifically assigned to these processed submolecules, which are "activated" or "inactivated" by specific endopeptidases.

In light of the existence of these subfragments, one of the next questions is what cleaves CCN2. According to a previous report, matrix metalloproteinases (MMPs) were capable of cleaving CCN2 at particular sites to yield subfragments (Hashimoto *et al.*, 2002) (Fig. 4). Considering the collaborative roles of MMPs and CCN2 in ECM remodeling, multiple molecular interplay between them is highly suspected.

4.2. Extracellular Factors that Regulate CCN2 Gene Expression

4.2.1. *Hormones, cytokines, and other small molecules*

Currently, several growth factors and cytokines, such as VEGF, bFGF, endothelin-1, TGF-β and even CCN2 itself, are known to modulate *ccn2* expression, a significant part of which information was obtained by recently-established genome-wide transcriptome analysis technology (Grotendorst *et al.*, 1996; Shimo *et al.*, 2001b; Xu *et al.*, 2004) (Fig. 4). Among those factors, the best-known upstream regulator of *ccn2* is TGF-β. Indeed, CCN2 was described as an immediate-early gene that was induced by TGF-β in early studies. Consistent with this initial proposal, a number of reports have described the involvement of TGF-β in the induction of *ccn2* gene expression in different types of cells and tissues (Eguchi *et al.*, 2001; Igarashi *et al.*, 1993; Kikuchi *et al.*, 1995). It is widely recognized that both TGF-β and CCN2 stimulate certain cells to deposit ECM components; hence CCN2 is considered to act directly on cells to accelerate ECM deposition in an autocrine/paracrine manner, under the control of TGF-β (Igarashi *et al.*, 1993).

Effects of inflammatory cytokines on *ccn2* expression are also of note. It was revealed that tumor necrosis factor-α (TNF-α), a quite common inflammatory cytokine, altered *ccn2* gene expression. That is, TNF-α repressed *ccn2* gene expression in skin fibroblasts, smooth muscle cells, and vascular endothelial cells (Abraham *et al.*, 2000; Lin *et al.*, 1998). CCN2 is required in the last phase of inflammation, which is, in other words, wound healing. Therefore, while TNF-α is actively promoting inflammation, *ccn2* expression may be still suppressed by this cytokine, awaiting the command to end the inflammatory events.

Steroid hormones comprise a major group of hormones that interact with intracellular receptors. Among them, glucocorticoids are also potent stimulators of *ccn2* expression at physiological doses. This effect was first described for fibroblasts (Dammeier *et al.*, 1998a). Recently, we found that dexamethasone strongly induced *ccn2* gene expression in chondrocytic cells as well. Moreover, it also promoted the maturation of chondrocytes at the same dose (Kubota *et al.*, 2003). Since the anti-inflammatory doses are far higher than those at which *ccn2* expression is induced, we may well discriminate this effect from the anti-inflammatory action of the hormone. At present, therefore, glucocorticoid ought to be regarded as one of the physiologic regulators in fibrous and cartilaginous tissues.

Not only hormones and cytokines, but also other local extracellular messengers and a variety of natural and artificial products are involved in the modulation of the *ccn2* gene expression. One of the autacoids, prostaglandin F2α, was found to drastically upregulate *ccn2* gene expression in a human embryonic kidney cell line (Liang *et al.*, 2003). In contrast, nitric oxide (NO) was reported to act as a negative regulator of *ccn2* expression (Keil *et al.*, 2002).

Effects of several artificial molecules on *ccn2* gene expression have been evaluated. The statins, which are 3-hydroxy-3-methyl glutaryl coenzyme A (HMG-CoA) reductase inhibitors and reduce *de novo* cholesterol synthesis, were shown to be an inducer of *ccn2* (Heusinger–Ribeiro *et al.*, 2004). Statins also act as osteogenic agents by interfering with isoprenylation of signaling molecules, which may account for such effects of statins on *ccn2*. A well-known and widely-used antioxidant, α-tocophenol, was described as an inducer of CCN2 in human smooth muscle cells as well (Villacorta *et al.*, 2003). Some of these molecules or their antagonists are already being applied clinically, and are promising candidates for CCN2-mediated regenerative therapeutics.

4.2.2. *Hypoxic and mechanical types of stress*

Under a stress load, specific groups of genes are induced to improve the intra-cellular and extracellular conditions of the stress-bearing region. In general, solid tumors undergo hypoxia along with tumor growth, which condition stimulates the tumor cells to produce angiogenic factors to obtain new blood vessels for fulfillment of their oxygen demand. When we exposed certain cell lines derived from solid tumors to hypoxia, both *ccn2* gene expression and CCN2 secretion were enhanced (Kondo *et al.*, 2002; Shimo *et al.*, 2001a) (Fig. 4). An increased mRNA level was distinctly observed relatively early after the beginning of hypoxic exposure, and then the effect subsided after 12 h. The observed increase in the steady-state mRNA level was realized by a post-transcriptional alteration causing stabilization of the *ccn2* mRNA. After this effect had subsided, post-translational control of CCN2 production supported the induction of CCN2 production during prolonged hypoxic exposure instead. Without an increase in the mRNA level, the cells continued to secrete more CCN2 molecules by releasing those stored, even after 72 h of continuous hypoxic exposure. The mechanism underlying the enhancement of CCN2 secretion is not clear; however, it is considered as a matricrine action mediated by certain proteases that degrade the ECM. In fact, hypoxic induction of certain MMPs was also described. It should also be noted that hypoxic induction of CCN2 was reported to occur in primary bovine chondrocytes (Grimshaw and Mason, 2001).

A mechanical load is another stressor that can provoke *ccn2* expression (Fig. 4). According to previous studies, tension loading of fibroblasts conferred reversible induction of *ccn2* (Schild and Trueb, 2004). Also, osteoblasts and osteocytes were found to express *ccn2* upon mechanical stress loads during experimental tooth movement (Yamashiro *et al.*, 2001). Such effects of mechanical stress on *ccn2* have been observed in chondrocytes under cyclic tensile strain as well. The signaling pathway from mechanical stress to *ccn2* gene is not exactly clear, except for the fact that a certain phosphorylation cascade is involved therein. Regarding this issue, it is also noteworthy that modification of the actin cytoskeleton strongly alters *ccn2* expression through Rho family signaling pathways (Ott *et al.*, 2003).

4.3. Genetic Elements and Their Intracellular Counterparts Controlling CCN2 Gene Expression

4.3.1. *Enhancers, silencers, and transcriptional factors*

The regulatory mechanism of the *ccn2* gene among the CCN family of genes has been relatively well characterized. To date, several *cis*-elements for the transcription control of *ccn2* have been described (Fig. 1). These elements are clustered within a 500-bp area upstream of the transcription initiation site. The most classical one is the basal control element 1 (BCE-1)/TGF-β response element (TbRE). BCE-1/TbRE was shown to be responsible for the basal promoter activity in mesangial cells, TGF-β responsiveness in fibroblasts, and endothelin-1 responsiveness in lung fibroblasts (Chen *et al.*, 2002; Grotendorst *et al.*, 1996). We also confirmed that BCE-1/TbRE also plays certain roles in the regulation of *ccn2* gene expression in chondrocytic cells. However, we found a more critical *cis*-regulatory element for the expression control in chondrocytes (Eguchi *et al.*, 2001). This element, designated TRENDIC, which stands for a <u>t</u>ranscriptional <u>en</u>hancer <u>d</u>ominant <u>i</u>n <u>c</u>hondrocytes, is located upstream of BCE-1/TbRE (Eguchi *et al.*, 2002) (Fig. 1). In most types of cells, TRENDIC enhances the gene expression, but the enhancing intensity is higher in chondrocytic cells. Between these 2 enhancers, a DNA segment with a Smad-binding consensus sequence (SBE) can be found. We confirmed that certain nuclear proteins bind specifically to SBE as well as to BCE-1/TbRE and TRENDIC. Also, the results of a recent study revealed that not only BCE-1/TbRE but also SBE played a role in supporting TGF-β responsiveness in mesangial cells (Chen *et al.*, 2002). Furthermore, apart from these juxtaposing enhancers, a tandem repeat of the TEF1-binding consensus sequence in the promoter area was recently investigated as well. Interestingly, this element was found to mediate the *ccn2* gene induction by TGF-β stimulation, despite its distinct identity from BCE-1/TbRE and SBE. Since TGF-β utilizes several distinct signal transduction pathways, involvement of different enhancers in its target gene might be plausible.

Characterization of most transcription factors that bind to the *cis*-elements mentioned above seems to be currently underway. At least, a functional contribution of Smads to the promoter activity has already been confirmed (Chen *et al.*, 2002). Association of AP-1 with the *ccn2* promoter segment was also indicated, albeit little information is available as to its binding target (Moritani *et al.*, 2003a) Contrarily, although a putative Sp-1-binding element can be found adjacent to the TATA box, this element appears to have no functional significance (Chen *et al.*, 2002).

The intracellular secondary messengers to transduce extracellular signals to *ccn2* are also under investigation. At present, Rho family members, MEK/ERK, and PKC have been shown to be involved in *ccn2* regulation (Ott *et al.*, 2003; Xu *et al.*, 2004). However, owing to the diversity of *ccn2*-modulating molecules, involvement of a vast number of signal transduction pathways is anticipated, and it may be quite difficult to summarize the entire events in a single illustration.

4.3.2. *post-transcriptional regulatory elements and their mechanism of action*

The *ccn2* mRNAs are characterized by their long 3′-untranslated regions (UTRs) (Fig. 4). It is widely recognized that 3′-UTRs of a variety of mRNAs contain critical genetic elements to determine the fate of their own mRNAs. When the nucleotide sequences of the 3′-UTRs in *ccn2* genes were compared among human, mouse, and chicken species, mutual homology was not high. However, surprisingly, when the 3′-UTRs were linked to the 3′-end of a reporter mRNA, all of them from different species remarkably repressed the reporter gene expression at similar levels (Kondo *et al.*, 2000; Kubota *et al.*, 1999; Mukudai *et al.*, 2003). Such functional conservation without structural conservation indicates crucial roles of the 3′-UTR for the regulation of *ccn2* gene expression among vertebrates. It should be noted that regulatory elements with different structures require different cell biological backgrounds to confer similar functionality. Indeed, the 3′-UTR of chicken *ccn2* displays strong *cis*-repressive effects on gene expression in chicken cells, whereas it shows absolutely no repressive effects in a monkey cell line. Similarly, human *ccn2* 3′-UTR acts as a potential repressive element, not in chicken cells, but in human and monkey cells. Subsequent studies to locate the critical post-transcriptional regulatory elements in the 3′-UTRs from different species have revealed several distinct elements therein (Mukudai *et al.*, 2003). The first post-transcriptional *cis*-element in the *ccn2* 3′-UTR was discovered in the human mRNA (Kubota *et al.*, 2000b). This element was originally located at the junction of the coding region and 3′-UTR and acted as a repressive element at either the 5′ or 3′ end of a reporter mRNA, whereas it displayed no such effect at the upstream of the promoter in a reporter construct. Moreover, the basal repressive effect was found to be realized without a change in the steady-state mRNA level. Therefore, this element is an RNA regulatory element acting at post-transcriptional stages. Of importance, this 84-base RNA driver of gene expression was observed to form a stable secondary structure, and subsequent mutational analyses clearly demonstrated the requirement of the RNA secondary structure in its repressive function. Based on the findings above, we designated this element as *cis*-acting element of structure-anchored repression (CAESAR) (Kubota *et al.*, 2000b) (Fig. 4).

Most recent studies suggest that the repressive effect of CAESAR may depend on the binding of a specific protein. Therefore, under particular conditions, the fate of CAESAR-containing *ccn2* mRNA may be modulated via alteration in the interaction between CAESAR and its protein counterpart. In fact, our recent study indicated that the 3′-UTR of chicken *ccn2* mRNA differentially regulated the *ccn2* expression at a post-transcriptional stage in such a manner (Mukudai *et al.*, 2005). However, in contrast to the human *ccn2*, the location of the post-transcriptional element was mapped in a small portion near the polyadenyl tail, representing the functional conservation without structural conservation. In the case of murine *ccn2*, the CAESAR orthologue was identified at the same position as in the human gene, which is consistent with closer evolutionary distance between the two (Kondo *et al.*, 2000; Kubota *et al.*, 2000b). Identification and characterization of CAESAR-binding proteins will provide us with more insight into the mechanism of the CAESAR-mediated gene regulatory system.

5. CONCLUSIONS AND OUTLOOK

In this chapter, we described CCN2 as a multifunctional growth/ differentiation factor for fibroblasts, chondrocytes, osteoblasts, and vascular endothelial cells, as well as for mesenchymal stem cells. Depending on the type of cell with which it interacts, it promotes chemotaxis, migration, adhesion, proliferation, differentiation and/or extracellular matrix formation. These cell/tissue specific functions may be due to interaction with other molecules such as growth factors and ECM components. Because of the very high expression of the CCN2 gene in hypertrophic chondrocytes in the physiological state, a major physiological role for CCN2 has been suggested to be the promotion of endochondral ossification. However, accumulating data that show its gene expression in various other types of cells including undifferentiated mesenchymal cells during embryonic development and during would healing of bone, cartilage, and skin, as well as its action on mesenchymal stem cells, indicate the involvement of CCN2 in embryonic development and tissue regeneration. In fact, recombinant CCN2 regenerated damaged cartilage by acting on chondrocytes and bone marrow-derived mesenchymal stem cells. From this point of view, CCN2 should be thought of as a regeneration factor, which may be called "regenerin."

In the case of organs/tissues composed of parenchymal cells and a connective tissue framework, the balance between regeneration of parenchymal cells and that of connective tissue framework is important for proper regeneration of the organs/tissues. Although CCN2 is known to be involved in fibrotic disorders and angiogenic diseases, these phenomena are characterized by overregeneration of connective tissue than that of parenchymal cells in the organs/tissues. Therefore, even taking into consideration these pathological findings, CCN2 should be defined as a regeneration factor rather than a pathological factor, although it should be controlled properly. Moreover, coordinate proliferation and differentiation of mesenchymal cells and parenchymal/epithelial cells is indispensable for proper development and growth of various organs/tissues. In this point of view, controling the action of CCN2 and its gene expression is critical for proper development and regeneration of these organs/tissues. As described in this chapter, much progress in understanding the molecular regulation of CCN2 has been made recently, but even more is still required for this purpose. The discovery of additional molecules that mediate, transduce and modify the actions and signaling pathways of CCN2 and control its gene expression will attract the interest of the scientific community at large as well as be beneficial to applied medicine. The delivery of these molecules, or their modifiers as well as CCN2 itself, would be therapeutically beneficial for the promotion of properly controlled tissue regeneration as well as for preventing pathological changes such as fibrotic disorders and diseases associated with defective CCN2 if such exist.

ACKNOWLEDGMENTS

This work was supported in part by grants from the program Grants-in-Aid for Scientific Research (S) (to M.T.) and (C) (to S.K.), for Exploratory Research (to M.T.) and for Young Scientists (B) (to T.N.) of the Ministry of Education, Science, Sports, and Culture of Japan. The authors wish to thank Drs. Tohru Nakanishi, Tomoichiro Yamaai, and other faculty members and graduate students who collaborate with us at Okayama University Graduate School of Medicine and Dentistry and Drs. Toshihiro Kushibiki and Yasuhiko Tabata at Kyoto University Graduate School of Medicine. We also give thanks to Ms. Yuki Nonami for technical and secretarial assistance.

REFERENCES

Abou–Shady M, Friess H, Zimmermann A, di Mola FF, Guo XZ, Baer HU and Buchler MW (2000) Connective tissue growth factor in human liver cirrhosis. *Liver* **20**:296–304.

Abraham DJ, Shiwen X, Black CM, Sa S, Xu Y and Leask A (2000) Tumor necrosis factor α suppresses the induction of connective tissue growth factor by transforming growth factor-β in normal and scleroderma fibroblasts. *J Biol Chem* **275**: 15220–15225.

Abreu JG, Ketpura NI, Reversade B and De Robertis EM (2002) Connective-tissue growth factor (CTGF) modulates cell signalling by BMP and TGF-β. *Nat Cell Biol* **4**:599–604.

Arikawa–Hirasawa E, Watanabe H, Takami H, Hassell J and Yamada Y (1999) Perlecan is essential for cartilage and cephalic development. *Nat Genet* **23**:354–358.

Babic AM, Chen CC and Lau LF (1999) Fisp12/mouse connective tissue growth factor mediates endothelial cell adhesion and migration through integrin $\alpha v\beta 3$, promotes endothelial cell survival, and induces angiogenesis *in vivo*. *Mol Cell Biol* **19**:2958–2966.

Blalock TD, Duncan MR, Varela JC, Goldstein MH, Tuli SS, Grotendorst GR and Schultz GS (2003) Connective tissue growth factor expression and action in human corneal fibroblast cultures and rat corneas after photorefractive keratectomy. *Invest Ophthalmol Vis Sci* **44**:1879–1887.

Bork P (1993) The modular architecture of a new family of growth regulators related to connective tissue growth factor. *FEBS Lett* **327**:125–130.

Bradham DM, Igarashi A, Potter RL and Grotendorst GR (1991) Connective tissue growth factor a cysteine-rich mitogen secreted by human vascular endothelial cells is related to the SRC-induced immediate early gene product CEF-10. *J Cell Biol* **114**:1285–1294.

Brigstock DR (1999) The connective tissue growth factor/cysteine-rich 61/nephroblastoma overexpressed (CCN) family. *Endocr Rev* **20**:189–206.

Brigstock DR (2003) The CCN family: a new stimulus package. *J Endocrinol* **178**: 169–175.

Brigstock DR, Steffen CL, Kim GY, Vegunta RK, Diehl JR and Harding PA (1997) Purification and characterization of novel heparin-binding growth factors in uterine secretory fluids. Identification as heparin-regulated Mr 10,000 forms of connective tissue growth factor. *J Biol Chem* **272**:20275–20282.

Brunner A, Chinn J, Neubauer M and Purchio AF (1991) Identification of a gene family regulated by transforming growth factor β. *DNA Cell Biol* **10**:293–300.

Buckwalter JA and Mankin H (1998) Articular cartilage: tissue design and chondrocyte-matrix interactions. *Instr Course Lect* **47**:477–486.

Cancedda R, Cancedda FD and Castagnola P (1995) Chondrocyte differentiation. *Int Rev Cyt* **159**:265–358.

Cash DE, Gates PB, Imokawa Y and Brockes JP (1998) Identification of newt connective tissue growth factor as a target of retinoid regulation in limb blastemal cells. *Gene* **222**:119–124.

Chen CC, Chen N and Lau LF (2001) The angiogenic factors Cyr61 and connective tissue growth factor induce adhesive signaling in primary human skin fibroblasts. *J Biol Chem* **276**:10443–10452.

Chen Y, Blom IE, Sa S, Goldschmeding R, Abraham DJ and Leask A (2002) CTGF expression in mesangial cells: involvement of SMADs, MAP kinase, and PKC. *Kidney Int* **62**:1149–1159.

Cicha I, Garlichs CD, Daniel WG and Goppelt–Struebe M (2004) Activated human platelets release connective tissue growth factor. *Thromb Haemost* **91**:755–760.

Croci S, Landuzzi L, Astolfi A, Nicoletti G, Rosolen A, Sartori F, Follo MY, Oliver N, De Giovanni C, Nanni P and Lollini PL (2004) Inhibition of connective tissue growth factor (CTGF/CCN2) expression decreases the survival and myogenic differentiation of human rhabdomyosarcoma cells. *Cancer Res* **64**:1730–1736.

Dammeier J, Beer HD, Brauchle M and Werner S (1998a) Dexamethasone is a novel potent inducer of connective tissue growth factor expression. Implications for glucocorticoid therapy. *J Biol Chem* **273**:18185–18190.

Dammeier J, Brauchle M, Falk W, Grotendorst GR and Werner S (1998b) Connective tissue growth factor: a novel regulator of mucosal repair and fibrosis in inflammatory bowel disease? *Int J Biochem Cell Biol* **30**:909–922.

Eguchi T, Kubota S, Kondo S, Kuboki T, Yatani H and Takigawa M (2002) A novel *cis*-element that enhances connective tissue growth factor gene expression in chondrocytic cells. *Biochem Biophys Res Commun* **295**:445–451.

Eguchi T, Kubota S, Kondo S, Shimo T, Hattori T, Nakanishi T, Kuboki T, Yatani H and Takigawa M (2001) Regulatory mechanism of human connective tissue growth factor (CTGF/Hcs24) gene expression in a human chondrocytic cell line, HCS-2/8. *J Biochem (Tokyo)* **130**:79–87.

Enomoto H, Enomoto–Iwamoto M, Iwamoto M, Nomura S, Himeno M, Kitamura Y, Kishimoto T and Komori T (2000) Cbfa1 is a positive regulatory factor in chondrocyte maturation. *J Biol Chem* **275**:8695–8702.

Enomoto M and Takigawa M (1992) Regulation of tumor derived and immortalized chondrocytes, in *Biological Regulation of the Chondrocytes*, Ed. M. Adolphe (CRC Press, Boca Raton), pp. 321–338.

Frazier K, Williams S, Kothapalli D, Klapper H and Grotendorst GR (1996) Stimulation of fibroblast cell growth, matrix production, and granulation tissue formation by connective tissue growth factor. *J Invest Dermatol* **107**:404–411.

Frazier KS and Grotendorst GR (1997) Expression of connective tissue growth factor mRNA in the fibrous stroma of mammary tumors. *Int J Biochem Cell Biol* **29**: 153–161.

Gao R and Brigstock DR (2004) Connective tissue growth factor (CCN2) induces adhesion of rat activated hepatic stellate cells by binding of its C-terminal domain to integrin $\alpha v\beta 3$ and heparan sulfate proteoglycan. *J Biol Chem* **279**:8848–8855.

Grimshaw MJ and Mason RM (2001) Modulation of bovine articular chondrocyte gene expression *in vitro* by oxygen tension. *Osteoarthritis Cartilage* **9**:357–364.

Grotendorst GR, Okochi H and Hayashi N (1996) A novel transforming growth factor β response element controls the expression of the connective tissue growth factor gene. *Cell Growth Differ* **7**:469–480.

Gupta N, Wang H, McLeod TL, Naus CC, Kyurkchiev S, Advani S, Yu J, Perbal B and Weichselbaum RR (2001) Inhibition of glioma cell growth and tumorigenic potential by CCN3 (NOV). *Mol Pathol* **54**:293–299.

Hashimoto G, Inoki I, Fujii Y, Aoki T, Ikeda E and Okada Y (2002) Matrix metalloproteinases cleave connective tissue growth factor and reactivate angiogenic activity of vascular endothelial growth factor 165. *J Biol Chem* **277**:36288–36295.

Herz J and Strickland DK (2001) LRP: a multifunctional scavenger and signaling receptor. *J Clin Invest* **108**:779–784.

Heuer H, Christ S, Friedrichsen S, Brauer D, Winckler M, Bauer K and Raivich G (2003) Connective tissue growth factor: a novel marker of layer VII neurons in the rat cerebral cortex. *Neuroscience* **119**:43–52.

Heusinger–Ribeiro J, Fischer B and Goppelt–Struebe M (2004) Differential effects of simvastatin on mesangial cells. *Kidney Int* **66**:187–195.

Hinton DR, Spee C, He S, Weitz S, Usinger W, LaBree L, Oliver N and Lim JI (2004) Accumulation of NH2-terminal fragment of connective tissue growth factor in the vitreous of patients with proliferative diabetic retinopathy. *Diabetes Care* **27**: 758–764.

Hishikawa K, Oemar BS, Tanner FC, Nakaki T, Fujii T and Luscher TF (1999a) Overexpression of connective tissue growth factor gene induces apoptosis in human aortic smooth muscle cells. *Circulation* **100**:2108–2112.

Hishikawa K, Oemar BS, Tanner FC, Nakaki T, Luscher TF and Fujii T (1999b) Connective tissue growth factor induces apoptosis in human breast cancer cell line MCF-7. *J Biol Chem* **274**:37461–37466.

Hunziker EB (1994) Mechanism of longitudinal bone growth and its regulation by growth plate chondrocytes. *Microsc Res Tech* **28**:505–519.

Igarashi A, Hayashi N, Nashiro K and Takehara K (1998) Differential expression of connective tissue growth factor gene in cutaneous fibrohistiocytic and vascular tumors. *J Cutan Pathol* **25**:143–148.

Igarashi A, Nashiro K, Kikuchi K, Sato S, Ihn H, Fujimoto M, Grotendorst GR and Takehara K (1996) Connective tissue growth factor gene expression in tissue sections from localized scleroderma, keloid, and other fibrotic skin disorders. *J Invest Dermatol* **106**:729–733.

Igarashi A, Nashiro K, Kikuchi K, Sato S, Ihn H, Grotendorst GR and Takehara K (1995) Significant correlation between connective tissue growth factor gene expression and skin sclerosis in tissue sections from patients with systemic sclerosis. *J Invest Dermatol* **105**:280–284.

Igarashi A, Okochi H, Bradham DM and Grotendorst GR (1993) Regulation of connective tissue growth factor gene expression in human skin fibroblasts and during wound repair. *Mol Biol Cell* **4**:637–645.

Inada M, Yasui T, Nomura S, Miyake S, Deguchi K, Himeno M, Sato M, Yamagiwa H, Kimura T, Yasui N, Ochi T, Endo N, Kitamura Y, Kishimoto T and Komori T (1999)

Maturational disturbance of chondrocytes in Cbfa1-deficient mice. *Dev Dynam* **214**:279–290.

Inoki I, Shiomi T, Hashimoto G, Enomoto H, Nakamura H, Makino K, Ikeda E, Takata S, Kobayashi K and Okada Y (2002) Connective tissue growth factor binds vascular endothelial growth factor (VEGF) and inhibits VEGF-induced angiogenesis. *Faseb J* **16**:219–221.

Ito Y, Aten J, Bende RJ, Oemar BS, Rabelink TJ, Weening JJ and Goldschmeding R (1998) Expression of connective tissue growth factor in human renal fibrosis. *Kidney Int* **53**:853–861.

Ivkovic S, Yoon BS, Popoff SN, Safadi FF, Libuda DE, Stephenson RC, Daluiski A and Lyons KM (2003) Connective tissue growth factor coordinates chondrogenesis and angiogenesis during skeletal development. *Development* **130**:2779–2791.

Jedsadayanmata A, Chen CC, Kireeva ML, Lau LF and Lam SC (1999) Activation-dependent adhesion of human platelets to Cyr61 and Fisp12/mouse connective tissue growth factor is mediated through integrin $\alpha IIb\beta 3$. *J Biol Chem* **274**:24321–24327.

Jin M, Chen Y, He S, Ryan SJ and Hinton DR (2004) Hepatocyte growth factor and its role in the pathogenesis of retinal detachment. *Invest Ophthalmol Vis Sci* **45**:323–329.

Kadota H, Nakanishi T, Asaumi K, Yamaai T, Nakata E, Mitani S, Aoki K, Aiga A, Inoue H and Takigawa M (2003) Expression of connective tissue growth factor/hypertrophic chondrocyte-specific gene product 24 (CTGF/Hcs24/CCN2) during distraction osteogenesis. *J Bone Miner Metab* **22**:293–302.

Kang Y, Siegel PM, Shu W, Drobnjak M, Kakonen SM, Cordon–Cardo C, Guise TA and Massague J (2003) A multigenic program mediating breast cancer metastasis to bone. *Cancer Cell* **3**:537–549.

Kanyama M, Kuboki T, Akiyama K, Nawachi K, Miyauchi FM, Yatani H, Kubota S, Nakanishi T and Takigawa M (2003) Connective tissue growth factor expressed in rat alveolar bone regeneration sites after tooth extraction. *Arch Oral Biol* **48**:723–730.

Kawaki H, Kubota S, Minato M, Moritani NH, Hattori T, Hanagata H, Kubota M, Miyauchi A, Nakanishi T and Takigawa M (2003) Novel enzyme-linked immunosorbent assay systems for the quantitative analysis of connective tissue growth factor (CTGF/Hcs24/CCN2): detection of HTLV-I tax-induced CTGF from a human carcinoma cell line. *DNA Cell Biol* **22**:641–648.

Keil A, Blom IE, Goldschmeding R and Rupprecht HD (2002) Nitric oxide down-regulates connective tissue growth factor in rat mesangial cells. *Kidney Int* **62**:401–411.

Kikuchi K, Kadono T, Ihn H, Sato S, Igarashi A, Nakagawa H, Tamaki K and Takehara K (1995) Growth regulation in scleroderma fibroblasts: increased response to transforming growth factor-β 1. *J Invest Dermatol* **105**:128–132.

Kireeva ML, Latinkic BV, Kolesnikova TV, Chen CC, Yang GP, Abler AS and Lau LF (1997) Cyr61 and Fisp12 are both ECM-associated signaling molecules: activities, metabolism, and localization during development. *Exp Cell Res* **233**:63–77.

Komori T, Yagi H, Nomura S, Yamaguchi A, Sasaki K, Deguchi K, Shimizu Y, Bronson RT, Gao Y-H, Inada M, Sato M, Okamoto R, Kitamura Y, Yoshiki S and Kishimoto T (1997) Target disruption of Cbfa1 results in a complete lack of bone formation owing to maturational arrest of osteoblasts. *Cell* **89**:755–764.

Kondo S, Kubota S, Eguchi T, Hattori T, Nakanishi T, Sugahara T and Takigawa M (2000) Characterization of a mouse ctgf 3'-UTR segment that mediates repressive regulation of gene expression. *Biochem Biophys Res Commun* **278**:119–124.

Kondo S, Kubota S, Shimo T, Nishida T, Yosimichi G, Eguchi T, Sugahara T and Takigawa M (2002) Connective tissue growth factor increased by hypoxia may initiate angiogenesis in collaboration with matrix metalloproteinases. *Carcinogenesis* **23**:769–776.

Kondo Y, Nakanishi T, Takigawa M and Ogawa N (1999) Immunohistochemical localization of connective tissue growth factor in the rat central nervous system. *Brain Res* **834**:146–151.

Kubo M, Kikuchi K, Nashiro K, Kakinuma T, Hayashi N, Nanko H and Tamaki K (1998) Expression of fibrogenic cytokines in desmoplastic malignant melanoma. *Br J Dermatol* **139**:192–197.

Kubota S, Eguchi T, Shimo T, Nishida T, Hattori T, Kondo S, Nakanishi T and Takigawa M (2001) Novel mode of processing and secretion of connective tissue growth factor/ecogenin (CTGF/Hcs24) in chondrocytic HCS-2/8 cells. *Bone* **29**:155–161.

Kubota S, Hattori T, Nakanishi T and Takigawa M (1999) Involvement of *cis*-acting repressive element(s) in the 3'-untranslated region of human connective tissue growth factor gene. *FEBS Lett* **450**:84–88.

Kubota S, Hattori T, Shimo T, Nakanishi T and Takigawa M (2000a) Novel intracellular effects of human connective tissue growth factor expressed in Cos-7 cells. *FEBS Lett* **474**:58–62.

Kubota S, Kawata K, Yanagita T, Doi H, Kitoh T and Takigawa M (2004) Abundant retention and release of connective tissue growth factor (CTGF/CCN2) by platelets, which promotes wound healing and tissue regeneration. *J Biochem* **136**:279–282.

Kubota S, Kondo S, Eguchi T, Hattori T, Nakanishi T, Pomerantz RJ and Takigawa M (2000b) Identification of an RNA element that confers post-transcriptional repression of connective tissue growth factor/hypertrophic chondrocyte specific 24 *(ctgf/hcs24)* gene: similarities to retroviral RNA-protein interactions. *Oncogene* **19**:4773–4785.

Kubota S, Moritani NH, Kawaki H, Mimura H, Minato M and Takigawa M (2003) Transcriptional induction of connective tissue growth factor/hypertrophic chondrocyte-specific 24 gene by dexamethasone in human chondrocytic cells. *Bone* **33**:694–702.

Lau LF and Lam SC (1999) The CCN family of angiogenic regulators: the integrin connection. *Exp Cell Res* **248**:44–57.

Leask A, Holmes A and Abraham DJ (2002) Connective tissue growth factor: a new and important player in the pathogenesis of fibrosis. *Curr Rheumatol Rep* **4**:136–142.

Liang Y, Li C, Guzman VM, Evinger AJ, 3rd, Protzman CE, Krauss AH and Woodward DF (2003) Comparison of prostaglandin F2α, bimatoprost (prostamide), and butaprost (EP2 agonist) on Cyr61 and connective tissue growth factor gene expression. *J Biol Chem* **278**:27267–27277.

Lin J, Liliensiek B, Kanitz M, Schimanski U, Bohrer H, Waldherr R, Martin E, Kauffmann G, Ziegler R and Nawroth PP (1998) Molecular cloning of genes differentially regulated by TNF-α in bovine aortic endothelial cells, fibroblasts and smooth muscle cells. *Cardiovasc Res* **38**:802–813.

Mercurio S, Latinkic B, Itasaki N, Krumlauf R and Smith JC (2004) Connective-tissue growth factor modulates WNT signalling and interacts with the WNT receptor complex. *Development* **131**:2137–2147.

Mo FE, Muntean AG, Chen CC, Stolz DB, Watkins SC and Lau LF (2002) CYR61 (CCN1) is essential for placental development and vascular integrity. *Mol Cell Biol* **22**:8709–8720.

Mori T, Kawara S, Shinozaki M, Hayashi N, Kakinuma T, Igarashi A, Takigawa M, Nakanishi T and Takehara K (1999) Role and interaction of connective tissue growth factor with transforming growth factor-β in persistent fibrosis: a mouse fibrosis model. *J Cell Physiol* **181**:153–159.

Moritani NH, Kubota S, Eguchi T, Fukunaga T, Yamashiro T, Takano–Yamamoto T, Tahara H, Ohyama K, Sugahara T and Takigawa M (2003a) Interaction of AP-1 and the *ctgf* gene: a possible driver of chondrocyte hypertrophy in growth cartilage. *J Bone Miner Metab* **21**:205–210.

Moritani NH, Kubota S, Nishida T, Kawaki H, Kondo S, Sugahara T and Takigawa M (2003b) Suppressive effect of overexpressed connective tissue growth factor on tumor cell growth in a human oral squamous cell carcinoma-derived cell line. *Cancer Lett* **192**:205–214.

Moussad EE, Rageh MA, Wilson AK, Geisert RD and Brigstock DR (2002) Temporal and spatial expression of connective tissue growth factor (CCN2: CTGF) and transforming growth factor β type 1 (TGF-β1) at the utero-placental interface during early pregnancy in the pig. *Mol Pathol* **55**:186–192.

Mukudai Y, Kubota S, Eguchi T, Kondo S and Takigawa M (2005) Regulation of chicken *ccn2* gene by interaction between RNA *cis*-element and *trans*-factor during differentiation of chondrocytes. *J Biol Chem* **280**:3166–3177.

Mukudai Y, Kubota S and Takigawa M (2003) Conserved repressive regulation of connective tissue growth factor/hypertrophic chondrocyte-specific gene 24 (*ctgf/hcs24*) enabled by different elements and factors among vertebrate species. *Biol Chem* **384**:1–9.

Murphy M, Godson C, Cannon S, Kato S, Mackenzie HS, Martin F and Brady HR (1999) Suppression subtractive hybridization identifies high glucose levels as a stimulus for expression of connective tissue growth factor and other genes in human mesangial cells. *J Biol Chem* **274**:5830–5834.

Nakanishi T, Kimura Y, Tamura T, Ichikawa H, Yamaai Y, Sugimoto T and Takigawa M (1997) Cloning of a mRNA preferentially expressed in chondrocytes by differential display-PCR from a human chondrocytic cell line that is identical with connective tissue growth factor (CTGF) mRNA. *Biochem Biophys Res Commun* **234**: 206–210.

Nakanishi T, Nishida T, Shimo T, Kobayashi K, Kubo T, Tamatani T, Tezuka K and Takigawa M (2000) Effects of CTGF/Hcs24, a product of a hypertrophic chondrocyte-specific gene, on the proliferation and differentiation of chondrocytes in culture. *Endocrinology* **141**:264–273.

Nakanishi T, Yamaai T, Asano M, Nawachi K, Suzuki M, Sugimoto T and Takigawa M (2001) Overexpression of connective tissue growth factor/hypertrophic chondrocyte-specific gene product 24 decreases bone density in adult mice and induces dwarfism. *Biochem Biophys Res Commun* **281**:678–681.

Nakata E, Nakanishi T, Kawai A, Asaumi K, Yamaai T, Asano M, Nishida T, Mitani S, Inoue H and Takigawa M (2002) Expression of connective tissue growth factor/hypertrophic chondrocyte-specific gene product 24 (CTGF/Hcs24) during fracture healing. *Bone* **31**:441–447.

Nawachi K, Inoue M, Kubota S, Nishida T, Yosimichi G, Nakanishi T, Kanyama M, Kuboki T, Yatani H, Yamaai T and Takigawa M (2002) Tyrosine kinase-type receptor ErbB4 in chondrocytes: interaction with connective tissue growth factor and distribution in cartilage. *FEBS Lett* **528**:109–113.

Nishida T, Kubota S, Fukunaga T, Kondo S, Yosimichi G, Nakanishi T, Takano–Yamamoto T and Takigawa M (2003) CTGF/Hcs24, hypertrophic chondrocyte-specific gene product, interacts with perlecan in regulating the proliferation and differentiation of chondrocytes. *J Cell Physiol* **196**:265–275.

Nishida T, Kubota S, Kojima S, Kuboki T, Nakao K, Kushibiki T, Tabata Y and Takigawa M (2004) Regeneration of defects in articular cartilage in rat knee joints by CCN2 (connective tissue growth factor). *J Bone Miner Res* **19**:1308–1319.

Nishida T, Kubota S, Nakanishi T, Kuboki T, Yosimichi G, Kondo S and Takigawa M (2002) CTGF/Hcs24, a hypertrophic chondrocyte-specific gene product, stimulates proliferation and differentiation, but not hypertrophy of cultured articular chondrocytes. *J Cell Physiol* **192**:55–63.

Nishida T, Nakanishi T, Asano M, Shimo T and Takigawa M (2000) Effects of CTGF/Hcs24, a hypertrophic chondrocyte-specific gene product, on the proliferation and differentiation of osteoblastic cells *in vitro*. *J Cell Physiol* **184**: 197–206.

Nishida T, Nakanishi T, Shimo T, Asano M, Hattori T, Tamatani T, Tezuka K and Takigawa M (1998) Demonstration of receptors specific for connective tissue growth factor on a human chondrocytic cell line (HCS-2/8). *Biochem Biophys Res Commun* **247**:905–909.

Oemar BS and Luscher TF (1997) Connective tissue growth factor. Friend or foe? *Arterioscler Thromb Vasc Biol* **17**:1483–1489.

Oemar BS, Werner A, Garnier JM, Do DD, Godoy N, Nauck M, Marz W, Rupp J, Pech M and Luscher TF (1997) Human connective tissue growth factor is expressed in advanced atherosclerotic lesions. *Circulation* **95**:831–839.

Omoto S, Nishida K, Yamaai Y, Shibahara M, Nishida T, Doi T, Asahara H, Nakanishi T, Inoue H and Takigawa M (2004) Expression and localization of connective tissue growth factor (CTGF/Hcs24/CCN2) in osteoarthritic cartilage. *Osteoarthritis Cartilage* **12**:771–778.

Ott C, Iwanciw D, Graness A, Giehl K and Goppelt–Struebe M (2003) Modulation of the expression of connective tissue growth factor by alterations of the cytoskeleton. *J Biol Chem* **278**:44305–44311.

Pan LH, Beppu T, Kurose A, Yamauchi K, Sugawara A, Suzuki M, Ogawa A and Sawai T (2002) Neoplastic cells and proliferating endothelial cells express connective tissue growth factor (CTGF) in glioblastoma. *Neurol Res* **24**:677–683.

Paradis V, Dargere D, Vidaud M, De Gouville AC, Huet S, Martinez V, Gauthier JM, Ba N, Sobesky R, Ratziu V and Bedossa P (1999) Expression of connective tissue growth factor in experimental rat and human liver fibrosis. *Hepatology* **30**:968–976.

Pawar S, Kartha S and Toback FG (1995) Differential gene expression in migrating renal epithelial cells after wounding. *J Cell Physiol* **165**:556–565.

Perbal B (2004) CCN proteins: multifunctional signalling regulators. *Lancet* **363**:62–64.

Riser BL and Cortes P (2001) Connective tissue growth factor and its regulation: a new element in diabetic glomerulosclerosis. *Ren Fail* **23**:459–470.

Riser BL, Denichilo M, Cortes P, Baker C, Grondin JM, Yee J and Narins RG (2000) Regulation of connective tissue growth factor activity in cultured rat mesangial cells and its expression in experimental diabetic glomerulosclerosis. *J Am Soc Nephrol* **11**:25–38.

Ryseck RP, Macdonald–Bravo H, Mattei MG and Bravo R (1991) Structure, mapping, and expression of *fisp-12*, a growth factor-inducible gene encoding a secreted cysteine-rich protein. *Cell Growth Differ* **2**:225–233.

Safadi FF, Xu J, Smock SL, Kanaan RA, Selim AH, Odgren PR, Marks SC, Jr., Owen TA and Popoff SN (2003) Expression of connective tissue growth factor in bone: its role in osteoblast proliferation and differentiation *in vitro* and bone formation *in vivo*. *J Cell Physiol* **196**:51–62.

Sato S, Nagaoka T, Hasegawa M, Tamatani T, Nakanishi T, Takigawa M and Takehara K (2000) Serum levels of connective tissue growth factor are elevated in patients with systemic sclerosis: association with extent of skin sclerosis and severity of pulmonary fibrosis. *J Rheumatol* **27**:149–154.

Schild C and Trueb B (2004) Three members of the connective tissue growth factor family CCN are differentially regulated by mechanical stress. *Biochim Biophys Acta* **1691**:33–40.

Schober JM, Chen N, Grzeszkiewicz TM, Jovanovic I, Emeson EE, Ugarova TP, Ye RD, Lau LF and Lam SC (2002) Identification of integrin $\alpha_M\beta_2$ as an adhesion receptor on peripheral blood monocytes for Cyr61 (CCN1) and connective tissue growth factor (CCN2): immediate-early gene products expressed in atherosclerotic lesions. *Blood* **99**:4457–4465.

Segarini PR, Nesbitt JE, Li D, Hays LG and Yates JR, 3rd and Carmichael DF (2001) The low density lipoprotein receptor-related protein/α2-macroglobulin receptor is a receptor for connective tissue growth factor. *J Biol Chem* **276**:40659–40667.

Shakunaga T, Ozaki T, Ohara N, Asaumi K, Doi T, Nishida K, Kawai A, Nakanishi T, Takigawa M and Inoue H (2000) Expression of connective tissue growth factor in cartilaginous tumors. *Cancer* **89**:1466–1473.

Shimo T, Kubota S, Kondo S, Nakanishi T, Sasaki A, Mese H, Matsumura T and Takigawa M (2001a) Connective tissue growth factor as a major angiogenic agent that is induced by hypoxia in a human breast cancer cell line. *Cancer Lett* **174**:57–64.

Shimo T, Nakanishi T, Kimura Y, Nishida T, Ishizeki K, Matsumura T and Takigawa M (1998) Inhibition of endogenous expression of connective tissue growth factor by its antisense oligonucleotide and antisense RNA suppresses proliferation and migration of vascular endothelial cells. *J Biochem (Tokyo)* **124**:130–140.

Shimo T, Nakanishi T, Nishida T, Asano M, Kanyama M, Kuboki T, Tamatani T, Tezuka K, Takemura M, Matsumura T and Takigawa M (1999) Connective tissue growth factor induces the proliferation, migration, and tube formation of vascular endothelial cells *in vitro*, and angiogenesis *in vivo*. *J Biochem (Tokyo)* **126**: 137–145.

Shimo T, Nakanishi T, Nishida T, Asano M, Sasaki A, Kanyama M, Kuboki T, Matsumura T and Takigawa M (2001b) Involvement of CTGF, a hypertrophic chondrocyte-specific gene product, in tumor angiogenesis. *Oncology* **61**:315–322.

Shimo T, Wu C, Billings PC, Piddington R, Rosenbloom J, Pacifici M and Koyama E (2002) Expression, gene regulation, and roles of Fisp12/CTGF in developing tooth germs. *Dev Dyn* **224**:267–278.

Steffen CL, Ball–Mirth DK, Harding PA, Bhattacharyya N, Pillai S and Brigstock DR (1998) Characterization of cell-associated and soluble forms of connective tissue growth factor (CTGF) produced by fibroblast cells *in vitro*. *Growth Factors* **15**: 199–213.

Surveyor GA and Brigstock DR (1999) Immunohistochemical localization of connective tissue growth factor (CTGF) in the mouse embryo between days 7.5 and 14.5 of gestation. *Growth Factors* **17**:115–124.

Surveyor GA, Wilson AK and Brigstock DR (1998) Localization of connective tissue growth factor during the period of embryo implantation in the mouse. *Biol Reprod* **59**:1207–1213.

Takigawa M (2003) CTGF/Hcs24 as a multifunctional growth factor for fibroblasts, chondrocytes and vascular endothelial cells. *Drug News Perspect* **16**:11–21.

Takigawa M, Nakanishi T, Kubota S and Nishida T (2003) Role of CTGF/Hcs24/ ecogenin in skeletal growth control. *J Cell Physiol* **194**:256–266.

Takigawa M, Pan H-O, Kinoshita A, Tajima K and Takano Y (1991) Establishment from a human chondrosarcoma of a new immortal cell line with high tumorigenicity *in vivo*, which is able to form proteoglycan-rich cartilage-like nodules and to respond to insulin *in vitro*. *Int J Cancer* **48**:717–725.

Takigawa M, Tajima K, Pan H-O, Enomoto M, Kinoshita A, Suzuki F, Takano Y and Mori Y (1989) Establishment of a clonal human chondrosarcoma cell line with cartilage phenotypes. *Cancer Res* **49**:3996–4002.

Twigg SM, Cao Z, SV MC, Burns WC, Brammar G, Forbes JM and Cooper ME (2002) Renal connective tissue growth factor induction in experimental diabetes is prevented by aminoguanidine. *Endocrinology* **143**:4907–4915.

Villacorta L, Graca–Souza AV, Ricciarelli R, Zingg JM and Azzi A (2003) α-tocopherol induces expression of connective tissue growth factor and antagonizes tumor necrosis factor-α-mediated downregulation in human smooth muscle cells. *Circ Res* **92**:104–110.

Wahab NA, Brinkman H and Mason RM (2001a) Uptake and intracellular transport of the connective tissue growth factor: a potential mode of action. *Biochem J* **359**:89–97.

Wahab NA, Yevdokimova N, Weston BS, Roberts T, Li XJ, Brinkman H and Mason RM (2001b) Role of connective tissue growth factor in the pathogenesis of diabetic nephropathy. *Biochem J* **359**:77–87.

Wandji SA, Gadsby JE, Barber JA and Hammond JM (2000) Messenger ribonucleic acids for MAC25 and connective tissue growth factor (CTGF) are inversely regulated during folliculogenesis and early luteogenesis. *Endocrinology* **141**:2648–2657.

Wenger C, Ellenrieder V, Alber B, Lacher U, Menke A, Hameister H, Wilda M, Iwamura T, Beger HG, Adler G and Gress TM (1999) Expression and differential regulation of connective tissue growth factor in pancreatic cancer cells. *Oncogene* **18**:1073–1080.

Wong M, Kireeva ML, Kolesnikova TV and Lau LF (1997) Cyr61, product of a growth factor-inducible immediate-early gene, regulates chondrogenesis in mouse limb bud mesenchymal cells. *Dev Biol* **192**:492–508.

Xie D, Nakachi K, Wang H, Elashoff R and Koeffler HP (2001) Elevated levels of connective tissue growth factor, WISP-1, and CYR61 in primary breast cancers associated with more advanced features. *Cancer Res* **61**:8917–8923.

Xu J, Smock SL, Safadi FF, Rosenzweig AB, Odgren PR, Marks SC, Jr., Owen TA and Popoff SN (2000) Cloning the full-length cDNA for rat connective tissue growth factor: implications for skeletal development. *J Cell Biochem* **77**:103–115.

Xu SW, Howat SL, Renzoni EA, Holmes A, Pearson JD, Dashwood MR, Bou–Gharios G, Denton CP, du Bois RM, Black CM, Leask A and Abraham DJ (2004) Endothelin-1 induces expression of matrix-associated genes in lung fibroblasts through MEK/ERK. *J Biol Chem* **279**:23098–23103.

Yamaai T, Asano M, Nakanishi T, Nishida T, Yosimichi G, Hattori T, Sugimoto T and Takigawa M (1999) Gene expression pattern of cartilage-derived growth factor CTGF/Hcs24 during mouse development. *Jpn J Oral Biol* **41**:464 (in Japanese).

Yamaai T, Nakanishi T, Asano M, Nawachi K, Yosimichi G, Ohyama K, Komori T, Sugimoto T and Takigawa M (2005) Gene expression of connective tissue growth factor (CTGF/CCN2) in calcifying tissues of normal and *cbfa1*-null mutant in late stage of mouse embryonic development. *J Bone Miner Metab*, in press.

Yamashiro T, Fukunaga T, Kobashi N, Kamioka H, Nakanishi T, Takigawa M and Takano–Yamamoto T (2001) Mechanical stimulation induces CTGF expression in rat osteocytes. *J Dent Res* **80**:461–465.

Yosimichi G, Kubota S, Hattori T, Nishida T, Nawachi K, Nakanishi T, Kamada M, Takano–Yamamoto T and Takigawa M (2002) CTGF/Hcs24 interacts with the cytoskeletal protein actin in chondrocytes. *Biochem Biophys Res Commun* **299**: 755–761.

Yosimichi G, Kubota S, Nishida T, Kondo S, Nakao K, Takano–Yamamoto T and Takigawa M (2005) Differential roles of PKC and PI3K in multiple transduction of CCN2/CTGF signals to downstream MAP kinases in chondrocytes. *J Cell Physiol*, submitted.

Yosimichi G, Nakanishi T, Nishida T, Hattori T, Takano–Yamamoto T and Takigawa M (2001) CTGF/Hcs24 induces chondrocyte differentiation through a p38 mitogen-activated protein kinase (p38MAPK), and proliferation through a p44/42 MAPK/extracellular-signal regulated kinase (ERK). *Eur J Biochem* **268**:6058–6065.

CHAPTER 3

INTEGRIN-MEDIATED CCN FUNCTIONS

Lester F. Lau* and Stephen C.-T. Lam[†]

*Department of Biochemistry and Molecular Genetics

and

[†]Department of Pharmacology

University of Illinois at Chicago College of Medicine

900 South Ashland Avenue, Chicago, IL 60607

Known activities of CCN proteins in isolated cell systems can be attributed to their interaction with integrin receptors, with cell surface heparan sulfate proteoglycans (HSPGs) serving as coreceptors in some contexts. Recent studies have identified specific binding sites on CCN proteins for integrin $\alpha_V\beta_3$ and the $\alpha_6\beta_1$-HSPG coreceptors, and mutations at these binding sites impair activities mediated through their cognate receptors. These findings provide compelling evidence that CCNs act through their binding to integrins and HSPGs, and provide a strategy for dissecting their integrin-mediated functions *in vivo*

1. CCN PROTEINS ARE MATRICELLULAR SIGNALING MOLECULES

CCN proteins are capable of mediating diverse cellular and biological functions. In isolated cell systems, they regulate cell adhesion, migration, proliferation, survival, differentiation, and gene expression. In the organism, CCNs function in angiogenesis and embryonic vascular development, as well as skeletal development and homeostasis. Recent studies suggest that the regulation of angiogenesis and matrix remodeling by CCNs may underlie or contribute to their roles in tissue repair and disease. Through what mechanism do they act to regulate such a multitude of biological processes? Critical to understanding how CCNs function mechanistically is the identification of the cell surface receptors that transduce their signals and mediate their activities. Although CCNs were once thought to act as growth factors, no classical growth factor receptors have been identified for any of them to date.

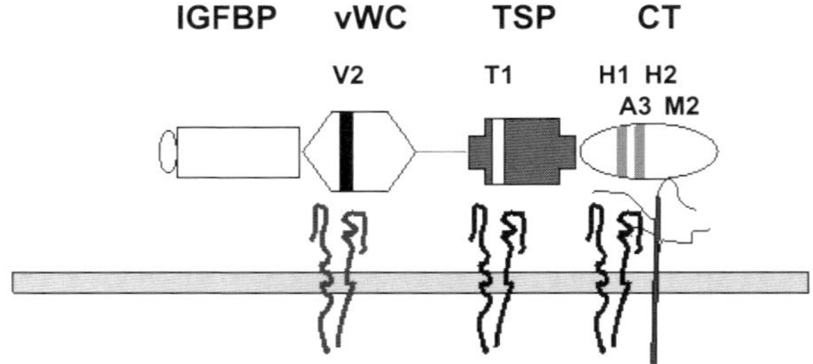

Fig. 1.   Receptor binding sites of CCN1. A schematic diagram represents the domain structure of CCNs, showing the locations and sequences of the binding sites for various receptors as indicated. All sequences are for CCN1 except A3, an $\alpha_V \beta_3$ binding site in CCN2. See text for references.

Bornstein and colleagues first recognized a novel class of proteins characterized by the presence of conserved domains from extracellular matrix (ECM) proteins (Bornstein, 1995; Bornstein and Sage, 2002). These "matricellular" proteins have features intermediate between growth factors, cytokines, and ECM molecules, such that they regulate diverse cellular functions without serving as structural components of the ECM. Archetypical examples of matricellular proteins include thrombospondins, SPARC, and osteopontin. CCNs can be considered matricellular proteins for several reasons. First, CCNs are ECM-associated proteins comprised of conserved domains from ECM molecules including the von Willebrand factor type C repeat (vWC), the thrombospondin type I repeat (TSP), and a C-terminus (CT) with similarity to Slit and mucins (Bork, 1993; Fig. 1). Second, as discussed below, CCNs are direct ligands of integrin receptors, which mediate ECM protein functions. Third, the dynamic expression of CCNs makes them unlikely to serve in a structural capacity, whereas their regulatory functions in modulating cell-cell and cell-matrix interactions are consistent with those of matricellular proteins. The recognition of CCNs as matricellular signaling molecules has provided a conceptual framework in which testable hypotheses regarding their mechanism of action can be developed (Lau and Lam, 1999).

2. INTEGRINS AS BIDIRECTIONAL SIGNALING RECEPTORS

Integrins form a widely distributed family of heterodimeric cell surface receptors consisting of non-covalently linked α and β subunits, both of which are type I transmembrane proteins. The extracellular domains of both subunits consist of large globular structures containing the ligand binding sites. A single transmembrane domain is present in each subunit, followed by a short cytoplasmic sequence. The non-random pairing of at least 18 α and 8 β subunits constitutes 24 known integrin $\alpha\beta$ heterodimers. Among the many ligands of integrins are large adhesive proteins in the ECM or in the plasma (e.g. collagen, fibronectin, laminin, thrombospondin, von Willebrand factor, fibrinogen and vitronectin), reflecting their roles in cell adhesion and migration. Aside from large structural ECM proteins, an increasing number of small matricellular proteins, including CCNs, are found to regulate cellular functions through interaction with integrins (Lau and Lam, 1999; Jain *et al.*, 2002). The prototypic integrin recognition sequence, Arg-Gly-Asp (RGD), is present in a number of adhesive proteins (e.g. fibronectin, fibrinogen, vitronectin, and von Willebrand factor), and RGD-containing peptides competitively inhibit ligand binding to some integrins (reviewed in Ruoslahti, 1996). However, it should be noted that not all integrins are RGD-sensitive and not all integrin ligands contain the RGD sequence. In addition to cell adhesion and migration, ligation of integrins to appropriate matrix proteins offers protection from apoptosis, and provides a permissive context in which cell proliferation, migration, and differentiation can occur. These cellular activities form the basis of integrin functions in diverse biological processes, such as embryogenesis, angiogenesis, wound healing, immune response, and hemostasis (reviewed in Bouvard *et al.*, 2001; Hood and Cheresh, 2002; Bökel and Brown, 2002; Giancotti and Tarone, 2003).

Extensive studies over the last two decades have established integrins as bidirectional signaling receptors capable of receiving and transducing environmental cues, activating signaling cascades, and evoking diverse cellular responses (reviewed in Hynes, 2002; Miranti and Brugge, 2002). Most integrins are present in an inactive state on resting cells with a low ligand binding affinity. Upon cellular activation (e.g. stimulation by growth factors or phorbol esters), "inside-out" signaling induces a conformational change in the integrin extracellular domain, converting it to a high affinity receptor for extracellular ligands. Although the precise regulatory mechanism for this activation process is not known, signaling enzymes including phospholipase C, phosphatidylinositol kinases, members of the Ras superfamily of GTPase (reviewed in Shattil, 1999; Xiong *et al.*, 2003), and physical separation of the integrin cytoplasmic domains have been implicated in this integrin "inside-out" signaling (Kim *et al.*, 2003a). Following binding to extracellular ligands, integrins initiate "outside-in" signaling processes that regulate myriad cellular functions, including cytoskeletal reorganization and morphological changes, alteration of gene expression, and activation of intracellular kinase cascades that converge with growth factor receptor signaling (reviewed in Giancotti and Ruoslahti, 1999; Miranti and Brugge, 2002; Schwartz and Ginsberg, 2002; Juliano, 2002). In this regard, the Src-family kinases are rapidly activated upon integrin-ligand interaction, resulting in the phosphorylation and activation of focal adhesion kinase (FAK). These events lead to the activation of many downstream pathways, including: Ras/Raf/Erk, which regulate cell differentiation and proliferation; PI3K/Akt, which promotes cell survival; and Rho/Rac/cdc42, which control cytoskeletal changes that mediate cell adhesion and motility. Thus, integrin "outside-in" signaling can act independently or converge with growth factor receptor signaling to elicit synergistic activation of cell function.

3. ACTIVITIES OF CCNS AND RECEPTORS THAT MEDIATE THEM

A number of CCN activities have been described in isolated cell systems, where analysis of signaling receptors is most tractable. In most cases, these activities can be attributed to the direct binding of CCNs to specific integrin receptors. To date, purified CCN1 and CCN2 have been shown to bind at least five distinct integrins, including $\alpha_6\beta_1$, $\alpha_v\beta_3$, $\alpha_v\beta_5$, $\alpha_{IIb}\beta_3$, and $\alpha_M\beta_2$ (Kireeva *et al.*, 1998; Chen *et al.*, 2000; Grzeszkiewicz *et al.*, 2001; Jedsadayanmata *et al.*, 1999; Babic *et al.*, 1999; Chen *et al.*, 2001a; Schober *et al.*, 2002; Leu *et al.*, 2003; Grzeszkiewicz *et al.*, 2002; Gao and Brigstock, 2004; Table 1). CCN3 binds $\alpha_6\beta_1$, $\alpha_v\beta_3$, $\alpha_v\beta_5$, as well as $\alpha_5\beta_1$, although its potential binding to $\alpha_{IIb}\beta_3$ and $\alpha_M\beta_2$ has not been tested (Lin *et al.*, 2003; Lin *et al.*, 2005; Ellis *et al.*, 2003). In each case, direct binding of the above integrins to CCNs has been established by solid phase binding assays, and their functional roles in mediating distinct CCN activities have been demonstrated by inhibition with

Table 1. Integrins shown to interact with CCNs

Protein	Integrin receptors
CCN1	$\alpha_6\beta_1$, $\alpha_v\beta_3$, $\alpha_v\beta_5$, $\alpha_{IIb}\beta_3$, $\alpha_M\beta_2$
CCN2	$\alpha_6\beta_1$, $\alpha_v\beta_3$, $\alpha_v\beta_5$, $\alpha_{IIb}\beta_3$, $\alpha_M\beta_2$
CCN3	$\alpha_6\beta_1$, $\alpha_v\beta_3$, $\alpha_v\beta_5$, $\alpha_5\beta_1$

specific integrin antagonists. In some contexts, cell surface heparan sulfate proteoglycans (HSPGs) serve as coreceptors with integrins (Chen *et al.*, 2000; Chen *et al.*, 2001a; Grzeszkiewicz *et al.*, 2002; Schober *et al.*, 2002; Gao and Brigstock, 2004). The utilization of distinct integrins by CCNs is both cell type- and context-dependent. For example, the expression and utilization of $\alpha_{IIb}\beta_3$ and $\alpha_M\beta_2$ are specific for platelets and leukocytes, respectively. However, CCN1 utilization of integrins in fibroblasts is function-dependent, and it supports cell adhesion, induces cell migration, and promotes cell proliferation through $\alpha_6\beta_1$, $\alpha_v\beta_5$, and $\alpha_v\beta_3$, respectively (Grzeszkiewicz *et al.*, 2001).

3.1. Cell Adhesion

Early observation showed that upon synthesis, CCN1 is secreted and associates with the ECM and the cell surface, largely through interaction with HSPGs (Yang and Lau, 1991). Likewise, secreted CCN2 is partitioned between the culture medium and the ECM/cell surface (Kireeva *et al.*, 1997). These studies suggested that CCNs may modulate cell-matrix interaction. Indeed, CCN1, CCN2, CCN3 and CCN5 have been shown to support adhesion of myriad cell types, including vascular endothelial cells and smooth muscle cells, fibroblasts, platelets, monocytes, osteoblasts, lung epithelial cells, and hepatic stellate cells (Kireeva *et al.*, 1996; Kireeva *et al.*, 1997; Jedsadayanmata *et al.*, 1999; Kumar *et al.*, 1999; Chen *et al.*, 2001a; Grzeszkiewicz *et al.*, 2002; Schober *et al.*, 2002; Ellis *et al.*, 2003; Gao and Brigstock, 2004; Table 2). The conclusion that cell adhesion to CCNs occurs through integrins is supported by several lines of

Table 2. Summary of integrin receptors that mediate distinct CCN1 activities in different cell types. ECs, endothelial cells; HSPGs, heparan sulfate proteoglycans; NA, not applicable

	Adhesion	Migration	Proliferation	Survival	Tubule formation
Fibroblasts	$\alpha_6\beta_1$ + HSPGs	$\alpha_v\beta_5$	$\alpha_v\beta_3$		NA
Unactivated ECs	$\alpha_6\beta_1$ + HSPGs				$\alpha_6\beta_1$
Activated ECs	$\alpha_v\beta_3$	$\alpha_v\beta_3$	$\alpha_v\beta_3$	$\alpha_v\beta_3$	$\alpha_v\beta_3$
Smooth muscle cells	$\alpha_6\beta_1$ + HSPGs	$\alpha_6\beta_1$ + HSPGs			NA
Platelets	$\alpha_{IIb}\beta_3$	NA	NA		NA
Monocytes	$\alpha_M\beta_2$				NA

evidence. First, adhesion to CCNs is divalent cation-dependent and enhanced by Mn^{2+}, which activates integrins directly by binding to the extracellular domain. Second, inhibitory monoclonal antibodies specific for the integrins in question abolished cell adhesion to CCNs. Furthermore, even though CCNs do not contain the RGD motif, CCN-supported cell adhesion is blocked by RGD-containing peptides where RGD-sensitive integrins are involved. Third, in solid phase binding assays, purified CCN proteins bind directly to the specific integrins that mediate their function.

In fibroblasts and aortic smooth muscle cells, cell adhesion to CCN1, CCN2, and CCN3 is mediated through integrin $\alpha_6\beta_1$ with HSPGs as core-ceptors (Chen *et al.*, 2000; Chen *et al.*, 2001a; Grzeszkiewicz *et al.*, 2002; Lin *et al.*, 2004a). The requirement for HSPGs is established by the observation that cell adhesion is obliterated upon either: i) saturation of the heparin binding site in CCNs with soluble heparin; ii) destruction of cell surface HSPGs by heparinase or inhibition of proteoglycan sulfation; or iii) mutation of the heparin binding sites (Chen *et al.*, 2000). The $\alpha_6\beta_1$-HSPG coreceptors also support the adhesion of unactivated human umbilical vein endothelial cells (HUVECs) to CCN1, whereas activated HUVECs adhere primarily through $\alpha_v\beta_3$ (Leu *et al.*, 2002). Thus, CCN1 is an activation-dependent ligand of $\alpha_v\beta_3$ and an activation-independent ligand of $\alpha_6\beta_1$. In hepatic stellate cells, CCN2 supports adhesion through integrin $\alpha_v\beta_3$ (Gao and Brigstock, 2004). In addition, peripheral blood monocytes and platelets also adhere to CCN proteins in an activation-dependent manner through integrin $\alpha_M\beta_2$ and $\alpha_{IIb}\beta_3$, respectively (Jedsadayanmata *et al.*, 1999; Schober *et al.*, 2002).

3.2. Cell Signaling

Consistent with their adhesive functions through integrins, CCN proteins induce adhesive signaling in cells adherent to them. Fibroblast adhesion to CCN1 or CCN2 occurs through $\alpha_6\beta_1$-HSPGs, resulting in the formation of $\alpha_6\beta_1$-containing focal adhesion complexes. Adherent cells spread and form filopodia and lamellipodia, pseudopods that allow the cells to sense chemotactic agents and to migrate. Concomitantly, outside-in integrin signaling results in the activation of FAK, paxillin, Rac, and Erk1/Erk2, culminating in the upregulation of matrix metalloproteinase (MMP)-1 and -3 (Chen *et al.*, 2001a). Alteration of gene expression is also observed in fibroblasts treated with soluble CCN1 (Chen *et al.*, 2001b), which upregulates the expression of genes perti-nent to processes in wound repair, including: i) angiogenesis and lymphogenesis (VEGF-A and VEGF-C); ii) inflammation (interleukin-1β); iii) extracellular matrix remodeling (MMP-1, MMP-3, TIMP1, uPA, and PAI-1); and

iv) cell-matrix interaction (integrin α_3 and α_5 subunits). CCN1 also downregulates type 1 collagen $\alpha 1$ expression. Likewise, CCN2 has been shown to upregulate types I and III collagen, fibromodulin, and basic fibroblast growth factor, and TIMP-1, -2. -3 and -4 in porcine fibroblasts (Wang *et al.*, 2003). Additionally, CCN2 causes an AP-2-dependent upregulation of MMP-2 in vascular smooth muscle cells (Fan and Karnovsky, 2002). In human mesangial cells, CCN2 induces a transient activation of the Erk1/2 and PI3K/Akt pathways through $\alpha_v\beta_3$, leading to upregulation of fibronectin and integrin α_1 (Crean *et al.*, 2002). Together, these studies suggest that CCNs may exert their biological function in cell proliferation and extracellular matrix remodeling through induction of integrin outside-in signaling and regulation of gene expression.

3.3. Cell Migration

Purified CCN1, CCN2, and CCN3 stimulate the migration of fibroblasts, vascular endothelial cells, vascular smooth muscle cells and mesangial cells (Kireeva *et al.*, 1996; Babic *et al.*, 1998; Babic *et al.*, 1999; Shimo *et al.*, 1999; Lin *et al.*, 2003; Crean *et al.*, 2004), whereas overexpression of CCN4 and CCN5 inhibits cell migration (Soon *et al.*, 2003; Lake *et al.*, 2003). Where examined, CCNs function as chemotactic factors (inducing directional cell movement), although CCN2 is both chemotactic and chemokinetic (inducing random cell movement) in microvascular endothelial cells (Babic *et al.*, 1999). Whereas $\alpha_v\beta_3$ mediates the migration of vascular endothelial cells to CCN1, CCN2, and CCN3, $\alpha_5\beta_1$ is also involved in endothelial cell migration to CCN3 (Lin *et al.*, 2003). In contrast, CCN1 stimulates fibroblast and vascular smooth muscle cell migration through $\alpha_v\beta_5$ and $\alpha_6\beta_1$-HSPGs, respectively (Grzeszkiewicz *et al.*, 2001; Grzeszkiewicz *et al.*, 2002). Migration of human mesangial cells to CCN2 is inhibited by anti-α_1 and anti-β_3 antibodies, suggesting the involvement of both integrins $\alpha_1\beta_1$ and $\alpha_v\beta_3$ (Crean *et al.*, 2002).

3.4. Cell Proliferation

Earlier studies of human CCN2 suggested that it may act as a mitogenic growth factor (Bradham *et al.*, 1991; Igarashi *et al.*, 1996), although work on purified mouse CCN2, mouse and human CCN1, and human CCN3 showed that these proteins enhance DNA synthesis induced by mitogenic growth factors but are unable to induce DNA synthesis on their own (Kireeva *et al.*, 1996; Kireeva *et al.*, 1997; Kolesnikova and Lau, 1998; Leu *et al.*, 2002; Lin *et al.*, 2004a). Recent studies show that human CCN2 also requires the presence of EGF to induce DNA synthesis, having no mitogenic activity by itself (Grotendorst *et al.*, 2004). In fibroblasts and endothelial cells, CCNs act through $\alpha_v\beta_3$ to promote growth factor-induced cell proliferation (Grzeszkiewicz *et al.*, 2001; Leu *et al.*, 2002; Lin *et al.*, 2005), consistent with their ability to induce integrin-mediated signals that converge with growth factor receptor signaling pathways. A truncated CCN1 mutant lacking the CT domain retains the ability to enhance bFGF-induced DNA synthesis in human skin fibroblasts (Grzeszkiewicz *et al.*, 2001), whereas the C-terminal 16–20 kDa fragments of CCN2 are unable to augment DNA synthesis in porcine fibroblasts (Wang *et al.*, 2003). These findings suggest the presence of an $\alpha_v\beta_3$ binding site within the first three domains of CCN proteins that is critical for enhancing DNA synthesis (see Section 4.2 below).

3.5. Cell Survival

Cell adhesion to ECM molecules, through α_v integrins in particular, can promote cell survival in many cell types via activation of the PI3K/Akt pathway, whereas cells detached from their matrices are prone to death by anoikis (reviewed in Frisch and Screaton, 2001; Frisch and Ruoslahti, 1997; Stupack and Cheresh, 2002). Activated endothelial cells adhered to CCN1 or CCN2 through integrin $\alpha_v\beta_3$ are protected from cell death upon growth factor withdrawal (Babic *et al.*, 1999; Leu *et al.*, 2002). In contrast, adhesion of endothelial cells to ECM proteins (e.g. laminin) through $\alpha_6\beta_1$ results in a high rate of apoptotic death when depleted of growth factors (Wary *et al.*, 1996; Leu *et al.*, 2002). CCN1 confers resistance to apoptosis in MCF-7 breast cancer cells via upregulation of the anti-apoptotic protein XIAP in a NF-κB-dependent manner (Lin *et al.*, 2004). Expression of CCN1 in glioma cells activates the PI3K/Akt pathway, resulting in phosphorylation and inhibition of Bad (Xie *et al.*, 2004). Likewise, CCN2 also activates the PI3K/Akt pathway in mesangial cells through integrin $\alpha_v\beta_3$ (Crean *et al.*, 2002). CCN4 protects cancer cells from p53-dependent apoptosis following DNA damage (Su *et al.*, 2002). In this context, CCN4 activates Akt and upregulates the anti-apoptotic protein Bcl-X_L. Thus, CCN proteins can promote cell survival in a variety of cell types and conditions. Surprisingly, CCN2 has been reported to induce or mediate apoptosis in MCF-7 breast cancer cells, vascular smooth muscle cells, and human mesangial cells (Hishikawa *et al.*, 1999b; Hishikawa *et al.*, 1999a; Hishikawa *et al.*, 2001). In MCF-7 cells, CCN2 induces a reduction of Bcl-2 expression without affecting Bax expression (Hishikawa *et al.*, 1999b). In addition, CCN1 expression has been associated with neuronal cell death (Kim *et al.*, 2003b). The receptor and signaling mechanism responsible for the pro-apoptotic activity of CCNs are currently unknown. While it is apparent that CCNs have both pro- and anti-apoptotic activities, their roles in regulating cell survival under various physiological conditions are still poorly understood.

3.6. Differentiation

Both CCN1 and CCN2 have been shown to promote cell differentiation *in vitro*. In a collagen gel assay, CCN1 and CCN2 can induce vascular endothelial cell differentiation into vessel-like tubular structures (Shimo *et al.*, 1999; Leu *et al.*, 2002). In activated and unactivated HUVECs, CCN1 induces tubule formation through $\alpha_v\beta_3$ and $\alpha_6\beta_1$, respectively, although a much higher concentration of CCN1 is required in unactivated cells (Leu *et al.*, 2002). CCN1 also induces chondrogenic differentiation in mouse limb bud mesenchymal cells in micromass culture, although the receptor requirement for this process has not been examined (Wong *et al.*, 1997). CCN2 promotes the proliferation and differentiation of HCS-2/8 chondrocytic cells, and induces matrix calcification of rabbit growth cartilage cells in culture (Nakanishi *et al.*, 2000; Yosimichi *et al.*, 2001). Interestingly, CCN2 can be cross linked to a protein complex on HCS-2/8 cells with a combined molecular size of integrin α and β subunits (Nishida *et al.*, 1998); however, the identity of this putative CCN2 receptor on HCS-2/8 cells remains to be defined.

3.7. Angiogenesis

CCN1, CCN2, and CCN3 induce angiogenesis *in vivo*, as demonstrated by the corneal micropocket implant assay (Babic *et al.*, 1998; Babic *et al.*, 1999; Lin *et al.*, 2003). In addition, CCN1 and CCN2 have been found to induce neovascularization in the rabbit ischemic hind limb assay and the chick chorioallantoic membrane assay, respectively (Shimo *et al.*, 1999; Fataccioli *et al.*, 2002). Overexpression of CCN1 in cancer cells enhances tumor growth and tumor vascular density (Babic *et al.*, 1998; Menendez *et al.*, 2003). Gene targeting studies revealed defects in embryonic vessel formation in *Ccn1* and *Ccn2* deficient mice, thus establishing their biological roles in developmental angiogenesis (Mo *et al.*, 2002; Ivkovic *et al.*, 2003). CCNs exert their pro-angiogenic activities in vascular endothelial cells largely through integrin receptors. In activated endothelial cells, CCN1 supports adhesion, stimulates migration, enhances DNA synthesis, promotes survival, and induces tubule formation, all mediated through integrin $\alpha_v\beta_3$ (Table 2: Leu *et al.*, 2002). In unactivated endothelial cells, CCN1-induced tubule formation is inhibited by the anti-α_6 monoclonal antibody GoH3 and by the $\alpha_6\beta_1$-binding T1 peptide (Leu *et al.*,

2002; Leu *et al.*, 2003). Likewise, CCN2 also supports microvascular endothelial cell adhesion, stimulates migration, and promotes DNA synthesis through integrin $\alpha_v\beta_3$ (Babic *et al.*, 1999). CCN3 supports endothelial cell adhesion through integrins $\alpha_6\beta_1$, $\alpha_v\beta_3$ and $\alpha_5\beta_1$, and stimulates migration through integrins $\alpha_v\beta_3$ and $\alpha_5\beta_1$ (Lin *et al.*, 2003). Inasmuch as $\alpha_v\beta_3$, $\alpha_v\beta_5$, and β_1 integrins have been shown to be involved in angiogenesis (reviewed in Hynes *et al.*, 2002), the proangiogenic activities of CCN proteins through integrin receptors are consistent with established role of integrins in this process.

4. RECEPTOR BINDING SITES AND SPECIFIC INTEGRIN-BINDING DEFECTIVE CCN1 MUTANTS

As summarized above, most, if not all, CCN activities in isolated cell systems are mediated through integrins, which bind CCNs directly. Recent studies have begun to identify binding sites for specific integrins in CCNs. Targeted mutagenesis demonstrates that disruption of these sites impairs activities mediated through the cognate integrin receptor specifically, providing compelling evidence that these CCN activities are indeed mediated through direct interaction with specific integrins.

4.1. Binding Sites for $\alpha_6\beta_1$ and HSPGs

Integrin $\alpha_6\beta_1$ and cell surface HSPGs act as coreceptors to mediate some CCN functions. Two consensus glycosaminoglycan binding sequences have been identified in the CT domain (Brigstock *et al.*, 1997; Chen *et al.*, 2000). Substitution of the basic residues in these sequences with alanines abolished the heparin binding activity in CCN1, confirming that these basic residues constitute the heparin binding site (Chen *et al.*, 2000). A novel binding site for $\alpha_6\beta_1$, T1 (GQKCIVQTTSWSQCSKS), has been identified in the TSP domain of CCN1 (Leu *et al.*, 2003; Fig. 1). Synthetic T1 peptide supports $\alpha_6\beta_1$-mediated cell adhesion as an immobilized substrate and inhibits $\alpha_6\beta_1$-dependent cell function as a soluble competitor. Further, T1 peptide-coupled matrix affinity purifies $\alpha_6\beta_1$ from an octylglucoside extract of fibroblasts, demonstrating that T1 is a *bona fide* $\alpha_6\beta_1$ binding site (Leu *et al.*, 2003). In addition, two other peptides, H1 (KGKKCSKTKKSPEPVR) and H2 (FTYAGCSSVKKYRPKY), in the CT domain of CCN1 also support $\alpha_6\beta_1$-HSPG-dependent cell adhesion, which can be inhibited by the anti-α_6 monoclonal antibody GoH3, soluble heparin, or heparinase treatment of cells (Leu *et al.*, 2004). The H1 and H2 sequences subsume the basic residues shown to be critical for heparin binding, suggesting that the sites for interaction with $\alpha_6\beta_1$ and HSPGs in these sequences are closely juxtaposed or overlapping.

Thus, three sites are involved in $\alpha_6\beta_1$-HSPG interaction in CCN1: T1, H1, and H2 (Fig. 1). However, these sites are functionally nonequivalent. Cells adhered to T1 (binding only $\alpha_6\beta_1$) results in rapid and transient activation of Erk1/Erk2, whereas cells adhered to H1 or H2 peptides (binding both $\alpha_6\beta_1$ and HSPGs) induce prolonged activation of Erk1/Erk2, similar to CCN1 (Leu *et al.*, 2004). To dissect their functions, each of these binding sites has been mutated in the context of full-length CCN1. As described above, alanine substitutions of the basic residues in the consensus heparin binding sites abolished heparin binding activity (Chen *et al.*, 2000). Surprisingly, disruption of the T1 sequence did not affect $\alpha_6\beta_1$-HSPG-mediated activities significantly, indicating that T1 is not the primary binding site mediating these activities. By contrast, disruption of H1 and H2 rendered CCN1 largely inactive in $\alpha_6\beta_1$-HSPG-mediated activities, and disruption of all these sites (T1, H1, and H2) completely abolished $\alpha_6\beta_1$-HSPG-dependent functions (Leu *et al.*, 2004). Importantly, all mutants affecting T1, H1/H2, or all three sites are fully active in all $\alpha_v\beta_3$-mediated pro-angiogenic activities. These findings show that the mutants are biologically active and are deficient only in $\alpha_6\beta_1$-HSPG-specific functions, and that $\alpha_v\beta_3$-mediated pro-angiogenic activities are independent of binding to heparin or $\alpha_6\beta_1$.

4.2. Binding Sites for Integrin $\alpha_v\beta_3$

CCNs do not contain the RGD sequence motif recognized by some integrins, including $\alpha_v\beta_3$. Nevertheless, CCN1, CCN2, and CCN3 bind $\alpha_v\beta_3$ directly as demonstrated by solid phase binding assays (Kireeva *et al.*, 1998; Lin *et al.*, 2003; Gao and Brigstock, 2004), and $\alpha_v\beta_3$ can be affinity purified from a CCN1-coupled matrix (Kireeva *et al.*, 1998). Analysis of the CT domain of CCN2 identified a novel, non-canonical $\alpha_v\beta_3$ binding site (IRTPKISKPIK-FELSG) that supports adhesion of rat hepatic stellate cells (Gao and Brigstock, 2004). This sequence is not well conserved among CCNs, although sequences of similar charge are found. A truncation mutant of CCN1 lacking the CT domain can enhance DNA synthesis through $\alpha_v\beta_3$ (Grzeszkiewicz *et al.*, 2001), and CCN5, which lacks the CT domain, competes with fibrinogen for binding to $\alpha_v\beta_3$ (Kumar *et al.*, 1999). These findings suggest that other $\alpha_v\beta_3$ binding sites are present within the first three domains.

Recent studies uncovered another novel $\alpha_v\beta_3$ binding site, V2 (NCK-HQCTCIDGAVGCIPLCP), in the vWC domain of CCN1 (Chen *et al.*, 2004).

The V2 sequence is well conserved in all CCNs except CCN6. Synthetic V2 peptide supports $\alpha_v\beta_3$-mediated cell adhesion when immobilized, and in soluble form inhibits endothelial cell adhesion to $\alpha_v\beta_3$ ligands such as CCN1, CCN2, and CCN3 but not to collagen, a ligand of β_1 integrins. An aspartate is present in the recognition of many $\alpha_v\beta_3$ ligands, and consistently, mutation of the aspartate in V2 obliterated the ability of the peptide to support cell adhesion (Chen *et al.*, 2004).

A single a.a. substitution CCN1 mutant (D125A) defective in the V2 binding site for $\alpha_v\beta_3$ has been constructed and analyzed (Chen *et al.*, 2004). Interestingly, as shown by direct solid phase binding assays, full-length D125A CCN1 mutant is defective in binding $\alpha_v\beta_3$ when it is in the soluble form, whereas the immobilized protein exposes a cryptic site for interaction with $\alpha_v\beta_3$. Consistently, this D125A mutant shows impaired $\alpha_v\beta_3$-dependent activity when assayed in the soluble form (induction of endothelial cell migration and enhancement of growth factor-induced DNA synthesis), but nevertheless supports $\alpha_v\beta_3$-dependent cell adhesion when immobilized. Again, this mutant is fully active in mediating $\alpha_6\beta_1$-HSPG-dependent CCN1 activities, showing that the $\alpha_v\beta_3$- and $\alpha_6\beta_1$-mediated activities can be dissociated.

4.3. Binding Site for Integrin $\alpha_M\beta_2$

Activated monocytes adhere to both CCN1 and CCN2 through the leukocyte-specific integrin $\alpha_M\beta_2$ (Schober *et al.*, 2002). A binding site for $\alpha_M\beta_2$ (SSVKKYRPKYCGS) has been identified in the CT domain of CCN1 (Schober *et al.*, 2003; Fig. 1). Mutation of the basic residues in this sequence in full-length CCN1 did not abolish the ability to support monocyte adhesion or binding to the integrin α_M ligand recognizing I domain, indicating that the basic residues in this sequence are not critical for $\alpha_M\beta_2$ binding (Schober *et al.*, 2002).

5. FUTURE QUESTIONS

Although a number of integrins have been identified to be signaling receptors for CCNs, several major questions remain. First, the detailed signaling pathways that emanate from CCN-integrin interactions have not been clearly defined. A few of the intracellular signaling molecules activated through this interaction have been identified, yet the signaling mechanisms that lead to various biological outcomes have not been fully delineated. In particular, what distinguishes CCN signaling from that of other ligands that bind the same integrins, and thereby conferring biological specificity, is unknown. Elucidating these signaling mechanisms will be an important line of inquiry.

While CCN activities in isolated cell systems are demonstrably mediated through integrins, the specific roles of CCN-integrin interaction in the organismal context, such as vascular and skeletal development, tissue repair, and disease conditions are yet unestablished. It is likely that other mechanisms also contribute to CCN functions *in vivo*. The recent identification of integrin binding sites and construction of specific integrin binding-defective CCN mutants will allow the physiological roles integrin- and non-integrin mediated actions to be dissected *in vivo* using gene replacement or transgenic approaches (Chen *et al.*, 2004; Leu *et al.*, 2004).

In addition to integrins and cell surface HSPGs, other cell surface binding proteins have been observed to interact with CCNs. Thus, CCNs may modulate other signaling pathways by interacting with their components, although how these interactions function and the biological responses they evoke are still poorly understood. For example, CCN2 has been shown to interact with the low density lipoprotein receptor related protein (LRP), which may mediate the internalization and degradation of CCN2 (Segarini *et al.*, 2001; Gao and Brigstock, 2003). Both CCN1 and CCN2 can modulate Wnt signaling in *Xenopus*, and in this context, CCN2 has been shown to interact with LRP-6, a coreceptor for Wnt (Latinkic *et al.*, 2003; Mercurio *et al.*, 2004). CCN3 binds Notch and inhibits myoblast differentiation through the Notch pathway (Sakamoto *et al.*, 2002). Understanding the biological functions of these interactions and their mechanisms of actions will be a major challenge in future research.

Other mechanisms of CCN actions are still being explored. For example, CCNs can interact with other growth factors or cytokines, and may modulate their activities. CCN2 has been shown to bind BMP4 and TGF-β1, thereby inhibiting BMP4 and enhancing TGF-β1 activities (Abreu *et al.*, 2002). Likewise, CCN2 also binds vascular endothelial growth factor and inhibits its angiogenic activities in some assays (Inoki *et al.*, 2002). CCN1 has been shown to mobilize fibroblast growth factor from the ECM, potentially enhancing the bioavailability of this growth factor (Kolesnikova and Lau, 1998). How these various mechanisms contribute to CCN functions physiologically remains to be addressed.

Finally, the question of whether there are specific, high-affinity cell surface receptors for CCNs that act in a manner similar to classical growth factor receptors is yet unanswered. To date, no such receptor has been identified for any CCN protein. Inasmuch as the activities of CCNs in isolated cell systems can be attributed to actions through integrins and HSPGs, it is unclear whether such high affinity receptors for CCNs exist. Nevertheless, this possibility cannot be excluded, and future analysis of CCN-induced signaling should help illuminate this important issue.

REFERENCES

Abreu JG, Ketpura NI, Reversade B and De Robertis EM (2002) Connective-tissue growth factor (CTGF) modulates cell signalling by BMP and TGF-β. *Nat Cell Biol* **4**:599–604.

Babic AM, Chen C-C and Lau LF (1999) Fisp12/mouse connective tissue growth factor mediates endothelial cell adhesion and migration through integrin $\alpha v\beta 3$, promotes endothelial cell survival, and induces angiogenesis *in vivo*. *Mol Cell Biol* **19**:2958–2966.

Babic AM, Kireeva ML, Kolesnikova TV and Lau LF (1998) CYR61, product of a growth factor-inducible immediate-early gene, promotes angiogenesis and tumor growth. *Proc Natl Acad Sci USA* **95**:6355–6360.

Bökel C and Brown NH (2002) Integrins in development: moving on, responding to, and sticking to the extracellular matrix. *Dev Cell* **3**:311–321.

Bork P (1993) The modular architecture of a new family of growth regulators related to connective tissue growth factor. *FEBS Lett* **327**:125–130.

Bornstein P (1995) Diversity of function is inherent in matricellular proteins: an appraisal of thrombospondin 1. *J Cell Biol* **130**:503–506.

Bornstein P and Sage EH (2002) Matricellular proteins: extracellular modulators of cell function. *Curr Opin Cell Biol* **14**:608–616.

Bouvard D, Brakebusch C, Gustafsson E, Aszodi A, Bengtsson T, Berna A and Fassler R (2001) Functional consequences of integrin gene mutations in mice. *Circ Res* **89**:211–223.

Bradham DM, Igarashi A, Potter RL and Grotendorst GR (1991) Connective tissue growth factor: a cysteine-rich mitogen secreted by human vascular endothelial cells is related to the SRC-induced immediate early gene product CEF-10. *J Cell Biol* **114**:1285–1294.

Brigstock DR, Steffen CL, Kim GY, Vegunta RK, Diehl JR and Harding PA (1997) Purification and characterization of novel heparin-binding growth factors in uterine-secretory fluids. *J Biol Chem* **272**:20275–20282.

Chen C-C, Chen N and Lau LF (2001a) The angiogenic factors Cyr61 and CTGF induce adhesive signaling in primary human skin fibroblasts. *J Biol Chem* **276**:10443–10452.

Chen C-C, Mo F-E and Lau LF (2001b) The angiogenic inducer Cyr61 induces a genetic program for wound healing in human skin fibroblasts. *J Biol Chem* **276**: 47329–47337.

Chen N, Chen C-C and Lau LF (2000) Adhesion of human skin fibroblasts to Cyr61 is mediated through integrin $\alpha 6\beta 1$ and cell surface heparan sulfate proteoglycans. *J Biol Chem* **275**:24953–24961.

Chen N, Leu S-J, Todorovic V, Lam SCT and Lau LF (2004) Identification of a novel integrin $\alpha v \beta 3$ binding site in CCN1 (CYR61) critical for pro-angiogenic activities in vascular endothelial cells. *J Biol Chem* **279**:44166–44176.

Crean JK, Finlay D, Murphy M, Moss C, Godson C, Martin F and Brady HR (2002) The role of p42/44 MAPK and protein kinase B in connective tissue growth factor induced extracellular matrix protein production, cell migration, and actin cytoskeletal rearrangement in human mesangial cells. *J Biol Chem* **277**:44187–44194.

Crean JK, Furlong F, Finlay D, Mitchell D, Conway B, Brady HR, Godson C and Martin F (2004) Connective tissue growth factor [CTGF]/CCN2 stimulates mesangial cell migration through integrated dissolution of focal adhesion complexes and activation of cell polarization. *FASEB J* **18**:1541–1543.

Ellis PD, Metcalfe JC, Hyvonen M and Kemp PR (2003) Adhesion of endothelial cells to NOV is mediated by the integrins $\alpha v \beta 3$ and $\alpha 5 \beta 1$. *J Vasc Res* **40**:234–243.

Fan WH and Karnovsky MJ (2002) Increased MMP-2 expression in CTGF over-expression vascular smooth muscle cells. *J Biol Chem* **277**:9800–9805.

Fataccioli V, Abergel V, Wingertsmann L, Neuville P, Spitz E, Adnot S, Calenda V and Teiger E (2002) Stimulation of angiogenesis by Cyr61 gene: a new therapeutic candidate. *Hum Gene Ther* **13**:1461–1470.

Frisch SM and Ruoslahti E (1997) Integrins and anoikis. *Curr Opin Cell Biol* **9**:701–706.

Frisch SM and Screaton RA (2001) Anoikis mechanisms. *Curr Opin Cell Biol* **13**: 555–562.

Gao R and Brigstock DR (2003) Low density lipoprotein receptor-related protein (LRP) is a heparin-dependent adhesion receptor for connective tissue growth factor (CTGF) in rat activated hepatic stellate cells. *Hepatol Res* **27**:214–220.

Gao R and Brigstock DR (2004) Connective tissue growth factor (CCN2) induces adhesion of rat activated hepatic stellate cells by binding of its C-terminal domain to integrin $\alpha v \beta 3$ and heparan sulfate proteoglycan. *J Biol Chem* **279**:8848–8855.

Giancotti FG and Ruoslahti E (1999) Integrin signaling. *Science* **285**:1028–1032.

Giancotti FG and Tarone G (2003) Positional control of cell fate through joint integrin/receptor protein kinase signaling. *Annu Rev Cell Dev Biol* **19**:173–206.

Grotendorst GR, Rahmanie H and Duncan MR (2004) Combinatorial signaling pathways determine fibroblast proliferation and myofibroblast differentiation. *FASEB J* **18**:469–479.

Grzeszkiewicz TM, Kirschling DJ, Chen N and Lau LF (2001) CYR61 stimulates human skin fibroblasts migration through integrin $\alpha v \beta 5$ and enhances mitogenesis through integrin $\alpha v \beta 3$, independent of its carboxy-terminal domain. *J Biol Chem* **276**:21943–21950.

Grzeszkiewicz TM, Lindner V, Chen N, Lam SCT and Lau LF (2002) The angiogenic factor CYR61 supports vascular smooth muscle cell adhesion and stimulates chemotaxis through integrin $\alpha 6 \beta 1$ and cell surface heparan sulfate proteoglycans. *Endocrinology* **143**:1441–1450.

Hishikawa K, Oemar BS and Nakaki T (2001) Static pressure regulates connective tissue growth factor expression in human mesangial cells. *J Biol Chem* **276**:16797–16803.

Hishikawa K, Oemar BS, Tanner FC, Nakaki T, Fujii T and Luscher TF (1999a) Overexpression of connective tissue growth factor gene induces apoptosis in human aortic smooth muscle cells. *Circulation* **100**:2108–2112.

Hishikawa K, Oemar BS, Tanner FC, Nakaki T, Luscher TF and Fujii T (1999b) Connective tissue growth factor induces apoptosis in human breast cancer cell line MCF-7. *J Biol Chem* **274**:37461–37466.

Hood JD and Cheresh DA (2002) Role of integrins in cell invasion and migration. *Nat Rev Cancer* **2**:91–100.

Hynes RO (2002) Integrins: bidirectional, allosteric signaling machines. *Cell* **110**: 673–687.

Hynes RO, Lively JC, McCarty JH, Taverna D, Francis SE, Hodivala–Dilke K and Xiao Q (2002) The diverse roles of integrins and their ligands in angiogenesis. *Cold Spring Harb Symp Quant Biol* **67**:143–153.

Igarashi A, Nashiro K, Kikuchi K, Sato S, Ihn H, Fujimoto M, Grotendorst GR and Takehara K (1996) Connective tissue growth factor gene expression in tissue sections from localized scleroderma, keloid, and other fibrotic skin disorders. *J Invest Dermatol* **106**:729–733.

Inoki I, Shiomi T, Hashimoto G, Enomoto H, Nakamura H, Makino K, Ikeda E, Takata S, Kobayashi K and Okada Y (2002) Connective tissue growth factor binds vascular endothelial growth factor (VEGF) and inhibits VEGF-induced angiogenesis. *FASEB J* **16**:219–221.

Ivkovic S, Yoon BS, Popoff SN, Safadi FF, Libuda DE, Stephenson RC, Daluiski A and Lyons KM (2003) Connective tissue growth factor coordinates chondrogenesis and angiogenesis during skeletal development. *Development* **130**:2779–2791.

Jain A, Karadag A, Fohr B, Fisher LW and Fedarko NS (2002) Three SIBLINGs (small integrin-binding ligand, N-linked glycoproteins) enhance factor H's cofactor activity enabling MCP-like cellular evasion of complement-mediated attack. *J Biol Chem* **277**:13700–13708.

Jedsadayanmata A, Chen CC, Kireeva ML, Lau LF and Lam SC (1999) Activation-dependent adhesion of human platelets to Cyr61 and Fisp12/Mouse connective tissue growth factor is mediated through integrin αIIbβ3. *J Biol Chem* **274**: 24321–24327.

Juliano RL (2002) Signal transduction by cell adhesion receptors and the cytoskeleton: functions of integrins, cadherins, selectins, and immunoglobulin-superfamily members. *Annu Rev Pharmacol Toxicol* **42**:283–323.

Kim M, Carman CV and Springer TA (2003a) Bidirectional transmembrane signaling by cytoplasmic domain separation in integrins. *Science* **301**:1720–1725.

Kim KH, Min YK, Baik JH, Lau LF, Chaqour B and Chung KC (2003b) Expression of angiogenic factor Cyr61 during neuronal cell death via the activation of c-Jun N-terminal kinase and serum response factor. *J Biol Chem* **278**:13847–13854.

Kireeva ML, Lam SCT and Lau LF (1998) Adhesion of human umbilical vein endothelial cells to the immediate-early gene product Cyr61 is mediated through integrin $\alpha v\beta 3$. *J Biol Chem* **273**:3090–3096.

Kireeva ML, Latinkic BV, Kolesnikova TV, Chen C-C, Yang GP, Abler AS and Lau LF (1997) Cyr61 and Fisp12 are both signaling cell adhesion molecules: comparison of activities, metabolism, and localization during development. *Exp Cell Res* **233**:63–77.

Kireeva ML, Mo F-E, Yang GP and Lau LF (1996) Cyr61, product of a growth factor-inducible immediate-early gene, promotes cell proliferation, migration, and adhesion. *Mol Cell Biol* **16**:1326–1334.

Kolesnikova TV and Lau LF (1998) Human CYR61-mediated enhancement of bFGF-induced DNA synthesis in human umbilical vein endothelial cells. *Oncogene* **16**:747–754.

Kumar S, Hand AT, Connor JR, Dodds RA, Ryan PJ, Trill JJ, Fisher SM, Nuttall ME, Lipshutz DB, Zou C, Hwang SM, Votta BJ, James IE, Rieman DJ, Gowen M and Lee JC (1999) Identification and cloning of a connective tissue growth factor-like cDNA from human osteoblasts encoding a novel regulator of osteoblast functions. *J Biol Chem* **274**:17123–17131.

Lake AC, Bialik A, Walsh K and Castellot JJ Jr. (2003) CCN5 is a growth arrest-specific gene that regulates smooth muscle cell proliferation and motility. *Am J Pathol* **162**:219–231.

Latinkic BV, Mercurio S, Bennett B, Hirst EM, Xu Q, Lau LF, Mohun TJ and Smith JC (2003) Xenopus Cyr61 regulates gastrulation movements and modulates Wnt signalling. *Development* **130**:2429–2441.

Lau LF and Lam SC (1999) The CCN family of angiogenic regulators: the integrin connection. *Exp Cell Res* **248**:44–57.

Leu S-J, Chen N, Chen C-C, Todorovic V, Bai T, Juric V, Liu Y, Yan G, Lam SCT and Lau LF (2004) Targeted mutagenesis of the matricellular protein CCN1 (CYR61): selective inactivation of integrin $\alpha 6\beta 1$-heparan sulfate proteoglycan coreceptor-mediated cellular activities. *J Biol Chem* **279**:44177–44187.

Leu S-J, Lam SCT and Lau LF (2002) Proangiogenic activities of CYR61 (CCN1) mediated through integrins $\alpha v\beta 3$ and $\alpha 6\beta 1$ in human umbilical vein endothelial cells. *J Biol Chem* **277**:46248–46255.

Leu S-J, Liu Y, Chen N, Chen CC, Lam SC and Lau LF (2003) Identification of a novel integrin $\alpha 6\beta 1$ binding site in the angiogenic inducer CCN1 (CYR61). *J Biol Chem* **278**:33801–33808.

Lin CG, Chen C-C, Leu S-J, Grzeszkiewicz TM and Lau LF (2005) Integrin-dependent functions of the angiogenic inducer CCN3 (NOV) in fibroblasts: implications in wound healing. In press.

Lin CG, Leu SJ, Chen N, Tebeau CM, Lin SX, Yeung CY and Lau LF (2003) CCN3 (NOV) is a novel angiogenic regulator of the CCN protein family. *J Biol Chem* **278**:24200–24208.

Lin MT, Chang CC, Chen ST, Chang HL, Su JL, Chau YP and Kuo ML (2004b) Cyr61 expression confers resistance to apoptosis in breast cancer MCF-7 cells by a mechanism of NF-kappaB-dependent XIAP up-regulation. *J Biol Chem* **279**: 24015–24023.

Menendez JA, Mehmi I, Griggs DW and Lupu R (2003) International Congress on Hormonal Steroids and Hormones and Cancer: The angiogenic factor CYR61 in breast cancer: molecular pathology and therapeutic perspectives. *Endocr Relat Cancer* **10**:141–152.

Mercurio S, Latinkic B, Itasaki N, Krumlauf R and Smith JC (2004) Connective-tissue growth factor modulates WNT signalling and interacts with the WNT receptor complex. *Development* **131**:2137–2147.

Miranti CK and Brugge JS (2002) Sensing the environment: a historical perspective on integrin signal transduction. *Nat Cell Biol* **4**:E83–E90.

Mo FE, Muntean AG, Chen CC, Stolz DB, Watkins SC and Lau LF (2002) CYR61 (CCN1) is essential for placental development and vascular integrity. *Mol Cell Biol* **22**:8709–8720.

Nakanishi T, Nishida T, Shimo T, Kobayashi K, Kubo T, Tamatani T, Tezuka K and Takigawa M (2000) Effects of CTGF/Hcs24, a product of a hypertrophic chondrocyte-specific gene, on the proliferation and differentiation of chondrocytes in culture. *Endocrinology* **141**:264–273.

Nishida T, Nakanishi T, Shimo T, Asano M, Hattori T, Tamatani T, Tezuka K and Takigawa M (1998) Demonstration of receptors specific for connective tissue growth factor on a human chondrocytic cell line (HCS-2/8). *Biochem Biophys Res Commun* **247**:905–909.

Ruoslahti E (1996) RGD and other recognition sequences for integrins. *Ann Rev Cell Dev Biol* **12**:697–715.

Sakamoto K, Yamaguchi S, Ando R, Miyawaki A, Kabasawa Y, Takagi M, Li CL, Perbal B and Katsube K (2002) The nephroblastoma overexpressed gene (NOV/ccn3) protein associates with Notch1 extracellular domain and inhibits myoblast differentiation via Notch signaling pathway. *J Biol Chem* **277**:29399–29405.

Schober JM, Chen N, Grzeszkiewicz TM, Emeson EE, Ugarova TP, Ye RD, Lau LF and Lam SCT (2002) Identification of integrin $\alpha M\beta 2$ as an adhesion receptor on peripheral blood monocytes for Cyr61 (CCN1) and connective tissue growth factor (CCN2), immediate-early gene products expressed in atherosclerotic lesions. *Blood* **99**:4457–4465.

Schober JM, Lau LF, Ugarova TP and Lam SC (2003) Identification of a novel integrin $\alpha M\beta 2$ binding site in CCN1 (CYR61), a matricellular protein expressed in healing wounds and atherosclerotic lesions. *J Biol Chem* **278**:25808–25815.

Schwartz MA and Ginsberg MH (2002) Networks and crosstalk: integrin signalling spreads. *Nat Cell Biol* **4**:E65–E68.

Segarini PR, Nesbitt JE, Li D, Hayes LG, Yates JR III and Carmichael DF (2001) The low density lipoprotein receptor-related protein/alpha 2-Macroglobulin receptor is a receptor for connective tissue growth factor (CTGF). *J Biol Chem* **276**: 40659–40667.

Shattil SJ (1999) Signaling through platelet integrin alpha IIb beta 3: inside-out, outside-in, and sideways. *Thromb Haemost* **82**:318–325.

Shimo T, Nakanishi T, Nishida T, Asano M, Kanyama M, Kuboki T, Tamatani T, Tezuka K, Takemura M, Matsumura T and Takigawa M (1999) Connective tissue growth factor induces the proliferation, migration, and tube formation of vascular endothelial cells *in vitro*, and angiogenesis *in vivo*. *J Biochem (Tokyo)* **126**:137–145.

Soon LL, Yie TA, Shvarts A, Levine AJ, Su F and Tchou-Wong KM (2003) Over-expression of WISP-1 down-regulated motility and invasion of lung cancer cells through inhibition of Rac activation. *J Biol Chem* **278**:11465–11470.

Stupack DG and Cheresh DA (2002) Get a ligand, get a life: integrins, signaling and cell survival. *J Cell Sci* **115**:3729–3738.

Su F, Overholtzer M, Besser D and Levine AJ (2002) WISP-1 attenuates p53-mediated apoptosis in response to DNA damage through activation of the Akt kinase. *Genes Dev* **16**:46–57.

Wang JF, Olson ME, Ball DK, Brigstock DR and Hart DA (2003) Recombinant connective tissue growth factor modulates porcine skin fibroblast gene expression. *Wound Repair Regen* **11**:220–229.

Wary KK, Mainiero F, Isakoff SJ, Marcantonio EE and Giancotti FG (1996) The adaptor protein Shc couples a class of integrins to the control of cell cycle progression. *Cell* **87**:733–743.

Wong M, Kireeva ML, Kolesnikova TV and Lau LF (1997) Cyr61, product of a growth factor-inducible immediate-early gene, regulates chondrogenesis in mouse limb bud mesenchymal cells. *Dev Biol* **192**:492–508.

Xie D, Yin D, Tong X, O'Kelly J, Mori A, Miller C, Black K, Gui D, Said JW and Koeffler HP (2004) Cyr61 is overexpressed in gliomas and involved in integrin-linked kinase-mediated Akt and beta-catenin-TCF/Lef signaling pathways. *Cancer Res* **64**:1987–1996.

Xiong JP, Stehle T, Goodman SL and Arnaout MA (2003) New insights into the structural basis of integrin activation. *Blood* **102**:1155–1159.

Yang GP and Lau LF (1991) Cyr61, product of a growth factor-inducible immediate early gene, is associated with the extracellular matrix and the cell surface. *Cell Growth Differentiation* **2**:351–357.

Yosimichi G, Nakanishi T, Nishida T, Hattori T, Takano–Yamamoto T and Takigawa M (2001) CTGF/Hcs24 induces chondrocyte differentiation through a p38 mitogen-activated protein kinase (p38MAPK), and proliferation through a p44/42 MAPK/extracellular-signal regulated kinase (ERK). *Eur J Biochem* **268**:6058–6065.

CHAPTER 4

EXPRESSION AND ROLES OF CCN2 DURING ODONTOGENESIS

Manabu Kanyama, Tsuyoshi Shimo, Changshan Wu,
Hiroki Sugito, Masahiro Iwamoto, Maurizio Pacifici and
Eiki Koyama

Department of Orthopaedic Surgery
Thomas Jefferson University Medical School
Philadelphia, PA 19107

Odontogenesis is a complex, multi-stage and fundamental process that involves epithelial-mesenchymal interactions, polarized growth, morphogenesis and differentiation of ameloblasts and odontoblasts. Studies have shown that these complex developmental events require continuous and reiterative action by signaling molecules at successive stages and multiple sites. However, the exact nature of these factors, the steps controlled by them, and the means by which they exert their roles at multiple sites and times are not fully known. Our recent work described here shows for the first time that CCN2/connective tissue growth factor (CTGF), a member of the powerful CCN effector family, is involved in, and regulates, odontogenesis. CCN2 is first expressed at the onset of odontogenesis along with transforming growth factor β_1 (TGF-β_1) and bone morphogenetic proteins 2 and 4 (BMP-2 and BMP-4), characterizes enamel knot and preameloblasts in conjunction with *Sonic hedgehog* (SHH) at subsequent stages, and then persists in differentiating and mature odontoblasts. CCN2 gene expression increases during differentiation of dental precursor cells in culture, and exogenous BMPs stimulate CCN2 expression in tooth germs explants. Exogenous recombinant CCN2 (rCCN2) stimulates proliferation in dental epithelium and mesenchyme. Interference with CCN2 action by neutralizing antibodies, however, inhibits cell proliferation and ameloblast and odontoblast differentiation. The results show that CCN2 is a powerful and developmentally controlled signaling molecules during tooth development, and has previously unsuspected roles such as stimulation of proliferation and promoting maturation in odontogenic cells. In this chapter, our previous studies are summarized and new data are presented. Taken together, our data demonstrate for the first time that CCN2 is expressed during, and regulates, odontogenesis. CCN2 expression is confined to specific sites and times in the developing tooth germ, is regulated by TGF-β and BMPs, and appears to be necessary for normal growth and cytodifferentiation of ameloblasts and odontoblasts.

1. ODONTOGENESIS

Mammalian tooth development remains at the center of much research activity, owing to the importance of normal dentition and the several abnormalities affecting it (Mina and Kollar, 1987; Lumsden, 1988; Maas and Bei, 1997; Peters and Balling, 1999; Salazar–Ciudad and Jernvall, 2002; Lisi *et al.*, 2003; Wu *et al.*, 2003). In mouse embryos, odontogenesis becomes appreciable around Day 11 (E11) with formation of dental laminas along the oral epithelium. Dental lamina cells located at prescribed locations proliferate, migrate and invade the underlying neural crest-derived mesenchyme; in so doing, they induce neighboring ectomesenchymal cells to undergo condensation, forming a dental papilla. The resulting tooth buds advance to the cap stage during which the cells at the tip of the invaginating and growing epithelium give rise to the primary enamel knot. The enamel knot is a small round-shaped structure made of a cluster of epithelial cells, protruding into the stellate reticulum and representing the site of future cusp formation. It is a transient structure that plays fundamental morphogenetic and signaling roles in tooth germ development (Vaahtokari *et al.*, 1996; Jernvall *et al.*, 1998; Pispa *et al.*, 1999; Coin *et al.*, 2000; Tucker *et al.*, 2000; Matalova *et al.*, 2004). Following the cap stage, the tooth germs advance to the bell stage, which is characterized by asymmetric growth and proliferation, a much more complex organization, and the presence of well-defined major and minor epithelial cell populations, including inner dental epithelium and stratum intermedium. Finally, during the crown stage in neonatal and postnatal mice, inner dental epithelial cells differentiate into ameloblasts and subadjacent mesenchymal cells differentiate into odontoblasts, which secrete enamel and dentin proteins, mineralize the extracellular matrix, and lead to formation of mature teeth in the growing organism.

Clearly, odontogenesis is a complex process that involves sequential and interdependent steps including: i) commitment of progenitor oral epithelial and mesenchymal cells to odontogenic lineages; ii) active and polarized phases of cell proliferation, cell migration and tissue invasion; iii) inductive events between epithelial and mesenchymal cells; iv) morphogenetic movements and tissue folding, such as those occurring from cap to bell stage; v) establishment of symmetrical structures such as molar cusps; and vi) cytodifferentiation of highly specialized and unique cells, ameloblasts and odontoblasts, which are responsible for the ultimate emergence of functional teeth. There have been major advances during the last few years in clarifying how odontogenesis is brought about during embryogenesis and early postnatal life, but we are far from having a detailed and thorough understanding of this process.

1.1. CCN2 Expression during Odontogenesis

We first analyzed CCN2 gene expression and distribution during odontogenesis of mouse embryonic and postnatal stages by *in situ* hybridization and immunohistochemistry. Indeed, CCN2 transcripts were detected at the onset of odontogenesis in thickening dental laminas of E11 embryos (Figs. 1A and 1B, arrows), an expression pattern closely resembling that of BMP-2 and BMP-4 (Vainio *et al.*, 1993; Åberg *et al.*, 1997). At the bud stage in E12.5 embryos, transcripts became more abundant and readily detectable in dental lamina

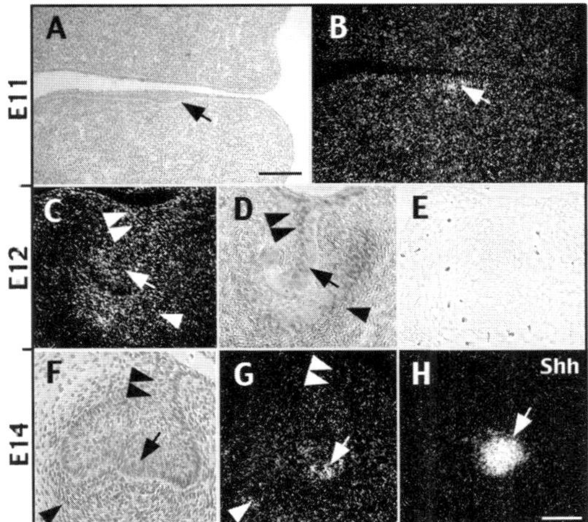

Fig. 1. **CCN2 expression and distribution in mouse embryo tooth germs.** (A, B) CCN2 transcripts in dental lamina (arrow). (C, D) Bud stage germs containing CCN2 transcripts (C) and protein (D) in dental lamina (double arrowhead), invaginating epithelial cells (arrow) and condensed mesenchyme (arrowhead). (E) Companion bud stage serial section reacted with preimmune IgGs and lacking staining. (F, G) Cap stage germs exhibiting CCN2 transcripts in enamel knot (arrow) and along perimeter of condensed mesenchyme (arrowhead) but lacking them in dental lamina (double arrowhead). (H) Companion cap stage section with abundant Shh transcripts in enamel knot. Bars in A and B, 150 μm and in C–H, 75 μm. (Shimo T, Wu C *et al.* (2002) Expression, gene regulation, and roles of Fisp12/CTGF in developing tooth germs. *Dev Dyn* **224**:267–278.)

(Fig. 1C, double arrowhead) and invaginating epithelium and adjacent condensing mesenchyme (Fig. 1C, arrow and arrowhead, respectively). Immunohistochemistry showed that CCN2 distribution paralleled that of its mRNA at these stages (Fig. 1D, arrow, arrowhead and double arrowhead). Pre-immune antibodies produced no staining (Fig. 1E). In E14 cap stage tooth germs, CCN2 transcripts were now restricted to the enamel knot (Figs. 1F and 1G, arrow) particularly near the underlying mesenchyme; the knot was identifiable by its characteristic morphology and location (Fig. 1F, arrow) and expression of *Sonic hedgehog* (SHH) (Fig. 1H, arrow) (Koyama *et al.*, 1996; Vaahtokari *et al.*, 1996). CCN2 transcripts were also present along the outer rim of the condensed mesenchyme (Figs. 1F and 1G, arrowhead), but were no longer detectable in dental lamina (Figs. 1F and 1G, double arrowhead).

In E16 early bell stage tooth germs, CCN2 expression was evident in inner dental epithelium (Figs. 2A and 2B, arrow), lingual side of the outer dental epithelium (Figs. 2A and 2B, double arrow) and outer rim of condensed mesenchyme (Figs. 2A and 2B, arrowhead). A sense probe produced no hybridization signal (Fig. 2C). These expression patterns were maintained in E18 bell stage tooth germs, where CCN2 transcripts were prominent in the secondary enamel knots (Figs. 2D and 2E, arrows) and very clear in inner dental epithelium connecting the developing cusps (Figs. 2D and 2E, double arrow) and outer dental epithelium/dental mesenchyme boundary (Figs. 2D and 2E, double arrowhead). Immunohistochemistry revealed that CCN2 distribution corresponded for the most part to that of its transcripts, with clear immunostaining in inner dental epithelium (Fig. 2F, double arrow) and along the outer epithelial/mesenchymal boundary (Fig. 2F, double arrowhead). In addition, there was reproducible staining of the mesenchyme immediately below the epithelium (Fig. 2F, arrowhead).

Interestingly, in E19 late bell stage tooth germs, CCN2 gene expression decreased in differentiating inner dental epithelial cells (Figs. 2G and 2H, double arrow), and was maintained only in cuspally located preameloblasts (Figs. 2G and 2H, arrow) that do not differentiate terminally and are eliminated from the enamel organ by apoptosis (Sutcliffe and Owens, 1980; Vaahtokari *et al.*, 1991). Lack of detectable transcripts also characterized fully differentiated ameloblasts in 2 day-old (D2) postnatal tooth germs (Figs. 2K and 2L, double arrow) that instead displayed strong amelogenin expression (Fig. 2M) (Snead *et al.*, 1988). In both E19 and D2 tooth germs, CCN2 expression continued to be present in outer dental epithelium and dental follicle (Figs. 2G, 2H, 2K and 2L, arrowhead), particularly in the outer layer of the dental follicle covering the base of the dental papilla (Figs. 2I and 2J, arrowheads) as well in

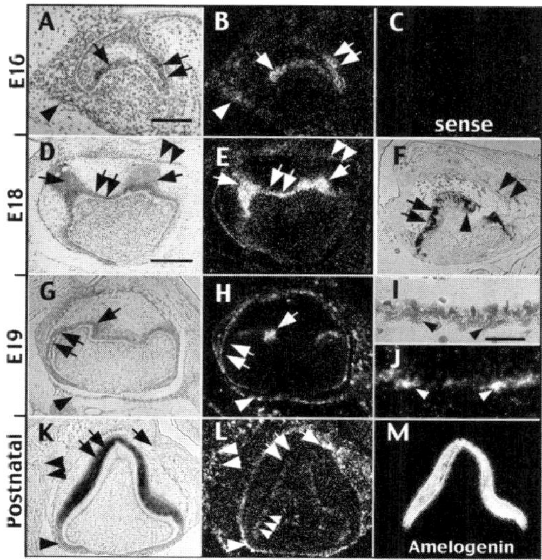

Fig. 2.   **CCN2 gene expression and distribution in early bell to crown stage mouse tooth germs.**
Sections from early bell (A–C), bell (D–F), late bell (G–J) and crown stages (K–M) were hybridized
with antisense (B, E, H, J and L) or sense (C) CCN2 riboprobes or with antisense amelogenin
riboprobe (M). (**A, B**) Inner dental epithelium (arrow), lingual side of outer dental epithelium (double arrow) and periphery of condensed mesenchyme (arrowhead). (**D, E**) Secondary enamel knot
(arrow), intercuspal inner dental epithelium (double arrow), and outer epithelial-mesenchymal
boundary (double arrowhead). (**F**) Immunostaining is present in mesenchyme below the positive
epithelium. (**G, H**) CCN2 positive preameloblasts (arrow) and CCN2 negative preameloblasts
(double arrow). (**I, J**) CCN2 positive cells in dental follicle (arrowheads). (**K, L**) Dental lamina (arrow), developing epithelial root sheath (arrowhead), ameloblasts (double arrow), odontoblasts (triple arrowhead) and osteoblasts (double arrowhead). Bars: A–C, 150 µm; D–H and K–M,
250 µm; and I–J, 30 µm. (Shimo T, Wu C *et al.* (2002) Expression, gene regulation, and roles of
Fisp12/CTGF in developing tooth germs. *Dev Dyn* **224**:267–278.)

developing epithelial root sheath (Figs. 2K and 2L, arrowhead). Lastly, we
observed fairly high hybridization signal in the involuting dental lamina and
alveolar bony crypts (Figs. 2K and 2L, arrow and double arrowhead) and persistent expression in differentiating and mature odontoblasts (Figs. 2K and
2L, triple arrow).

To verify our interesting finding that CCN2 expression ceases in differentiating ameloblasts but persists in differentiating and mature odontoblasts, we
examined postnatal teeth. As shown in Fig. 3, 4 day-old mouse incisor tooth
germs exhibited significant and clear CCN2 expression in differentiating and
secreting odontoblasts (Figs. 3A and 3B, arrows) but negligible expression in

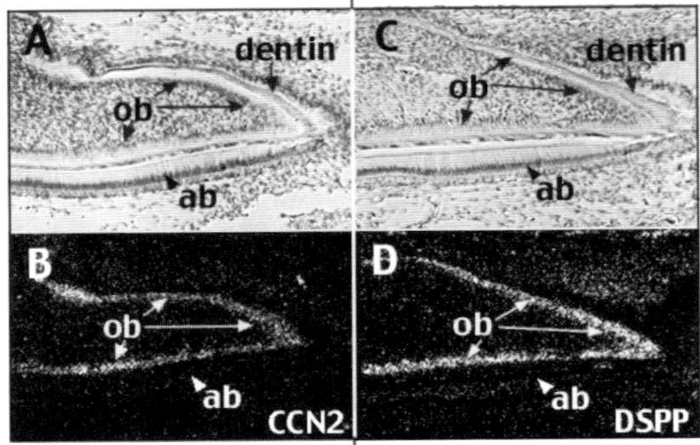

Fig. 3.   **CCN2 expression in postnatal odontoblasts.** Note presence of abundant CCN2 (A and B) and DSPP (C and D) transcripts in maturing odontoblasts (ob, arrows) and their absence in ameloblasts (ab, arrowhead).

ameloblasts (Figs. 3A and 3B, arrowhead). The differentiated state of odontoblasts was affirmed morphologically as well as by strong expression of Dentin Sialophosphoprotein (DSPP) (Figs. 3C and 3D, arrows).

1.2. Regulation of CCN2 Gene Expression during Odontogenesis

Expression and roles of several regulators of odontogenesis require epithelial-mesenchymal interactions (Mina and Kollar, 1987; Thesleff and Sharp, 1997; Wu *et al.*, 2003). Thus, it became of relevance to ask whether such interactions regulate CCN2 expression as well. This possibility was tested by tissue recombination studies. Mandibular explants containing bud stage tooth germs were isolated from Day 12 mouse embryos (E12) by microsurgical procedures. After enzymatic treatment, one side of dental epithelium was removed from the explant and was then grown in organ culture for up to 16–20 hrs in serum-free culture medium. Whole mount *in situ* hybridization revealed that strong CCN2 gene expression was lost in mesenchyme grown without epithelium (Fig. 4D).

Because BMPs, TGF-βs and other upstream molecules induce CCN2 gene expression in a variety of cell types (Kothapalli *et al.*, 1997; Shimo *et al.*, 1999; Folger *et al.*, 2001), we determined whether these signaling molecules regulate CCN2 gene expression in dental epithelium. Accordingly, agarose or heparin acrylic beads filled with recombinant BMP-4 were applied to the surface of E12 mandibular explants lacking dental epithelium, and explants were incubated for 16–20 hrs. BMP-4 (Fig. 4A) induced strong CCN2 expression, whereas

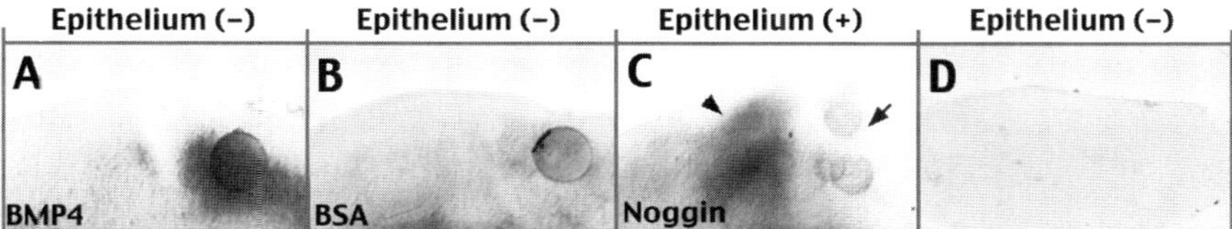

Fig. 4. **Analysis of CCN2 gene regulation.** E12 mandibular explants with (C) and without epithelium (A, B, and D) were maintained in organ culture for 16–20 hrs association with: **(A)** 100 ng/µL BMP-4 filled bead; **(B)** 100 ng/µL BSA-filled control bead; and **(C)** 200 ng/µL Noggin filled beads. Specimens were then processed for whole mount *in situ* hybridization analysis of CCN2 gene expression.

control BSA beads containing no factor did not (Fig. 4B). BMP-2 could also induce CCN2 expression, but much less than BMP-4 (data not shown).

To determine whether endogenous BMP signaling is needed for CCN2 expression, beads filled with the BMP antagonist *Noggin* were placed near the incisor tooth germs on the right side of E12 mouse embryo mandibular explants (Fig. 4C, arrow). Explants were processed for hybridization 20 hrs later. Obvious CCN2 expression was detectable in untreated tooth germs (Fig. 4C, arrowhead) but not in *Noggin*-treated ones (Fig. 4C, arrow).

2. CCN2 AND DENTAL CELL PROLIFERATION

As CCN2 is expressed intensively in dental epithelium and mesenchyme of growing tooth germs, one role of CCN2 may be to maintain and/or regulate proliferation of dental cell populations. To obtain direct evidence for such role, we tested the mitogenic response of tooth germ cell populations in culture to treatment with exogenous recombinant CCN2 (rCCN2). For these experiments, we used epithelial and mesenchymal cell populations isolated from bell stage molar bovine tooth germs. This is a unique experimental system that we developed previously and that allows efficient dissection of primary cell populations for *in vitro* studies. This is nearly impossible to do with murine specimens, given their small size.

Cells were reared in primary monolayer multiwell cultures for 3 days and then treated for 24 hrs with increasing amounts of rCCN2. Epithelial cultures were pulse-labeled with BrdU during the last 2 hrs and incorporation was determined by a colorimetric assay. The rCCN2 treatment resulted in a stimulation of proliferation ($p < 0.05$), which was maximal at 50–100 ng/mL (Fig. 5A). Companion day 4 epithelial (Fig. 5B) and mesenchymal (Fig. 5C) cultures were treated for 24 hrs with 50 ng/mL rCCN2 or with 50 ng/mL rCCN2 *plus*

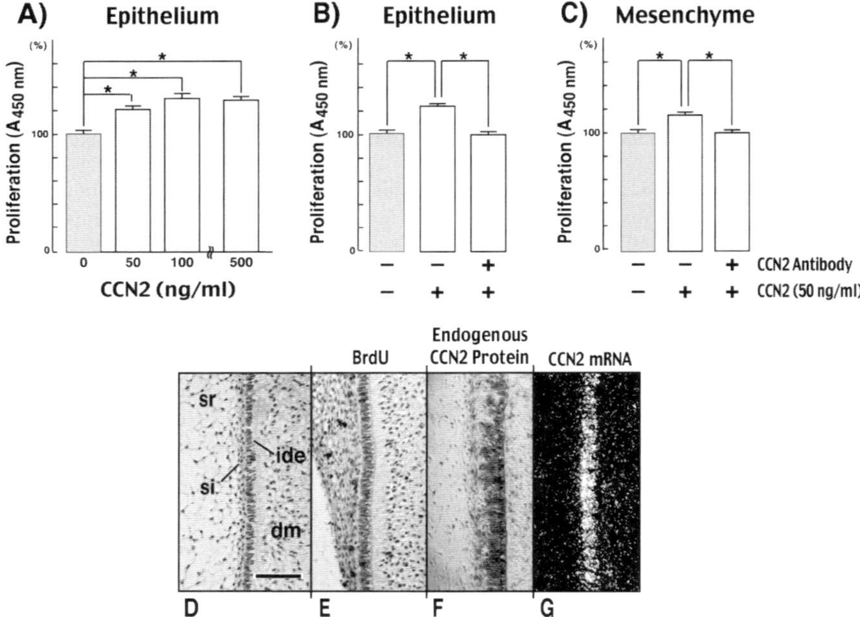

Fig. 5.   **Analysis of bell stage bovine tooth germs and cells.** (A) Proliferative responses of epithelial cells to increasing amounts of rCCN2. (**B, C**) Proliferative response of epithelial (B) and mesenchymal cells (C) to treatment with rCCN2 alone or a mixture of rCCN2 plus its antibodies. (**D**) Hematoxylin-eosin staining of tooth germ parasagittal section. (**E**) Immunohistochemical detection of BrdU-labeled proliferation cells. Note their presence in stratum intermedium (si), inner dental epithelium (ide) and dental mesenchyme (dm). (**F**) Immunostaining with CCN2 antibodies. Note specific staining of inner dental epithelium, stratum intermedium and dental mesenchyme. (**G**) *In situ* hybridization showing that CCN2 expression is limited to inner dental epithelium. $^*p < 0.5$; Bar, 50 μm. (Shimo T, Wu C *et al.* (2002) Expression, gene regulation, and roles of Fisp12/CTGF in developing tooth germs. *Dev Dyn* **224**:267–278.)

50 ng/mL of CCN2 neutralizing antibodies; parallel control cultures were left untreated. All cultures were labeled with BrdU during the last 2 hrs. The CCN2 antibody treatment clearly reversed the stimulation of proliferation induced by rCCN2 in both epithelial and mesenchymal cultures and brought it down to control level (Figs. 5B and 5C).

To verify these data, we determined whether there is a correlation between proliferating cells and CCN2 distribution *in vivo* at early cytodifferentiation stages. Sections from bell stage bovine tooth germs (pre-labeled with BrdU for 2 hrs) were processed for immunohistochemistry and histochemistry.

Indeed, inner dental epithelium, stratum intermedium and adjacent dental mesenchyme contained proliferating cells (Fig. 5E) and sizable amounts of CCN2 (Fig. 5F; see Fig. 2F for similar data in mouse). Interestingly, *in situ* hybridization of companion sections revealed that CCN2 transcripts were largely restricted to inner dental epithelium at this stage (Fig. 5G), indicating that some CCN2 can diffuse away from its source into adjacent tissues and exert effects (Shimo *et al.*, 2002).

3. CCN2 AND DENTAL CELL DIFFERENTIATION

To determine whether CCN2 regulates dental cell differentiation at later stages, we used primary cultures of epithelial and mesenchymal cells from bell stage bovine tooth germs. These cells undergo a spontaneous process of differentiation over time in culture (Fig. 6), as indicated by time-dependent increases in RNAs encoding APase (Fig. 6A), type I collagen (Fig. 6B), and DSPP (Fig. 6C). Similar day 3 cultures were treated with rCCN2 for different lengths of time (1.5, 3.0 or 6 days). All cultures including untreated cultures were harvested simultaneously at the end of treatment period and processed for gene expression analysis of markers of differentiation. rCCN2 treatment significantly stimulated APase activity (Fig. 7A) and gene expression of APase and type I collagen (Figs. 7B and 7C). The data strengthen the conclusion reached with explant cultures that CCN2 favors progression of dental cells towards differentiation.

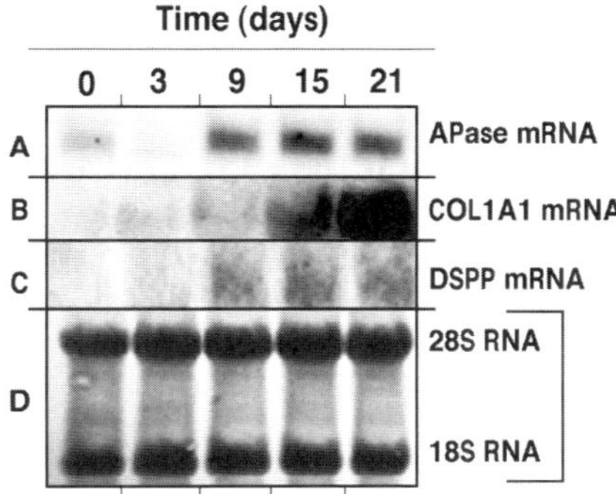

Fig. 6.   **Northern blot analysis of APase, type I collagen and DSPP expression in cultures of dental mesenchymal cells.** Note that expression of the genes increases with age, indicating progressive advance of the cells toward differentiation.

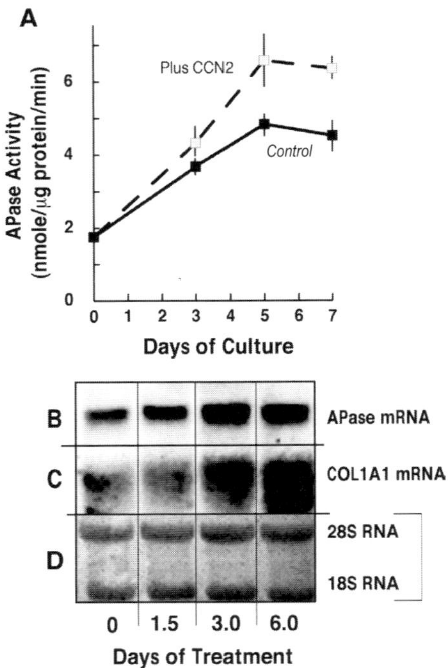

Fig. 7. **Stimulatory effects of CCN2 on cultured cells.** (A) APase activity in dental epithelial cells isolated from bell stage bovine tooth germs and grown for indicated days with or without exogenous rCCN2 (50 ng/mL). (**B–D**) Mesenchymal cells from the same tooth germs first grown for 3 days and then treated for up to 6 days with rCCN2 (50 ng/mL). All cultures were harvested simultaneously at end of treatment and process for gene expression analysis of APase (B) and type I collagen (C).

4. CCN FAMILY IN DEVELOPING TOOTH GERMS

As described in an earlier chapter of this book, CCN2 is a member of the CCN signaling and growth factor family of secreted proteins which currently include six members. Recent studies have shown that the CCN family of genes are co-expressed and have functions in developing organs (Brigstock, 2003; Ivkovic *et al.*, 2003; Takigawa, 2003; Perbal, 2004). Thus, it becomes interesting to ask whether other CCN family members are expressed in developing tooth germs. To begin exploring this possibility, we have completed a series of *in situ* hybridization analyses of representative stages in wild-type mouse tooth germs. Indeed, we found that both CCN1 and CCN3 are co-expressed in E13

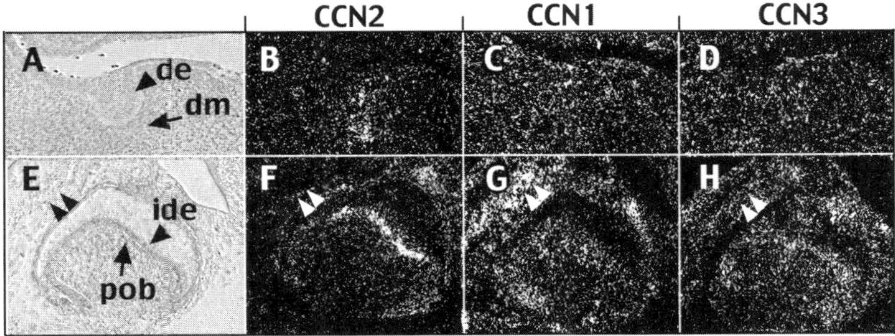

Fig. 8.   CCN1 and CCN3 expression in developing tooth germs and alveolar bone. E13 (A–D) and E17 (E–H) mouse embryo tooth germs were processed for *in situ* hybridization analysis of expression of: (**B** and **F**) CCN2; (**C** and **G**) CCN1; and (**D** and **H**) CCN3. The restricted and specific expression of CCN2 contrasts with the diffuse non-descript expression of CCN1 and CCN3 in dental epithelium (de), dental mesenchyme (dm), and inner dental epithelium (ide). Note the presence of CCN1, CCN2 and CCN3 transcripts in differentiating osteoblasts (double arrowhead).

and E17 tooth germs. While expression of these molecules was diffuse, non-descript and fairly low (Fig. 8C and 8D) in E13 tooth germs, those transcripts were markedly up-regulated and restricted to differentiating pre-odontoblasts in E17 tooth germs (Fig. 8G and 8H). Based on overlapping expression patterns in CCNs at multiple sites and stages during odontogenesis, it is conceivable that these proteins may influence multiple events, including commitment of dental cell lineage and odontogenic cell differentiation.

5. SUMMARY AND CONCLUSIONS

Our data provide evidence that CCN2 is initially expressed in dental lamina cells and transiently shifts to mesenchyme during the bud stage, a pattern closely resembling that reported for BMP-2 and BMP-4 (Åberg *et al.*, 1997). CCN2 expression then becomes a prominent feature of epithelial cells, with obvious and strong expression in primary and secondary enamel knots and differentiating inner dental epithelial cells; these patterns are reminiscent of those of SHH we and others reported previously (Koyama *et al.*, 1996; Vaahtokari *et al.*, 1996; Shimo *et al.*, 2002; Wu *et al.*, 2003). CCN2 expression is reduced in terminally differentiated ameloblasts, but is maintained in differentiating and mature odontoblasts. In addition, we show that i) CCN2 expression is regulated by epithelial-mesenchymal interactions and signaling by BMPs; ii) CCN2 stimulates dental cell proliferation and differentiation in culture while CCN2 neutralizing antibodies inhibited both proliferation and differentiation; and iii) CCN2 can diffuse away from its site of synthesis possibly reaching longer-range targets. Taken together, the results of several studies and the additional evidence presented here provide firm support for our central hypotheses that i) CCN2 is required for progression and completion of odontogenesis; and ii) CCN2 acts as an effector of upstream factors at multiple stages and sites in developing tooth germs.

ACKNOWLEDGMENTS

We thank our colleagues Drs. P. C. Billings, B. Abrams and J. Rosenbloom who participated in studies summarized here and reported previously. Original work was supported by NIH grant (E.K.).

REFERENCES

Åberg T, Wozney J and Thesleff I (1997) Expression patterns of bone morphogenetic proteins (Bmps) in the developing mouse tooth suggests roles in morphogenesis and differentiation. *Dev Dyn* **210**:383–396.

Brigstock DR (2003) The CCN family: a new stimulus package. *J Endocrinol* **178**:169–175.

Coin R, Schmitt R, Lesot H, Vonesch JL and Ruch JV (2000) Regeneration of halved embryonic lower first mouse molars: correlation with the distribution pattern of non dividing IDE cells, the putative organizers of morphogenetic units, the cusps. *Int J Dev Biol* **44**:289–295.

Folger P, Zekaria D, Grotendorst G and Masur S (2001) Transforming growth factor-beta-stimulated connective tissue growth factor expression during corneal myofibroblast differentiation. *Invest Ophthalmol Vis Sci* **42**:2534–2541.

Ivkovic S, Yoon BS, Popoff SN, Safadi FF, Libuda DE, Stephenson RC, Daluiski A and Lyons KM (2003) Connective tissue growth factor coordinates chondrogenesis and angiogenesis during skeletal development. *Development* **130**:2779–2791.

Jernvall J, Åberg T, Kettunen P, Keranen S and Thesleff I (1998) The life history of an embryonic signaling center: BMP-4 induces p21 and is associated with apoptosis in the mouse tooth enamel knot. *Development* **125**:161–169.

Kothapalli D, Frazier KS, Welply A, Segarini P and Grotendorst G (1997) Transforming growth factor beta induces anchorage-independent growth of NRK fibroblasts via a connective tissue growth factor-dependent signaling pathway. *Cell Growth Differ* **8**:61–68.

Koyama E, Yamaai T, Iseki S, Ohuchi H, Nohno T, Yoshioka H, Hayashi Y, Leatherman JL, Golden EB, Noji S and Pacifici M (1996) Polarizing activity, Sonic hedgehog, and tooth development in embryonic and postnatal mouse. *Dev Dyn* **206**:59–72.

Lisi S, Peterkova R, Peterka M, Vonesch JL, Ruch JV and Lesot H (2003) Tooth morphogenesis and pattern of odontoblast differentiation. *Connect Tissue Res* **44**:167–170.

Lumsden AG (1988) Spatial organization of the epithelium and the role of neural crest cells in the initiation of the mammalian tooth germ. *Development* **103**:155–169.

Maas R and Bei M (1997) The genetic control of early tooth development. *Crit Rev Oral Biol Med* **8**:4–39.

Matalova E, Tucker AS and Sharpe PT (2004) Death in the life of a tooth. *J Dent Res* **83**:11–16.

Mina M and Kollar EJ (1987) The induction of odontogenesis in non-dental mesenchyme combined with early murine mandibular arch epithelium. *Arch Oral Biol* **32**:123–127.

Perbal B (2004) CCN proteins: multifunctional signalling regulators. *Lancet* **363**:62–64.

Peters H and Balling R (1999) Teeth. Where and how to make them. *Trends Genet* **15**:59–65.

Pispa J, Jung HS, Jernvall J, Kettunen P, Mustonen T, Tabata MJ, Kere J and Thesleff I (1999) Cusp patterning defect in Tabby mouse teeth and its partial rescue by FGF. *Dev Biol* **216**:521–534.

Salazar–Ciudad I and Jernvall J (2002) A gene network model accounting for development and evolution of mammalian teeth. *Proc Natl Acad Sci USA* **99**:8116–8120.

Shimo T, Nakanishi T, Nishida T, Asano M, Kanyama M, Kuboki T, Tamatani T, Tezuka K, Takemura M, Matsumura T and Takigawa M (1999) Connective tissue growth factor induces the proliferation, migration, and tube formation of vascular endothelial cells *in vitro*, and angiogenesis *in vivo*. *J Biochem* **126**:137–145.

Shimo T, Wu C, Billings PC, Piddington R, Rosenbloom J, Pacifici M and Koyama E (2002) Expression, gene regulation, and roles of Fisp12/CTGF in developing tooth germs. *Dev Dyn* **224**:267–278.

Snead M, Luo W, Lau E and Slavkin H (1988) Spatial- and temporal-restricted pattern for amelogenin gene expression during mouse molar tooth organogenesis. *Development* **104**:77–85.

Sutcliffe JE and Owens PD (1980) A light and scanning electron microscopic study of the development of enamel-free areas on the molar teeth of the rat. *Arch Oral Biol* **25**:263–268.

Takigawa M (2003) CTGF/Hcs24 as a multifunctional growth factor for fibroblasts, chondrocytes and vascular endothelial cells. *Drug News Perspect* **16**:11–21.

Thesleff I and Sharp P (1997) Signalling networks regulating dental development. *Mach Dev* **67**:111–123.

Tucker AS, Headon DJ, Schneider P, Ferguson BM, Overbeek P, Tschopp J and Sharpe PT (2000) Edar/Eda interactions regulate enamel knot formation in tooth morphogenesis. *Development* **127**:4691–4700.

Vaahtokari A, Vainio S and Thesleff I (1991) Associations between transforming growth factor beta 1 RNA expression and epithelial-mesenchymal interactions during tooth morphogenesis. *Development* **113**:985–994.

Vaahtokari A, Åberg T, Jernvall J, Keranen S and Thesleff I (1996) The enamel knot as a signaling center in the developing mouse tooth. *Mech Dev* **54**:39–43.

Vainio S, Karavanova I, Jowett A and Thesleff I (1993) Identification of BMP-4 as a signal mediating secondary induction between epithelial and mesenchymal tissues during early tooth development. *Cell* **75**:45–58.

Wu C, Shimo T, Liu M, Pacifici M and Koyama E (2003) Sonic hedgehog functions as a mitogen during bell stage of odontogenesis. *Connect Tissue Res* **44 Suppl 1**:92–96.

CHAPTER 5

CCN GENES AND THE KIDNEY

Bruce L. Riser, Sujatha Karoor and Darryl R. Peterson

Baxter Healthcare, Renal Division, and
Rosalind Franklin University of Medicine and Science
Department of Physiology and Biophysics

Since their first description as CYR-61 (CCN1), CTGF (CCN2) and NOV (CCN3) proteins more than a decade ago, the number of publications and in turn our knowledge of the structure and function of these and other members of the CCN family of genes has increased in a logarithmic fashion. Their involvement has now been shown in multiple organ systems. One organ that has been the focus of a large number of these reports is the kidney. The majority of these renal studies have centered on, and have now documented, the role of CCN2 as an important pathogenic factor in fibrosis/sclerosis in a number of models of chronic kidney disease (CKD). This chapter will focus on these discoveries, the future direction, and the resulting possibilities for novel therapeutics and diagnostics, both targeting CCN. In addition, limited data indicates a possible role for CCN genes in embryonic development and in normal renal function. This area will also be briefly considered and the possible relationship to the pathological response discussed.

CCN2 IN RENAL DISEASE; EARLY REPORTS

With the initial description of CCN2 (CTGF) by Bradham and Grotendorst (Bradham *et al.*, 1991) and early reports suggesting a possible interactive role with TGF-β in skin fibrosis and scleroderma, our laboratory and several others around the world began simultaneous investigations of a possible role for CCN2 in renal pathology. The first published reports suggesting possible involvement of CCN2 in renal pathology came from Hammes, Pawar, Toback and colleagues (Hammes *et al.*, 1995; Pawar *et al.*, 1995). Monolayer cultures of monkey kidney epithelial cells (BSC-1 line) and calcium oxalate monohydrate (COM) crystals were used as a model to study cell responses to crystal

95

interactions that might occur in the nephrons of patients during periods of hyperoxaluria. They found that the immediate early genes c-myc, EGR-1, and Nur-77, along with CCN2 were induced at one hour, with the latter persisting for 24 hours. The genes encoding plasminogen activator inhibitor (PAI-1) and platelet-derived growth factor (PDGF)-A chain were induced soon after (Hammes *et al.*, 1995). In the same cell line, using a scrape model for migration in wound healing with acute tubular necrosis (ATN), they found that CCN2 gene expression was upregulated along with other genes thought to be involved in the wound healing response (Pawar, Kartha and Toback, 1995). This was followed by a report from Ito *et al.* (Ito *et al.*, 1998) in which kidney biopsies were examined for CCN2 mRNA expression by *in situ* hybridization. Renal expression in patients with a variety of different renal diseases were compared to controls consisting of the non-involved areas of kidneys removed due to tumors. CCN2 expression was weak and appeared limited to a few areas of the control kidneys. However, there was an upregulation of specific mRNA in cases of crescentic glomerulonephritis, IgA nephropathy, focal and segmental glomerulosclerosis, and in the 2 cases of diabetic nephropathy studied. This upregulation was observed predominately in the tubulointerstitial areas at sites of chronic damage.

CCN2 IN DIABETIC GLOMERULOSCLEROSIS

Our laboratory had begun investigating the role of CCN2 in chronic kidney disease (CKD), focusing primarily on diabetic nephropathy (DN). One of the primary lesions in DN, and in CKD in general, is sclerosis or fibrosis that develops in the mesangium of the glomerulus, the filtering unit of the kidney. The mesangial cell (MC) is located in the extracapillary space of the glomerular lobule and may influence glomerular function, at least in part by maintaining the structural integrity via foot-processes attached to the peri-mesangial basement membrane, as well as the production of mesangial extracellular matrix (ECM) (Riser, Cortes and Yee, 2000).

We, and others, have shown that the response to intraglomerular hypertension and thus capillary dilation is MC contraction and excessive cyclic stretch (Riser *et al.*, 1992). However, these events are not unique to DN. They occur under other conditions where intraglomerular hypertension is present. In diabetes, there is an additional element of exposure to hyperglycemia, a factor that has clearly been shown to be causal in the progression of disease. Our laboratory and others have shown by *in vitro* modeling that DN is likely to occur also as a result of MC response to both pathological cyclic stretch and elevated glucose exposure. For example, cultured MC when subjected to cyclic stretching approximating the stretch occurring during fluctuations in systemic blood pressure with impaired autoregulation at the afferent artriole, respond by increasing the production and accumulation of those mesangial ECM components that characterize the glomerular lesion (Riser *et al.*, 1992). It has been well established that the prosclerotic molecule TGF-β is a causal factor, not only in models of diabetic nephropathy, but numerous other non-diabetic models where the end result of a variety of different insults to the kidney is renal fibrosis or sclerosis. Indeed, evidence accumulated on this molecule has been one of the driving forces for the hypothesis that diabetic nephropathy and CKD are the result of an imbalance of cytokines created as a result of long-term or chronic pathological stimuli. Our laboratory has shown how these elements are interconnected by demonstrating that intraglomerular hypertension is translated to MC stretch, which results in the increased production and activation of TGF-β (Riser *et al.*, 2001). This production of TGF-β is also induced in MC by elevated glucose concentrations, but in the presence of both stimuli there is a differential, synergistic induction of ECM accumulation (Riser *et al.*, 1998). We therefore sought to next determine if the potential profibrotic factor CCN2 was implicated in this pathogenic process. To accomplish this we studied MC in culture as well as in the db/db mouse model of type II diabetes, the most prevalent form of the disease in humans.

When we examined the expression of CCN2 mRNA in cultured rat MC and compared the results with those from whole kidney and extra-renal tissue, we found that the transcript was weakly expressed in cultured MC and brain tissue, and was not detectable in cultured kidney fibroblasts (Riser *et al.*, 2000). In comparison, the message was strongly expressed in the heart and kidney tissue. To then determine if CCN2 could influence the synthesis or secretion of ECM, cultured MC were exposed to the recombinant human (rh) CCN2 protein, and the effects compared to those produced by TGF-β, or high glucose exposure. CCN2 produced a marked increase in key mesangial ECM components tested. For example, fibronectin increased 45% (compared to a 23 and 30% increase with TGF-β and the high glucose respectively), and type I collagen increased 64%, which was roughly equal, or greater, than that induced by TGF-β (50%) or high glucose (22%).

When TGF-β was examined as a possible regulatory factor in MC expression of CCN2 message, we found that exogenous TGF-β exposure increased the expression of CCN2 transcripts greater than 4-fold, whereas autoinduction of TGF-β1 mRNA was increased only 80% (Fig. 1).

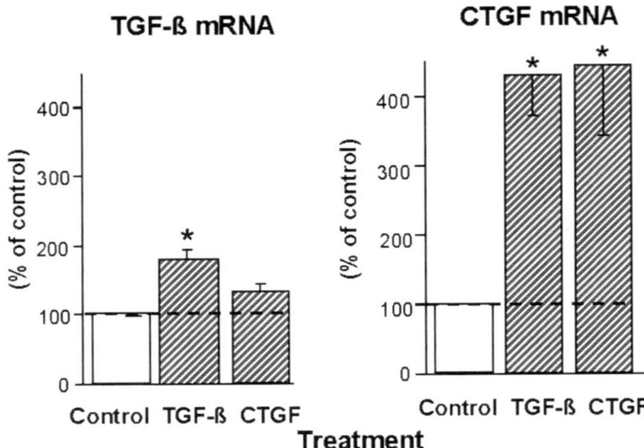

Fig. 1.   Regulation of TGF-β and CCN2 (CTGF) mRNA by exogenous TGF-β and CCN2 (CTGF). Serum-deprived mesangial cells were incubated for 24 hr in the presence of 2 ng/mL TGF-β or 20 ng/mL rhCTGF. RNA was extracted for Northern analyses and probed for, TGF-β1 or CTGF. The mRNA bands from replicate experiments were quantified by densitometric analysis, and the results normalized to the values of the ribosomal RNA. N = 4, *P < 0.05 vs. control. Reprinted courtesy of Renal Failure (Riser and Cortes, 2001).

Exposure to CCN2 failed to alter the level of TGF-β mRNA, but surprisingly auto-induced CCN2 message. When we examined the potential regulation of CCN2 by high glucose, we found that MC grown in media containing 5 mM glucose, when switched for 14 days to one containing 35 mM glucose, increased CCN2 message nearly sevenfold. The observed effects on CCN2 mRNA levels were confirmed in our studies of the protein, using Western blotting and ELISA. In unstimulated MC cultures, two faint CCN2 bands were detected at the level of the recombinant standard, approximately 36 and 39 kDa (Fig. 2A). The intensities of these full-length CCN2 bands increased markedly following treatment with either TGF-β or high glucose. Further, a smaller CCN2 molecule(s) (18–20 kDa, half-length CCN2) was strongly induced by TGF-β. These findings corresponded to the changes noted by a quantitative ELISA, i.e. unstimulated MC secreted 2.3 ng/10^6 cells/24 hr, but increased 2.5- and 2-fold following exposure to TGF-β or high glucose, respectively (Fig. 2B). The effect of high glucose levels was not due to an osmolar effect since exposure to 5 mM glucose plus 15 mM mannitol had no effect on the amount of CCN2, or the distribution of secreted CCN2 forms, as determined by immunoblotting. To determine if these effects of high glucose were mediated by TGF-β, we examined the response to elevated glucose in both the presence and absence of an

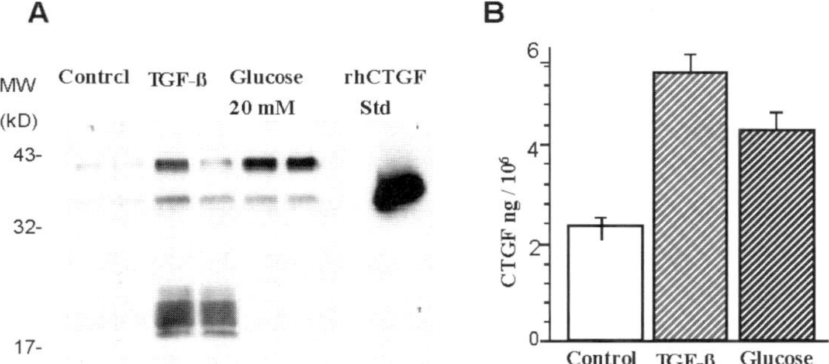

Fig. 2.   CCN2 (CTGF) mRNA (A, by Northern analysis) and protein (B, by ELISA) induction in MC. Cells grown in a medium containing 5 mM glucose were serum-depleted and then cultured for an additional 48 hr in the presence or absence of 2 ng/mL TGF-β or 20 mM glucose. Reprinted courtesy of *J Amer Soc Nephrol* (Riser *et al.*, 2000).

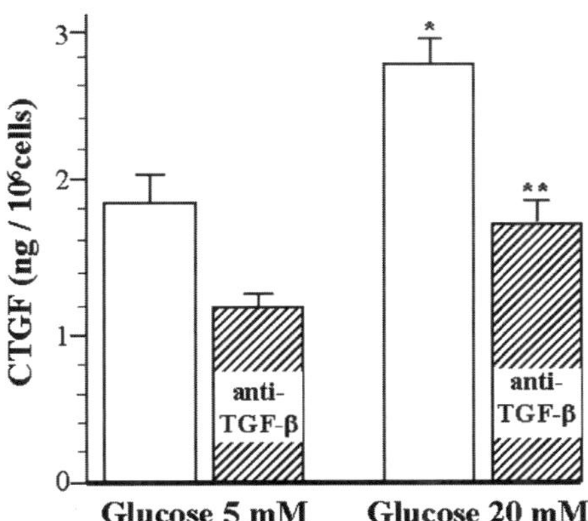

Fig. 3.   TGF-β blockade of high glucose-induced CCN2 (CTGF) production. MC cultured for 14 d in the presence of either 5 or 20 mM glucose were seeded and grown for 8 d. On day 4, the cultures were serum-deprived and half of the cultures received 20 µg/mL anti-TGF-β antibody. Fresh antibody was added daily, and the media was replaced 24 hr before collection. N = 4. *P < 0.05 *versus* 5 mM glucose; **P < 0.05 *versus* 20 mM glucose without TGF-β antibody. Reprinted courtesy of *J Amer Soc Nephrol* (Riser *et al.*, 2000).

antibody that neutralizes TGF-β. We found that neutralization of endogenous TGF-β had no effect on the constitutive secretion of CCN2 (i.e. in the presence of normal concentrations of glucose), but totally blocked the stimulation of CCN2 that occurred in response to high glucose (Fig. 3).

Last, we tested the effects of cyclic mechanical strain, mimicking conditions of MC stretch during possible low frequency oscillations in intraglomerular pressure. Since we, and others, had already shown that such mechanical strain upregulates TGF-β activity, we wanted to determine if the effects of stretch might also occur independently of TGF-β. The response of increased secreted TGF-β protein requires a minimum of 24 hours (Riser *et al.*, 1996). Therefore, for these experiments we tested the response at 4 and 8 hours. We found that the CCN2 mRNA levels were strongly increased as early as 4 hours, remaining elevated at 8 (Fig. 4), 24 and 48 hours (Riser *et al.*, 2000).

To last determine if CCN2 was mediating the effects of TGF-β on ECM, we cultured MC following stimulation with TGF-β in the presence or absence of an antisense oligonucleotide (AS-ODN) directed at the initiation codon of CCN2. We found that treatment with the ODN resulted in the blockade of elevated collagen type I production (Fig. 5). Taken as a whole, these results show that MC express low levels of CCN2 mRNA and protein under normal, non-simulated conditions. However, in the presence of elevated glucose, TGF-β, or increased mechanical strain, the cells respond by increasing their production and response to CCN2. Occurring downstream is the increased production of ECM and in turn the fibrosis or sclerosis that characterizes CKD associated with diabetes, hypertension, and reduction in renal mass. The finding of a CCN2 autoinduction *in vitro* suggests a possible *in vivo* mechanism for explaining the perpetual progression of ECM accumulation that occurs in this disease.

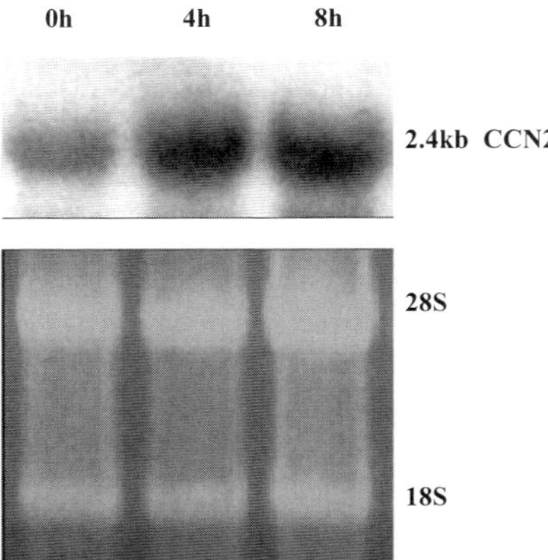

Fig. 4.   Effect of cyclic stretching on the MC expression of CCN2 transcripts. Cells cultured overnight on collagen-coated flexible-bottomed dishes were subjected to cyclic stretching (3 cycles/minute, with 11% average elongation) or control, static conditions. At the indicated periods, RNA was extracted and probed for CCN2 message. Each lane represents the results of samples pooled from 24 difference culture wells. 28 and 18S RNA levels were used to verify equal loading. Reprinted courtesy of *J Amer Soc Nephrol* (Riser *et al.*, 2000).

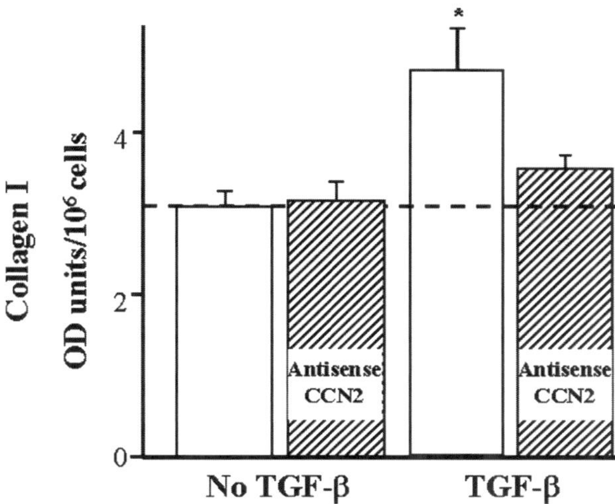

Fig. 5. Effects of CCN2 antisense oligonucleotide treatment on TGF-β stimulated collagen type I secretion in rat mesangial cells. Rat MC grown to approximately 90% confluency then maintained under serum-deprived conditions for 24 hr, were exposed to TGF-β in the presence or absence of a CCN2-specific antisense ODN. Collagen type I secretion was measured by ELISA. * = significantly (P < 0.01) different than all other groups.

To determine if CCN2 was upregulated in early DN, we studied diabetic *db/db* mice and nondiabetic *db/m* controls. The *db/db* mouse has a defective receptor gene for leptin, and therefore becomes obese at 3 to 4 weeks of age developing hyperglycemia. This is followed by a nephropathy that is primarily proteinuria and mesangial expansion with increased mesangial matrix that develops by 5–7 months. In our experiment, at 5 months of age, or approximately 3.5 months after the onset of diabetes, mean blood glucose levels, body weights, but not the level of proteinuria, were significantly elevated in the *db/db* animals (Fig. 6).

The diabetic animals exhibited notable glomerular changes, i.e. mild, but statistically significant increased mesangial matrix expansion without apparent tubulointerstitial disease. These changes are consistent with early diabetic glomerulosclerosis. Northern analysis of whole kidney RNA indicated that the CCN2 message levels increased 2-fold in diabetic mice with similar changes in fibronectin (Fig. 7A). In comparison, analysis of micro-dissected glomeruli by quantitative PCR showed a low transcript level of CCN2 in the glomeruli of control animals that were dramatically increased (27-fold) with diabetes (Fig. 7B). Fibronectin message was increased 5-fold. Our findings, in a model,

     Riser BL, Karoor S & Peterson DR

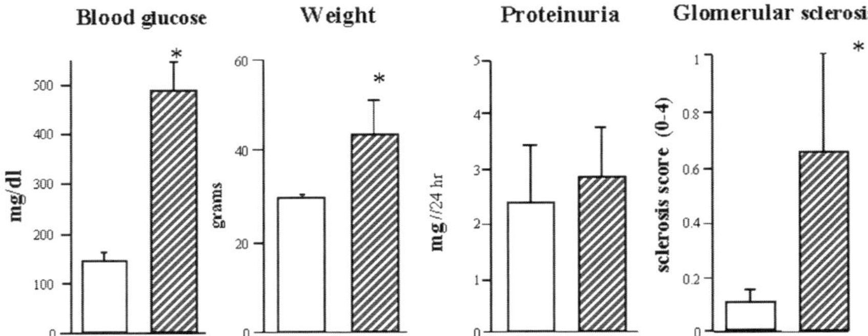

Fig. 6.   Measurement of clinical changes in *db/db* mice at 5 months. Protein excretion was the mean of 2 consecutive 24 hr urine collections. Mesangial sclerosis was scored on a scale of 0–4 (0 being no lesion to 4 consisting of diffuse collapse of capillary lumina, and sclerosis involving 75% or more of the tuft) from PAS stained kidney sections (a total of 100–150 glomeruli per kidney). *Values are significantly different from control, $P < 0.05$. Open bars are control *db/m* and striped bars are diabetic *db/db* mice. Reprinted courtesy of *Renal Failure* (Riser and Cortes, 2001).

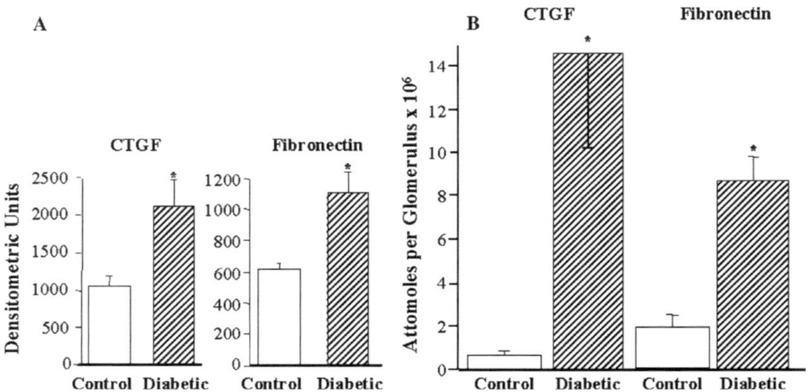

Fig. 7.   Effects of diabetes on whole kidney (A) or glomerular (B) expression of CCN2 (CTGF) and fibronectin transcript levels in *db/db* mice. At 5 months of age, analyses were carried out on whole kidneys (northern blot) or quantitative RT-PCR glomeruli microdissected (50 from each kidney) or five control (*db/m*) and five diabetic (*db/db*) mice. *P $<$ 0.05 *versus* control. Figure adapted, with courtesy, from *J Amer Soc Nephrol* (Riser *et al.*, 2000).

of type II diabetes, were supported by a report from Murphy *et al.* (1999) in a model of type I diabetes. They too demonstrated an upregulation in MC CCN2 transcript labels upon high glucose exposure, and an increase in glomerular expression of CCN2 mRNA as well.

The laboratories of Mason and Wahab have also investigated this issue. Using human MC, they also reported the production of a 36–38 kDa CCN2 with a minor band at < 30 kDa (Wahab *et al.*, 2001). However, they additionally observed a major band at approximately 56 kDa that they speculated could be full-length CCN2 complexed to one of the smaller cleavage products, or another protein. To our knowledge others have not reported this form. However, we have described a large, approximately 200 kDa form (see below) in the urine of patients with kidney disease that reacts with CCN2 specific antibody suggesting that the full-length molecule may either complex to other CCN2 molecules, or to other proteins in solution. Further support for *in vivo* findings described above comes from Wahab *et al.* (2001). They reported the detection of immunoreactive CCN2 protein in the glomeruli of non-obese diabetic mice 14 days after the onset of diabetes, prominent by 70 days. CCN2 was also present in the glomeruli of patients with DN, but no CCN2 could be detected in either normal murine or human glomeruli. Along these lines, Adler *et al.* (2001) have reported CCN2 transcript levels in glomeruli of diabetic patients that are significantly elevated over those from living donor kidneys (see further description below).

Wahab and colleagues have also transfected the MC to overexpress CCN2. They reported a resulting increase in fibronectin protein secretion as detected in a qualitative Western analysis. The same group has further shown (Yevdokimova *et al.*, 2001) that the molecule thrombospondin-1 (TSP-1), known to activate TGF-β *in vivo*, is involved. They showed that the blockade of TSP-1 was able to inhibit the *in vitro* MC response to high glucose. One of the recognized early changes in experimental DN is mesangial cell hypertrophy. How this event influences the later changes that occur in the mesangium is not clearly known at this time. However, the same laboratory has recently investigated the possibility that CCN2 might be involved. They reported that CCN2 stimulates human MC to actively enter G_1 phase from G_o, but apparently does not stimulate progression further through the cell cycle (Wahab *et al.*, 2002). This appeared to occur via the induction of cyclin-dependent kinase inhibitors, CDKI, $p15^{INK4}$, and $p27^{Kip1}$. Their use of CCN2 AS-ODN also indicated that the previously identified TGF-β induced hypertrophy in MC is CCN2-dependent.

POSSIBLE ROLE OF AGE IN CCN2 MEDIATED DIABETIC NEPHROPATHY

Advanced glycosylated end products (AGE), irreversibly formed biochemical end products of non-enzymatic glycosylation, have been implicated as causal in the expansion of ECM that occurs in DN. AGE are present in the serum and in many tissues of diabetic patients, and have the ability to covalently cross-link and biochemically modify protein structure and function. Twigg *et al.* (2002) examined a possible connection between CCN2 and AGE. They found that AGE exposure was able to upregulate, in dermal fibroblasts, CCN2 mRNA and protein levels in a specific manner. This suggests that AGE may augment the development of fibrosis in DN via its effects on CCN2. This hypothesis was furthered by the findings of Forbes *et al.* (2003). In the streptozotocin (STZ) model of diabetes, they found that treatment with an AGE cross-link breaker ALT711 resulted in a significant reduction in serum and renal AGE that was associated with a decrease in albumin excretion rate (AER), blood pressure, and renal hypertrophy, along with a decrease in gene expression of CCN2, TGF-β, and collagen type IV. These effects were seen following both early (long-term 16 weeks) and late (shorter, 8 weeks) treatments. However, early treatment was required to reduce the glomerulosclerotic index, the tubulointerstitial area, and the expression of collagen type IV and TGF-β proteins. In another treatment study, Twigg and colleagues (Twigg *et al.*, 2002) were able to show that long-term treatment with aminoguanidine, an advanced glycation inhibitor, blocked the increased CCN2 and fibronectin expression seen in experimental DN. Further, the exposure of cultured MC to AGE increased both CCN2 and fibronectin expression (Twigg *et al.*, 2002). These results indicate yet another role for CCN as a downstream mediator of the effects of a known causal factor in experimental and human renal disease, in this case DN. This supports the idea of CCN2 as a final common pathway to the development of sclerosis/fibrosis in the kidney.

CCN2 IN TUBULOINTERSTITIAL FIBROSIS

Further evidence for CCN2 upregulation as a downstream common element to the development of fibrosis comes from work on tubulointerstitial fibrosis (TIF). This form of renal pathology is itself regarded as a common sequel to renal injury and is considered a late stage final common pathway in the progression to end stage renal failure in numerous afflictions, including DN. One model that has been used to study TIF is the 5/6 nephrectomy. This model involves the complete removal of one kidney, and 2/3 nephrectomy on the contra-lateral side. The severe reduction in renal mass results in a marked increase in intraglomerular pressure, which appears to largely drive the resulting glomerulosclerosis, interstitial fibrosis, and eventual tubular atrophy. In addition to the degenerative and regenerative changes, there is tubular epithelial-myofibroblast transdifferentiation. The transition to myofibroblast-like cells has been the focus of many recent studies and appears to be critical to the development of fibrosis at this site. The rate and final stage of progression in the 5/6-nephrectomy model vary with the surgical technique used as well as the species and strain of animal.

Frazier *et al.* (2000) used the 5/6-nephrectomy model, in Sprague–Dawley rats, to examine CCN2 expression, and the relationship to other growth factors known to be involved in the fibrosis process. They reported that CCN2 mRNA expression was minimal in controls, mild at 2 weeks, and marked by 4 to 8 weeks in interstitial fibroblasts, coinciding with damage, regeneration, and fibrosis. TGF-β expression, now thought to at least in part drive the epithelial-myofibroblast transdifferentiation described above, on the other hand, was increased in many cell types at 2 weeks. CCN2 expression was additionally augmented by 4 weeks, and then remained constant. PDGF-β message was found in many stromal cells at 2–4 weeks, but expression decreased at 8 weeks. No significant IL-1 or TNF-α was detected. The proximity of these upregulated growth factors to regenerative epithelial cells and those transdifferentiating to myofibroblasts suggested a role for CCN2 in modulation of renal tubular epithelial differentiation.

Using the STZ-induced model of type I diabetes, Wang *et al.* (2001) found a doubling of the CCN2 mRNA levels in kidney cortex (identical to what we reported in the whole kidney of the *db/db* mouse model of type II diabetes). Further, they found that the glomerular ultra-filtrate from diabetic rats, TGF-β, and to a lesser degree hepatocyte growth factor (HGF), but not high glucose exposure, induced increased expression of CCN2 in cultured mouse proximal convoluted tubule cells (MCT) (Wang *et al.*, 2001). rhCCN2 exposure moderately increased fibronectin, but not collagen type I and III in MCT cells. However the latter was elevated by rhCCN2 in cultured normal rat kidney fibroblasts (NRK-49F). Yokoi and colleagues (Yokoi *et al.*, 2001) also investigated the role of CCN2 in renal tubulointerstitial fibrosis using the unilateral ureteral obstruction (UUO) model in rats. This model has the advantage of inducing a rapid, albeit transient, fibrosis. They found that following obstruction, there was an early upregulation in both TGF-β and CCN2 message, followed by an induction of fibronectin and collagen type I mRNA levels. In cultured NRK-49F cells CCN2 antisense oligonucleotide treatment appeared to attenuate TGF-β-induced increase in fibronectin and collagen type I mRNA levels (Yokoi *et al.*, 2001). Along these lines, Zang and colleagues recently investigated the role of CCN2 in renal tubular epithelial-myofibroblast transdifferentiation and ECM accumulation by examining the effects of TGF-β

on a cultured human proximal tubule epithelial cell line (Zhang *et al.*, 2004). They found that exogenous exposure upregulated the transcripts for CCN2, followed by α-smooth muscle actin, fibronectin, and plasminogen activator inhibitor-1 (PAI-1). These markers all appeared inhibited by CCN2 AS-ODN treatment. Yokoi *et al.* (2004) have very recently provided proof of a causal role for CCN2 in the development of TIF. In the UUO model they reported that treatment of mice with AS-ODN directed against CCN2 specific sequences, when injected into the renal artery were able to markedly attenuate the upregulation of CCN2, fibronectin and collagen type I, without affecting the expression of TGF-β. This treatment also reduced the number of myofibroblasts as determined by α-smooth muscle actin, but did not alter the number of proliferating tubule or interstitial cells. Very little is currently known about how CCN2 mediates its biological effects on target cells, including MC and fibroblasts. This includes cellular transformation and increased production of ECM. Heusinger–Ribeiro, Goppelt–Struebe, and colleagues have investigated (Heusinger–Ribeiro *et al.*, 2001) the potential roles of RhoA and the cytoskeleton in CCN2 mediated effects, providing evidence for such mechanisms. This area is the focus of another chapter in this text.

Taken as a whole, these results provide experimental evidence of a role for CCN2, also in TIF. Further, they suggest that the damage driven by CCN2 in CKD, including that associated with diabetes is not limited to the glomerulus. The pathological cellular response to CCN2 at the two different sites does not appear to be identical however, as seen for example by the apparent CCN2-induced myofibroblast transformation. The at least partial differential response of proximal tubule cells and interstitial fibroblasts as compared to MC may be the result of a cell specific response to CCN2, or alternatively a response to CCN2 seen in combination with other local cytokine/growth factors unique to the different sites. This could be an area for fruitful study.

CCN2 IN GLOMERULONEPHRITIS (GN)

The first evidence of a possible role for CCN2 in GN came from Ito *et al.* (1998). As described above, they reported an upregulation of specific mRNA in human cases of crescentic glomerulonephritis, IgA nephropathy, and focal and segmental glomerulosclerosis. This upregulation was observed predominately in the tubulointerstitial areas at sites of chronic damage. Exploration of the mechanisms and role of CCN2 in GN, although limited has included the anti-Thy-1 experimental model of proliferative GN. In this model, antibody directed against Thy-1 is typically injected as a single dose into rats or mice. In addition to expression on thymoyctes, for which it was named, the Thy-1 protein is also found on MC *in situ*. Anti-Thy-1 injection results at day 1 in a complement-dependent mesangiolysis, and the formation of aneurysms that are followed by a glomerular-proliferative phase from day 4 to 14. This is characterized by a replacement of MC accompanied by increased production of mesangial ECM. The proliferative response is followed by a relatively rapid resolution of pathology by day 14, unless a second administration of antibody is given. Using this model, Ito *et al.* (2001) reported that CCN2 was strongly increased in extracapillary and mesangial proliferative lesions and in areas of periglomerular fibrosis. Early glomerular CCN2 overexpression in glomerular visceral epithelial cells coincided temporally with the upregulation of TGF-β1 and to a lesser degree TGF-β3. Glomerular CCN2 and TGF-β mRNA expression was maximal at day 7. CCN2 expression by parietal epithelial cells preceded the periglomerular appearance of α-smooth muscle actin-positive fibroblasts. Further, cultured glomerular visceral epithelial cells responded by quickly upregulating CCN2 mRNA levels when exposed to treatment by TGF-β (Ito *et al.*, 2001).

Recently, Kanemoto *et al.* (2003) studied the expression of CCN2 and TGF-β in crescentric GN (CRGN). Crescents are a determinant of progression in some renal diseases, leading to scar formation, and are made up primarily of glomerular parietal epithelial cells (PEC). In a rat model of CRGN, they found that in the early acute phase, the majority of cells were macrophages that did not express CCN2 (Kanemoto *et al.*, 2003). However, in the advanced phase, crescents strongly expressed CCN2 mRNA, and demonstrated markers for epithelial cells, and not macrophages. PEC in culture were stimulated by TGF-β or PDGF-BB to upregulate both ECM and CCN2. The same group also showed relevance to human CRGN by examining biopsy specimens from 18 CRGN patients (Kanemoto *et al.*, 2004). CCN2 mRNA was expressed in the podocytes and PEC in unaffected glomeruli and was strongly expressed in cellular and fibrocellular crescents particularly in pseudotubule structures. Expression was co-localized with TGF-β, TGF-β receptor, collagen type I and fibronectin message.

These studies of experimental and human GN indicate that CCN2 is also involved in the renal response to immune- or inflammation-mediated damage, and may explain the resulting fibrosis/sclerosis.

OTHER CCN GENES IN RENAL DISEASE

There currently exists little data on the role of other CCN, i.e. non-CCN2, members in the kidney. However, Sawai (2003) interestingly found that CCN1 (Cyr-61), known particularly for its angiogenic activity, was present in proximal tubules and afferent and efferent arterioles of normal kidneys. However, in the glomeruli of animals with Thy-1 induced glomerulonephritis the protein was induced in podocytes, and appeared less broadly distributed than CCN2 upregulation in this model, suggesting perhaps distinct roles. This finding is very interesting. Future studies are justified, and will be necessary to determine which CCN members are involved in renal disease processes, and whether they play supportive, overlapping, or opposing roles. Our laboratory is currently investigating this area.

CCN MOLECULES AS TARGETS FOR DIAGNOSIS AND PREDICTION OF DISEASE

It is currently impossible to reliably predict which diabetic patients will be among the 30–40% that will develop nephropathy and progress to kidney failure. Determination of microalbuminuria is regularly sought in these patients, with positive results serving to signal renal involvement and the need to begin renal-focused therapy. Microalbuminuria, therefore, is not a predictor of overt DN, but rather an indicator of established damage. Our previous observations (described above) of CCN2 in MC, coupled with those of an early and marked expression of the cytokine in the glomerulus, led us to postulate that CCN2 might be excreted in the urine and that its early presence might serve as a predictor for the progression of DN. To test this, we first examined urine from healthy rats in comparison to those made diabetic by STZ. Low levels of urinary CCN2 were present in healthy, control rats, but were increased 7-fold overall in diabetic animals (Fig. 8A) (Riser *et al.*, 2003). Levels were highest at 3 weeks of diabetes, then decreased with time, but remained elevated over controls, even after 32 weeks. In contrast, 32 weeks was required in order to detect a significant increase in albumin levels (Fig. 8B).

By analogy, consistent low levels of CCN2 were also detected in healthy human volunteers. However, levels were elevated approximately 6-fold in the majority of diabetic patients with nephropathy (Fig. 9). As predicted, a small number of patients not yet exhibiting evidence of renal involvement showed urinary CCN2 levels 9-fold greater than controls, even though the remaining normoalbuminuric patients had CCN2 levels indistinguishable from the healthy controls (Fig. 9). A qualitative analysis by Western blotting demonstrated that while a band equivalent to the full-length CCN2 (37/39 kDa doublet) both healthy and diabetic persons, diabetic patients exhibited other multiple bands, including a high (approximate 200 kDa) molecular weight band, as well as a low weight (approximate 12 kDa) band equivalent to the

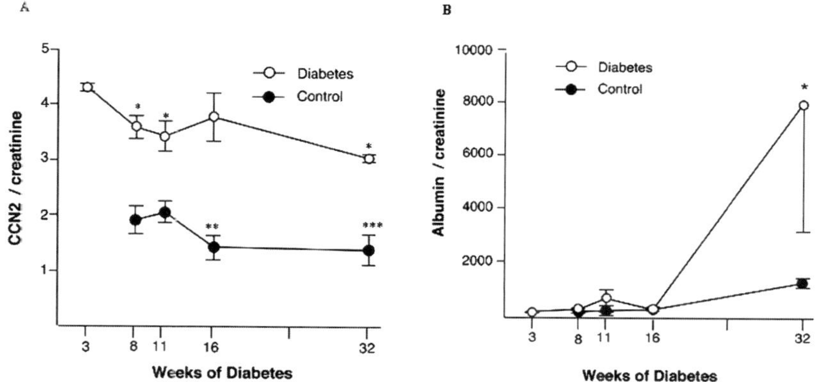

Fig. 8. Urinary CCN2 (A) and albumin levels (B) in normal and diabetic rats. Concentrations in two consecutive 24 hr urine samples were measured by ELISA and normalized to the amount of creatinine in the same sample. All CCN2 values (mean log ± SE) in the experimental group ($n = 12$) were significantly different than those in the control group ($n = 6$) at the corresponding periods ($P < 0.0001$). Within each group, CCN2 levels decreased significantly over time. *P < 0.0006 *versus* 3 wks; **P < 0.0001 *versus* 11 wks; ***P < 0.032 *versus* 8 wks and 11 wks. The value (mean ± SE) of albumin in diabetic animals ($n = 12$) was significantly greater at 32 wks, only, when compared to the corresponding control ($n = 6$) group ($P = 0.042$). Reprinted courtesy of *Kidney Int* (Riser *et al.*, 2003).

quarter fragment of full-length CCN2. Our results suggest that glomerular and urinary CCN2 may provide a unique target for, not only diagnosing the onset and stage of progression of renal disease including CKD, but also, in the case of diabetes may be able to predict which patients are destined for later progression to DN.

CCN MOLECULES AS TARGETS FOR THERAPEUTICS

We, and now many others, have proposed the use of CCN2 as a therapeutic target for various forms of renal disease, particularly those where progression leads ultimately to a fibrosis and/or sclerosis. To date experimental evidence supports this as a reasonable approach. One method for the control of CCN2 activity may be possible by the administration of neutralizing antibodies. Antibodies directed against CCN2 have been generated in a number of different species, and to different sites on the molecule, and might seem an obvious potential therapy. However, very few of these antibodies have been reported to demonstrate neutralizing activity. This seems quite unlike the case for other cytokines, e.g. TGF-β where effective blocking antibodies have been generated. The explanation for this observation with CCN2 is currently unexplained to

     Riser BL, Karoor S & Peterson DR

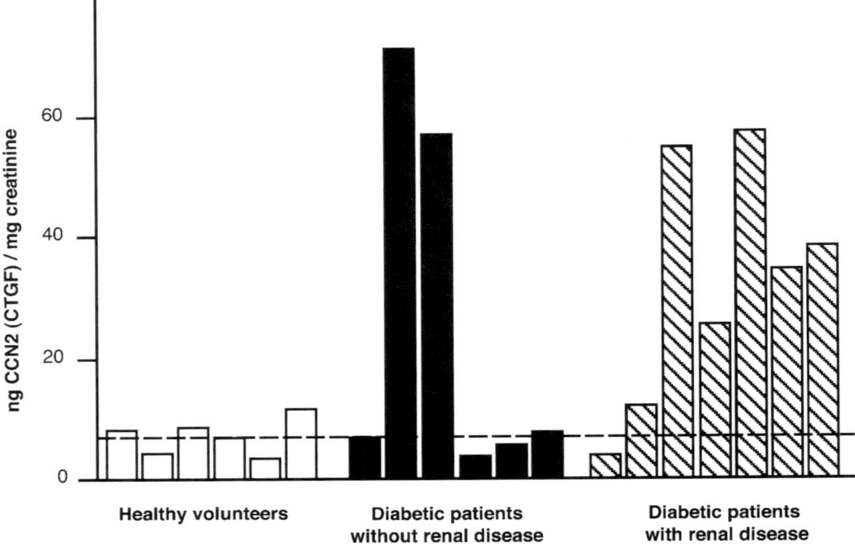

Fig. 9.   Measurement of urinary CCN2 from patients and healthy volunteers. Urinary CCN2 levels were measured by ELISA. Each bar represents the determination from an individual patient sample run in triplicate. The dashed line indicated the mean value for all healthy volunteers. Reprinted courtesy of *Kidney Int* (Riser *et al.*, 2003).

us. However, FibroGen, a company involved early in CCN2 biology, has developed an anti-CCN2 monoclonal antibody that is currently in an open-label, Phase I clinical trial (public announcement). This is the first step in evaluating the therapeutic potential of the antibody, with the indication of idiopathic pulmonary fibrosis.

Another potential area for anti-CCN2 based therapy may come from the utilization of endogenous factors that are part of the normal mechanism for countering, or down regulating, CCN2 production and/or activity. Along these lines, Cruzado *et al.* (2004) have reported that the gene delivery and expression of human hepatocyte growth factor (HGF) to STZ-diabetic rats reduced albuminuria, regressed mesangial expansion and the development of glomerulosclerosis. This was associated with the suppression of glomerular CCN2 upregulation. Although a number of recent reports have indicated a possible role for HGF as an endogenous factor reducing fibrosis development, the direct effects of HGF on CCN2 appear to be unknown. Another hormone-like molecule bone morphogenic peptide 7 (BMP-7) has been shown to ameliorate experimental forms of interstitial fibrosis (Wang and Hirschberg, 2003).

Treatment of MC with BMP-7 was shown to reduce the CCN2 and ECM levels stimulated by TGF-β (Wang and Hirschberg, 2003).

Agents currently on the market and used in the treatment of CKD, for example angiotensin-converting enzyme (ACE) inhibitors, may exert their effects, at least in part, by their effect on CCN2 activity. Understanding how this effect is mediated may lead to improvement within these classes of drugs. Along these lines, Goppelt–Struebe *et al.* (2001) reported that simvastatin an HMG CoA reductase inhibitor, inhibited CCN2 mRNA expression in cultured MC in a concentration-dependent manner, suggesting that statins may have a potential to reduce fibrosis development via CCN2. In another study, Makino *et al.* (2003) in order to examine the response of CCN2 to prostanoids treated STZ-diabetic rats with aspirin. They reported that chronic treatment significantly attenuated mesangial expansion and effectively suppressed CCN2. Last, new therapeutics targeted at AGEs, for example ALT-711 (Forbes *et al.*, 2003a; 2003b) may work through their ability to block stimulation of CCN2.

Our laboratory has focused attention of the development of inhibitors of CCN2 RNA regulation and utilization in the form of both AS-ODN and siRNAs. This area has the potential advantage of targeting specific effects. The greatest obstacles with this form of potential therapy involve delivery to the kidney, and proof of limited off target effects. As described above, CCN2 AS-ODN administered via the renal vein has been used to successfully demonstrate an attenuation of fibrosis in the UUO model. This is ostensibly a reasonable model to test such agents, because of the rapid development of fibrosis. However, the use of such an agent to treat CKD in patients will likely require a long-term stable agent with the acceptability of an oral, or infrequent IV parenteral administration providing targeted delivery to the appropriate renal site.

CCN IN KIDNEY DEVELOPMENT AND NORMAL PHYSIOLOGY

While little is currently known about the role of CCN molecules in normal renal physiology, a number of studies have shown a differential time-related expression of the CCN genes during development, thus indicating a role in embryogenesis. For example, Surveyor and Bridgestock (1999) identified in mouse embryos, by immunolocalization, CCN2 reactive protein in the developing kidney along with many other, although not all, major organs. Expression in renal tissue appeared at, but not before, day 14 of gestation. Chevalier *et al.* (1998) showed that during normal nephrogenesis CCN3 (NovH) protein was tightly associated with differentiation glomerular podocytes and was also detected in endothelium and neural tissue of the kidney. Abnormally high expression of CCN3 interestingly occurred in sporadic and heritable Wilms tumors suggesting how this particular renal pathology may result from the abnormal regulation of the CCN family member.

A role for CCN members in normal physiology, beyond this rudimentary information in development, has not been established. Studies employing immunostaining have largely failed to consistently, and precisely detect and map the expression of CCN protein in the tissues examined. Our laboratory, like others, have failed to detect marked staining for CCN2 protein in the normal glomerulus, in spite of the fact that specific mRNA can be measured in isolated glomeruli. However, we have observed reproducible, focal staining of CCN2 protein associated with tubules of normal mice (unpublished observation). The finding of CCN2 at low levels in the urine of both healthy humans and rodents, suggests a role in normal physiological function. However the exact source of this CCN2 is currently unknown. Of the members of the CCN family, CCN2 would be expected to also play a beneficial role in the normal wound healing process, although currently this has not been the target of may investigations. One can envision the circumstance where, like with TGF-β, it is only in the environment where a chronic insult or overwhelming injury occurs that the uncontrolled expression of this normally beneficial cytokine becomes a causal factor now driving the course of the disease.

SUMMARY AND CONCLUSIONS

The formation of sclerosis or fibrosis in the kidney is a common response to severe or chronic forms of injury. At least in CKD, there appears to be three predominant casual factors, metabolic, genetic, and homodynamic (Fig. 10). All of these factors can interact, particularly in DN, to drive progression. We envision CCN2 as a central, downstream mediator of the effects of these three elements. For example, pathological shear or stretching force resulting from intraglomerular hypertension appears to stimulate the production of cytokines including CCN2. This same force also appears to be responsible for increased vascular permeability leading to both proteinuria and an increased production of vasoactive hormones such as Ag II and endothelin, which in turn elevate CCN2 and further enhance the mechanical force. The abnormal accumulation of AGE that occurs with the altered metabolism of glucose in DN may also work to both directly increase ECM cross-linking and accumulation, as well as to increase CCN2. The genetic background of the individual can influence

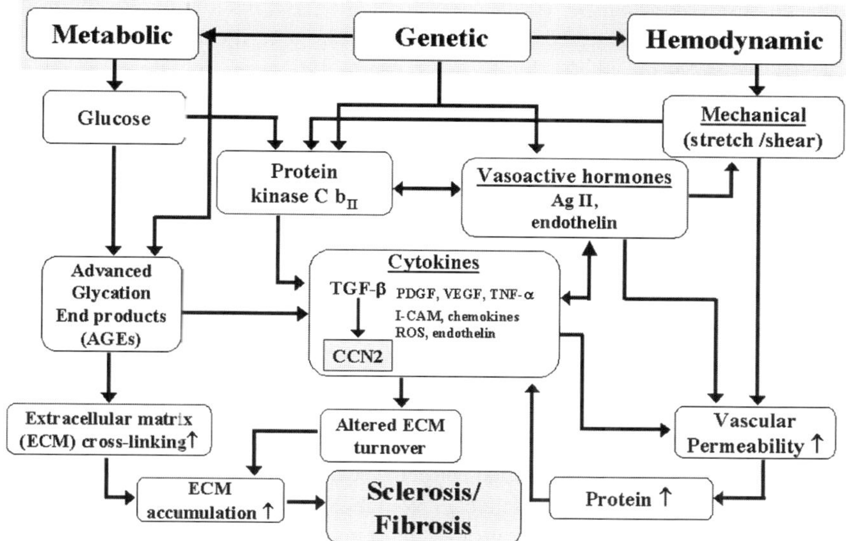

Fig. 10.   Schematic depicting the proposed central role of CCN2 in renal fibrosis and sclerosis.

the elements of hemodynamics and metabolism, and in turn the resulting pathways as described. Additionally there is a likely influence on PKC activity and production of vasoactive hormones. In all cases, the chronic upregulation of CCN2 activity is likely to result in altered ECM turnover and increasing ECM accumulation, producing fibrosis or sclerosis. These findings support the postulate that CCN2 is a central downstream element in the progression of renal fibrosis, and as such, provides a reasonable and novel target for both diagnostics and therapeutics. Clearly, understanding these pathways will be important in the development of both.

REFERENCES

Adler SG, Kang S-W, Feld S, *et al.* (2001) Glomerular mRNAs in human type 1 diabetes: biochemical evidence for microalbuminuria as a manifestation of diabetic nephropathy. *Kidney Int* **60**:2330–2336.

Bradham DM, Igarashi A, Potter RL and Grotendorst GR (1991) Connective tissue growth factor: a cysteine-rich mitogen secreted by human vascular endothelial cells is related to the SRC-induced immediate early gene product CEF-10. *J Cell Biol* **114**:1285–1294.

Chevalier G, Yeger H, Martinerie C, *et al.* (1998) novH: differential expression in developing kidney and Wilm's tumors. *Amer J Pathol* **152**:1563–1575.

Cruzado JM, Lloberas N, Torras J, *et al.* (2004) Regression of advanced diabetic nephropathy by hepatocyte growth factor gene therapy in rats. *Diabetes* **53**:1119–1127.

Forbes JM, Cooper ME, Oldfield MD and Thomas MC (2003b) Role of advanced glycation end products in diabetic nephropathy. *J Amer Soc Nephrol* **14**:S254–S258.

Forbes JM, Thallas V, Thomas MC, *et al.* (2003a) The breakdown of pre-existing advanced glycation end products is associated with reduced renal fibrosis in experimental diabetes. *FASEB* **17**:1762–1764, 1710.1096/fj.1702-1102fje.

Frazier KS, Paredes A, Dube P and Styer E (2000) Connective tissue growth factor expression in the rat remnant kidney model and association with tubular epithelial cells undergoing transdifferentiation. *Vet Pathol* **37**:328–335.

Goppelt–Struebe M, Hahn A, Iwanciw D, *et al.* (2001) Regulation of connective tissue growth factor (ccn2: ctgf) gene expression in human mesangial cells: modulation by HMG CoA reductase inhibitors (statins). *Mol Pathol* **54**:176–179.

Hammes MS, Lieske JC, Pawar S, *et al.* (1995) Calcium oxalate monohydrate crystals stimulate gene expression in renal epithelial cells. *Kidney Int* **48**:501–509.

Heusinger–Ribeiro J, Eberlein M, Abdel Wahab N and Goppelt–Struebe M (2001) Expression of connective tissue growth factor in human renal fibroblasts: regulatory roles of RhoA and cAMP. *J Amer Soc Nephrol* **12**:1853–1861.

Ito Y, Aten J, Bende RJ, *et al.* (1998) Expression of connective tissue growth factor in human renal fibrosis. *Kidney Int* **53**:853–861.

Ito Y, Goldschmeding R, Bende RJ, *et al.* (2001) Kinetics of connective tissue growth factor expression during experimental proliferative glomerulonephritis. *J Amer Soc Nephrol* **12**:472–484.

Kanemoto K, Usui J, Nitta K, *et al.* (2004) *In situ* expression of connective tissue growth factor in human crescentic glomerulonephritis. *Virchows Archiv* **444**:257–263.

Kanemoto K, Usui J, Tomari S, *et al.* (2003) Connective tissue growth factor participates in scar formation of crescentic glomerulonephritis. *Lab Invest* **83**:1615–1625.

Makino H, Mukoyama M, Sugawara A, *et al.* (2003) Roles of connective tissue growth factor and prostanoids in early streptozotocin-induced diabetic rat kidney: the effect of aspirin treatment. *Clin Exp Nephrol* **7**:33–40.

Murphy M, Godson C, Cannon S, *et al.* (1999) Suppression subtractive hybridization identifies high glucose levels as a stimulus for expression of connective tissue growth factor and other genes in human mesangial cells. *J Biol Chem* **274**:5830–5834.

Pawar S, Kartha S, Toback FG (1995) Differential gene expression in migrating renal epithelial cells after wounding. *J Cell Physiol* **165**:556–565.

Riser BL and Cortes P (2001) Connective tissue growth factor and its regulation: a new element in diabetic glomerulosclerosis. *Renal Failure* **23**:459–470.

Riser BL, Cortes P and Yee J (2000) Modelling the effects of vascular stress in mesangial cells. *Curr Opin Nephrol Hy* **9**:43–47.

Riser BL, Cortes P, DeNichilo M, *et al.* (2003) Urinary CCN2 (CTGF) as a possible predictor of diabetic nephropathy: preliminary report. *Kidney Int* **64**:451–458.

Riser BL, Cortes P. Heilig C, *et al.* (1996) Cyclic stretching force selectively up-regulates transforming growth factor-beta isoforms in cultured rat mesangial cells. *Amer J Pathol* **148**:1915–1923.

Riser BL, Cortes P, Yee J, *et al.* (1998) Mechanical strain- and high glucose-induced alterations in mesangial cell collagen metabolism: role of TGF-beta. *J Amer Soc Nephrol* **9**:827–836.

Riser BL, Cortes P, Zhao X, *et al.* (1992) Intraglomerular pressure and mesangial stretching stimulate extracellular matrix formation in the rat. *J Clin Invest* **90**:1932–1943.

Riser BL, Denichilo M, Cortes P, *et al.* (2000) Regulation of connective tissue growth factor activity in cultured rat mesangial cells and its expression in experimental diabetic glomerulosclerosis. *J Amer Soc Nephrol* **11**:25–38.

Riser BL, Varani J, Cortes P, *et al.* (2001) Cyclic stretching of mesangial cells up-regulates intercellular adhesion molecule-1 and leukocyte adherence: a possible new mechanism for glomerulosclerosis. *Amer J Pathol* **158**:11–17.

Sawai K, Mori K, Mukoyama M, *et al.* (2003) Angiogenic protein Cyr61 is expressed by podocytes in anti-Thy-1 glomerulonephritis. *J Amer Soc Nephrol* **14**:1154–1163.

Surveyor GA and Brigstock DR (1999) Immunohistochemical localization of connective tissue growth factor (CTGF) in the mouse embryo between days 7.5 and 14.5 of gestation. *Growth Factors* **17**:115–124.

Twigg SM, Cao Z, McLennan SV, *et al.* (2002) Renal connective tissue growth factor induction in experimental diabetes is prevented by aminoguanidine. *Endocrinol* **143**:4907–4915.

Wahab NA, Weston BS, Roberts T and Mason RM (2002) Connective tissue growth factor and regulation of the mesangial cell cycle: role in cellular hypertrophy. *J Amer Soc Nephrol* **13**:2437–2445.

Wahab NA, Yevdokimova N, Weston BS, *et al.* (2001) Role of connective tissue growth factor in the pathogenesis of diabetic nephropathy. *Biochem J* **359**:77–87.

Wang S, Denichilo M, Brubaker C and Hirschberg R (2001) Connective tissue growth factor in tubulointerstitial injury of diabetic nephropathy. *Kidney Int* **60**:96–105.

Wang S and Hirschberg R (2003) BMP7 antagonizes TGF-b-dependent fibrogenesis in mesangial cells. *Amer J Physiol* **284**:F1006–F1013.

Yevdokimova N, Abdel Wahab N and Mason RM (2001) Thrombospondin-1 is the key activator of TGF-b1 in human mesangial cells exposed to high glucose. *J Amer Soc Nephrol* **12**:703–712.

Yokoi H, Mukoyama M, Nagae T, *et al.* (2004) Reduction in connective tissue growth factor by antisense treatment ameliorates renal tubulointerstitial fibrosis. *J Amer Soc Nephrol* **15**:1430–1440.

Yokoi H, Sugawara A, Mukoyama M, *et al.* (2001) Role of connective tissue growth factor in profibrotic action of transforming growth factor-beta: a potential target for preventing renal fibrosis. *Amer J Kidney Dis* **38**.

Zhang C, Meng X, Zhu Z, *et al.* (2004) Role of connective tissue growth factor in renal tubular epithelial-myofibroblast transdifferentiation and extracellular matrix accumulation *in vitro. Life Sci* **75**:367–379.

CHAPTER 6

CCN PROTEINS IN LIVER INJURY AND DISEASE

Amy W. Rachfal and David R. Brigstock*

Center for Cell and Vascular Biology
Children's Research Institute
700 Children's Drive
Columbus Ohio 43205 USA
Tel. 614-722-2840; Fax. 614-722-5892
**e-mail: brigstod@pediatrics.ohio-state.edu*

CCN proteins regulate a broad spectrum of cellular functions including adhesion, migration, proliferation and survival. CCN family members are produced by various hepatic cells in response to injury or cellular transformation. Repair or regeneration following drug-induced hepatoxicity or hepatic resection is characterized by enhanced expression of CCN2 via pathways that are downstream of the actions of tumor necrosis factor-α and/or transforming growth factor-β. CCN2 and CCN3 levels are enhanced in primary and metastatic hepatocellular carcinoma, while CCN4 is over-produced in cholangiocarcimonas. In hepatic fibrosis, CCN2 is produced by all major cell types, often downstream of transforming growth factor-β. CCN2 stimulates a profibrotic phenotype in hepatic stellate cells, the principal fibrogenic cell type, through its regulation of proliferation, adhesion, migration, proliferation and survival as well as production of α-smooth muscle actin or collagen I. Binding of CCN2 to stellate cells occurs via cell surface integrin $\alpha v \beta 3$ and low density lipoprotein receptor-related protein, both of which utilize heparan sulfate proteoglycans as a co-receptor. CCN proteins are a novel class of cell regulatory molecules that may represent new therapeutic targets for a variety of liver diseases, especially fibrosis and cirrhosis.

INTRODUCTION

The CCN gene family comprises six 30–40 kDa cysteine-rich mosaic proteins (CCN1–6). These molecules are secreted modular proteins that become extracellular matrix (ECM)-associated and regulate diverse cellular activities such as adhesion, migration, mitogenesis, differentiation and survival. CCN proteins likely exert many of their biological effects by direct binding interactions with components of the ECM, integrins and other cell surface molecules including heparan sulfate proteoglycans (HSPG) and low density lipoprotein receptor related protein (LRP). Connective tissue growth factor (CCN2, also known as CTGF) has attracted particular interest in the context of liver disease because of its roles in fibrosis, injury and cancer. CCN2 is a multi-functional matricellular protein that is produced by a variety of cell types and acts via autocrine and paracrine circuits to regulate many cellular functions including growth, proliferation, apoptosis, adhesion, migration, extracellular matrix production, and differentiation (Brigstock, 1999; Lau and Lam, 1999; Moussad and Brigstock, 2000; Perbal, 2001). CCN2 appears to play a role in diverse biological processes including angiogenesis, chondrogenesis, embryogenesis, implantation, development, ovarian function, tumorigenesis and fibrosis. This article will focus on the involvement of the CCN gene family in liver injury and regeneration, carcinoma and fibrosis.

HEPATIC STRUCTURE AND FUNCTION

The liver is divided into lobes that are further subdivided into smaller anatomic units or lobules. These lobules serve as functional units within which many substances are metabolically processed by hepatocytes. A central vein is at the middle of the classic hexagonal lobule with portal triads at three of its six angles. The portal triads contain a portal vein, hepatic artery and bile duct. Between the central vein and portal triads are cords of hepatocytes surrounded by a fenestrated endothelium. Hepatic stellate cells (HSC) reside in the intervening space of Disse within the sinusoidal lumen while tissue macrophages, also known as Kupffer cells, lie in the blood sinusoid in close proximity to the endothelium (Fig. 1).

The liver plays an essential function in maintaining the metabolic homeostasis of the body. It performs a major role in the synthesis and metabolism of protein, carbohydrate and fats while also performing phagocytosis. Many foreign substances and hormones are detoxified as they pass through the liver. In addition, the liver is responsible for the synthesis and secretion of bile as well as storing many vitamins and minerals. Therefore, hepatic disorders have wide-ranging consequences and are often life-threatening.

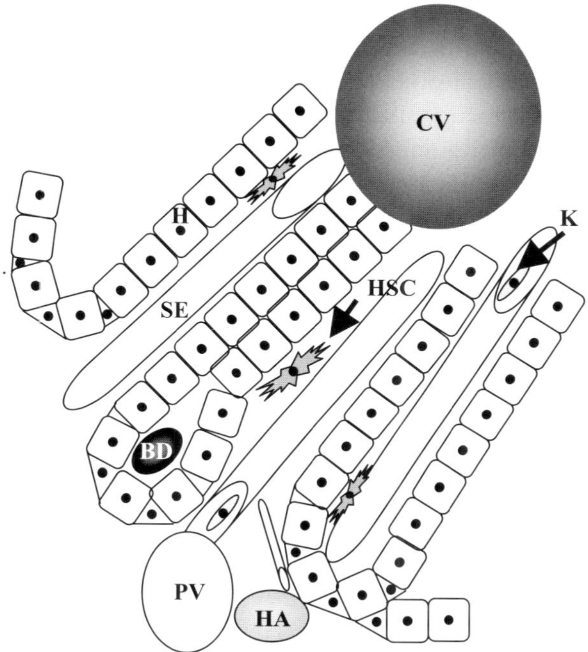

Fig. 1.   Organization of the principal hepatic cell types.

Between the central vein (CV) and portal triad (the latter of which comprises the bile ductile (BD), portal vein (PV), and hepatic arteriole (HA)) cords of hepatocytes (H) are surrounded by a fenestrated sinusoidal endothelium (SE). Hepatic stellate cells (HSC) are located within the space of Disse and when activated, proliferate and produce excess ECM. In healthy subjects, Kupffer cells (K) reside within the blood sinusoids along the endothelium.

DRUG-INDUCED HEPATOTOXICITY AND INJURY

Drug-induced liver disease is a relatively common, but often undiagnosed, cause of liver injury (Goodman, 2002). The unique metabolic processes within the liver coupled with its intimate relationship to the gastrointestinal tract make it a key target for drug cytotoxicity (Pineiro–Carrero and Pineiro, 2004). Many drugs are lipophilic and readily cross the gastrointestinal tract and subsequently undergo biotransformation in the liver (oxidation and conjugation) into hydrophilic derivatives that are excreted in the urine or bile. Adverse drug reactions involving the liver are the most common reason for an approved drug to be withdrawn from the market.

Overdoses of the common analgesic acetaminophen (Tyleonol^{TM}, Paracetemol^{TM}) can cause severe liver injury and acute liver failure. In the USA,

50% of acute liver failure cases are due to acetaminophen-induced liver injury (Lee, 2004). The drug is converted by cytochrome P4502E1 (CYP2E1)-mediated oxidative pathways in hepatocytes to the toxic metabolite N-acetyl-p-quinone imine which is subsequently detoxified by preferential binding to glutathione (GSH). Liver injury occurs upon depletion of cytosolic and mitochondrial GSH, though the progression and degree of injury is dependent on genetic and environmental factors (e.g. expression levels of CYP2E1 or GSH, alcohol consumption, etc). The bioenergetic consequences of overdoses of acetaminophen are manifested as a centrilobular necrosis that can be fatal. Liver pathogenesis is also associated with the production of tumor necrosis factor-α (TNF-α) by Kupffer cells (Decker, 1990). TNF-α acts as a stimulus for the production of other cytokines that activate inflammatory cells and initiate wound healing. In mice receiving acetaminophen, centrilobular necrosis was associated with increased transcription of a variety of inflammatory cytokines and repair molecules, including CCN2 (Gardner *et al.*, 2002; Gardner *et al.*, 2003). Although CCN2 is a transforming growth factor-β (TGF-β) immediate early gene in other systems (Leask and Abraham, 2003), TGF-β levels in the liver were unaffected by acetaminophen (Gardner *et al.*, 2003). The TNFR1 receptor, which is expressed predominantly by Kupffer cells and hepatocytes and plays a key role in hepatotoxicity, was implicated in mediating CCN2 production since the level of hepatic CCN2 mRNA was reduced in acetaminophen-treated TNFR1−/−mice as compared to wild type controls (Gardner *et al.*, 2003). TNFR1−/−mice exhibited delayed or reduced production of several other inflammatory cytokines and wound healing factors as well as an exaggerated hepatotoxicity to acetaminophen. These data show that TNF-α signaling is central to acetaminophen-induced inflammation and liver injury, and that CCN2 is a component of the signaling cascade.

Related studies in mice that were knockout for inducible nitric oxide synthase (NOS II) showed that they exhibit reduced hepatotoxicity in response to acetaminophen as compared to wild-type controls (Gardner *et al.*, 2002). The NOS II knockout mice demonstrated increased expression of both TNF-α and CCN2 highlighting the potential importance of CCN2 as a wound healing factor downstream of TNF-α. Increased CCN2 expression has thus been proposed to contribute to the mechanism whereby NOS II knockout mice exhibit reduced hepatotoxicity in response to acetaminophen (Gardner *et al.*, 2002). Although TNF-α was originally reported to suppress CCN2 production in endothelial cells, smooth muscle cells, and fibroblasts (Lin *et al.*, 1998), CCN2 expression was reportedly enhanced in HSC following prolonged exposure to TNF-α (Liu *et al.*, 2001). Thus, drug-induced liver injury results in complex inflammatory and wound healing processes that involve interacting cell types and multiple coordinated signaling pathways of which CCN2 is a component.

HEPATIC REGENERATION

Unlike other vital organs, the liver can regenerate after injury or resection. Resections of up to two-thirds of the liver are followed by a carefully orchestrated and fully integrated sequence of events that generally result in complete restitution of liver architecture and re-establishment of the specific functions of the liver (Minuk, 2003). Processes in the residual liver include the induction of cytoprotective mechanisms, deletion of mortally wounded cells, repair of less damaged survivors, liver cell proliferation to replace the cells that died, deposition of new matrix, and tissue remodeling to restore normal hepatic mass and architecture (Diehl, 2002). Regeneration involves the proliferation of the normally quiescent cell populations within the liver. Replication proceeds first with hepatocytes, followed by biliary, epithelial, Kupffer, stellate, and sinusoidal endothelial cells. Regeneration is halted once liver architecture, function and mass are restored, though abnormal regeneration can occur in the face of other confounding factors and thereby contribute to the pathogenesis of liver failure, cirrhosis, and primary liver cancers (Diehl, 2002).

A common experimental model is partial hepatectomy in which 70% of the rodent liver is removed. The remaining lobes are left intact and undergo profound compensatory growth over the next few days; in fact only two rounds of hepatocyte replication are required for complete regeneration, though the proliferative potential of hepatocytes is much greater. This model has become an important means of analyzing the molecular signals involved in hepatic regeneration and it is not surprising that the role of growth-regulatory molecules in liver regeneration have been studied extensively. Regeneration is initiated or primed by cytokines such as TNF-α or interleukin 6 and this is followed by the action of multiple growth factors, most notably hepatocyte growth factor, epidermal growth factor, and transforming growth factor-α. Activation of cytokine and growth factor signaling pathways leads to the induction of transcription factor complexes followed by DNA synthesis and cell division. TGF-β appears to be involved in terminating the proliferative response once liver mass is attained (Zimmermann, 2004).

A study of CCN2 production after 70% partial hepatectomy in rats revealed that it was induced very rapidly (within 2 hours) during the regeneration phase, with kinetics that were comparable to the immediate early gene, c-fos (Ujike *et al.*, 2000). These kinetics were fairly well correlated with those of TGF-β suggesting that, as in other systems, CCN2 was a TGF-β immediate early gene during liver regeneration. It was proposed that CCN2 and TGF-β may function cooperatively to regulate matrix remodeling prior to the G1-S transition of regeneration. This was supported by the kinetics of collagen production as well as the finding that both CCN2 and TGF-β were localized to fibroblasts and stellate-like cells in the partial hepatectomy model (Ujike *et al.*, 2000). Regeneration in response to the hepatotoxin D-galactosamine was also characterized by the production of TGF-β and CCN2, although their kinetics of induction were delayed as compared to the resection model. This was correlated with delayed onset of DNA synthesis, possibly due to toxin-induced block in the hepatocytes (Ujike *et al.*, 2000). Collectively these data show that TGF-β and CCN2 are coordinately produced during liver regeneration following injury by resection or toxin. However, the precise role played by each factor and the target cells involved requires additional investigation.

HEPATIC CARCINOMA

Hepatocellular carcinoma (HCC) is the fifth most common malignant disorder and results in nearly 1 million deaths a year worldwide (Burroughs *et al.*, 2004). HCC is the most common primary hepatic malignant tumor, with nearly 1 million new cases annually worldwide. The tumors are hard to treat, not usually detected early, and are rapidly fatal. HCC is the third cause of cancer-related death worldwide. Since cirrhosis is the major risk factor for HCC, Asian and subSaharan African populations have historically suffered enormously because of chronic infection with hepatitis C (Burroughs *et al.*, 2004). However, as a result of the high prevalence of hepatitis B in Europe and the USA, the incidence of HCC is increasing dramatically in Western nations (Burroughs *et al.*, 2004). HCC is also the leading cause of death amongst cirrhotic patients overcoming liver failure. The only curative treatments are surgical resection or liver transplantation, but few patients are eligible for these procedures (Burroughs *et al.*, 2004).

Progress in combating HCC will rely, at least in part, on the identification of key molecules that can be therapeutically targeted. Although insufficient data have so far accumulated to rule CCN proteins in or out as possible targets, they nonetheless have been implicated in the disease process. For example, 74% of human HCC samples were shown to express CCN1 (also known as CYR61) and CCN3 (also known as NOV) while 26% of the samples expressed

CCN2 (Hirasaki *et al.*, 2001). While CCN1 expression was not elevated relative to normal liver tissue, expression of both CCN2 and CCN3 was significantly higher than in surrounding non-tumor tissue. Nonetheless, there was no apparent relationship between expression of CCN2 or CCN3 and clinicopathogical parameters (Hirasaki *et al.*, 2001). CCN1-3 were expressed by metastatic liver tumors though their respective roles in metastasis as well as their pathological significance has yet to be addressed. Also, CCN2 was identified as an early response gene that was activated by TGF-β in FaO hepatoma cells and was proposed to play a role in TGF-β-mediated apoptosis (Coyle *et al.*, 2003). Expression of CCN4 (also known as WISP-1 or ELM-1) was not detected in normal livers yet it was apparent in 49% of cholangiocarcinomas studied (Tanaka *et al.*, 2003). Expression of CCN4 was significantly associated with lymphatic and perineural invasion of tumor cells and a poor clinical prognosis (Tanaka *et al.*, 2003). While further studies are required to understand their role in HCC, CCN2, CCN3 and CCN4 may play a role in cell proliferation, adhesion, or migration within the primary tumor or its metastatic lesions.

HEPATIC FIBROSIS

Hepatic fibrosis, caused by hepatitis and other chronic liver diseases, is a major cause of morbidity and mortality throughout the world. Liver fibrosis is a wound healing response in which damaged regions are surrounded by ECM or scar. Chronic liver disease often leads to scarring and is frequently accompanied by a progressive loss of liver function and cirrhosis, a combination of cell death, fibrosis and regeneration. Cirrhosis is usually initiated by consistent and persistent cell death over a long period. Fibrosis is the central pathological process and the disease is typified by the formation of nodules that arise due to hepatocyte regeneration within the confines of fibrous septa. Most cases (60–70%) of liver cirrhosis in the Western world are caused by alcoholic liver disease while cryptogenic cirrhosis, viral hepatitis, biliary diseases and hereditary haemochromatosis make up the remaining 30–40% of cases (Crawford *et al.*, 2002). Liver cirrhosis is the ninth leading cause of death in the West, with half of the cases due to alcohol abuse. Annually, it accounts for 27,000 deaths in the USA and 6000 deaths in the UK (Crawford *et al.*, 2002). Worldwide, millions of individuals suffer from cirrhosis of the liver. Since there are no FDA-approved anti-fibrotic drugs, treatment modalities for liver fibrosis and chronic liver disease are a huge untapped market for biotechnology and pharmaceutical companies.

Hepatic Stellate Cells in Liver Fibrosis

In healthy subjects, the ECM in the Space of Disse is of low density and functions in maintaining normal homeostasis of the surrounding cell types. However, during fibrosing liver injury, a high density ECM comprised mainly of fibrillar collagens becomes deposited that eventually severely limits cell and tissue function. The striking change in the amount and composition of ECM levels is principally a reflection of a functional change by HSC in response to tissue injury. HSC are the major fibrogenic cell type in the liver and signals from these cells drive the formation as well as degradation of ECM. A central event in liver fibrosis is the activation of HSC from quiescent vitamin A-rich cells to vitamin A-deficient, proliferative, fibrogenic and contractile myofibroblasts that express α-SMA, type 1 collagen and tissue inhibitors of metalloproteases (Britton and Bacon, 1999; Burt, 1999; Eng and Friedman, 2000; Friedman, 1999; 2000). Thereafter, activated HSC demonstrate perpetuation which involves changes in cell behavior, proliferation, chemotaxis, fibrogenesis, contractility, matrix degradation, retinoid loss, leukocyte chemotaxis, and cytokine release (Friedman, 1999; 2000). Collectively, these changes increase ECM accumulation such that HSC become the principal effector cells of the fibrotic response.

Although the notion has existed that fibrosis is non-reversible, this is not actually true. Reversal of hepatic fibrosis has, in fact, been documented in patients after venesection for hemochromatosis, immunosuppresive therapy of autoimmune chronic active hepatitis and primary biliary cirrhosis, antiviral treatment of chronic hepatitis B and chronic delta hepatitis, and biliary decompression for chronic pancreatitis and stenosis of the common bile duct (Dufour *et al.*, 1997; Hammel *et al.*, 2001; Kaplan *et al.*, 1997; Kweon *et al.*, 2001; Lau *et al.*, 1999; Powell and Kerr, 1970; Wanless, 2001). While these outcomes involved small patient numbers and have not been verified, a more compelling argument comes from a retrospective analysis of four clinical trials involving more than 3000 chronic hepatitis C patients randomized to treatment with interferon and/or rabivarin (Poynard *et al.*, 2002). Antiviral therapy had major beneficial effects on liver fibrosis and even cirrhosis was reversed in 49% of the patients (Poynard *et al.*, 2002). Thus, while the idea that severe fibrosis or cirrhosis inevitably leads to liver transplantation or death was once considered the dogma, these diseases are now believed to be largely reversible by attacking the primary disease or reducing scarring (Friedman and Arthur, 2002). The underlying basis for the latter is that each step fibrogenesis is potentially amenable to therapeutic targeting. Strategies that target the activation and survival of HSC in addition to the ability of HSC to deposit as well as degrade collagen and secrete matrix metalloproteinases (MMP) and tissue inhibitors of metalloproteases (TIMP) are promising starting points for treating and reversing advanced fibrosis and cirrhosis. As will be discussed below, CCN2 activates both fibrogenic and anti-apoptotic pathways in HSC suggesting that it is a rational target in an anti-fibrotic regime.

CCN2 and Hepatic Fibrosis

Human disease

Over the last few years, data reported have provided persuasive evidence that CCN2 mRNA and protein levels are correlated with the degree of hepatic fibrosis, irrespective of the underlying etiology. Northern blot analysis showed that, as compared to normal livers, CCN2 and TGF-β1 mRNA expression was 6- to 8-fold higher in cirrhotic livers from patients with chronic viral hepatitis, PBC, PSC, cryptogenic and alcoholic liver disease (ALD) (Abou–Shady *et al.*, 2000). Elevated levels of CCN2 mRNA were also present in livers from patients with non-alcoholic steatohepatitis (NASH) and the degree of fibrosis was correlated with CCN2 protein levels in the ECM (Paradis *et al.*, 2001). Similarly, ribonuclease protection assay analysis of cirrhotic livers from patients with primary biliary cirrhosis (PBC), primary sclerosing cholangitis (PSC) or biliary atresia showed that CCN2 mRNA in fibrotic livers was elevated 3–5-fold as compared to their normal counterparts (Williams *et al.*, 2000). Immunohistochemical staining of normal and hepatitis C livers showed that increased CCN2 protein levels were linked with a higher score of fibrosis (Paradis *et al.*, 1999).

In situ hybridization of cirrhotic livers resulting from PBC, PSC, cryptogenic, ALD and chronic viral hepatitis, contained marked CCN2 staining in fibroblast and myofibroblast-like cells within the fibrotic portal tracts and fibrous septa (Abou–Shady *et al.*, 2000; Hayashi *et al.*, 2002; Konishi *et al.*, 2001). CCN2 mRNA was also observed in HSC, myofibroblast-like spindle cells around proliferating ductules, a few duct and ductular epithelial cells, inflammatory cells, and sinusoidal and vascular endothelial cells (Abou–Shady *et al.*, 2000; Paradis *et al.*, 1999). CCN2 protein staining was observed in portal tracts, the ECM of the fibrous septa, sinusoidal lining and in the proliferating bile ducts in hepatitis C diseased livers (Hayashi *et al.*, 2002; Paradis *et al.*, 1999). Additionally, many cells in fibrotic liver that stained for α-smooth muscle actin (α-SMA) were also shown to contain CCN2 protein (Paradis *et al.*, 1999).

In congenital hepatic fibrosis, which is marked by profound hyperplasia of the bile ducts, CCN2 mRNA and protein are strongly expressed by bile duct epithelial cells (Rachfal and Brigstock, 2003). In ALD, hepatocytes may be an important source of CCN2 since it is produced in cultured HepG2 cells downstream of CYP2E1-mediated ethanol oxidation (Konishi *et al.*, 2001; Konishi *et al.*, 2002). Patients suffering from idiopathic portal hypertension demonstrated enhanced CCN2 expression in their periductal mononuclear cells and this was correlated with collagen and elastin deposition and the presence of activated HSC and pericellular fibrosis (Tsuneyama *et al.*, 2002). Hence, while its mode of action and timing of production may vary according to the initial hepatic insult and type of damage, CCN2 has the potential of being produced by virtually all major cell types in the liver. Furthermore, high intra-hepatic levels of CCN2 are associated with its entry into the circulation since, as compared to control subjects, serum CCN2 levels were higher in patients with biliary atresia and were correlated with the progression of liver fibrosis (Tamatani *et al.*, 1998).

Animal models of hepatic fibrosis

Examination of mRNA by real-time PCR or reverse transcriptase PCR showed that hepatic CCN2 and collagen expression was increased with the degree of fibrosis following bile duct ligation (BDL) or CCl_4 treatment in rats (Paradis *et al.*, 1999; Williams *et al.*, 2000). In the BDL model, TGF-β and CCN2 mRNA were increased 4- and 7-fold, respectively, as compared to normal liver (Sedlaczek *et al.*, 2001). Both CCN2 and TGF-β mRNA increased 6 hours post CCl_4 administration and levels peaked at 72 hours (Sedlaczek *et al.*, 2001). The main site of CCN2 mRNA in BDL livers was proliferating bile duct epithelial cells, though the protein was present in the fibrous septa as well as in areas of ductular proliferation (Paradis *et al.*, 1999; Sedlaczek *et al.*, 2001). TGF-β was only expressed at minimal levels in these cells but was highly expressed in the surrounding HSC (Sedlaczek *et al.*, 2001). In contrast, in the CCl_4 model, CCN2 mRNA was expressed at high levels in HSC and the protein was localized to fibrous septa and centrilobular regions (Paradis *et al.*, 1999; Sedlaczek *et al.*, 2001). Lastly, CCN2 mRNA and protein were up-regulated in the obesity and type II diabetes Zucker (fa/fa) rat as compared to the wild-type counterpart (Paradis *et al.*, 2001).

CCN2 biology in cultured HSC

Substantial data have shown that CTGF is produced by HSC as a function of activation or in response to stimulation of the cells by pro-fibrotic molecules or growth factors. Cultured activated primary HSC were found to contain CCN2 by Western blotting (Paradis *et al.*, 1999) and both Northern and Western blots demonstrated CCN2 was increasingly expressed during progressive activation of cultured primary rat HSC (Williams *et al.*, 2000). In primary HSC from a CCl$_4$-treated rat, CCN2 was localized exclusively to the Golgi apparatus and was quantitatively secreted into the medium (Chen *et al.*, 2001). Additionally, both intermediate and activated cultured HSC were induced by TGF-β1 to express CCN2 mRNA and protein (Gao and Brigstock, 2003; Williams *et al.*, 2000) while CCN2 promoter activity was enhanced in a Smad7-dependent fashion by TGF-β or PDGF in HSC transfected with a CCN2 promoter luciferase reporter construct (Gao *et al.*, 2004). CCN2 mRNA expression was also induced in HSC treated with VEGF, lipid peroxidation products, acetaldehyde or PDGF-BB (Paradis *et al.*, 2002). In the HSC-T6 cell line, CCN2 production was stimulated by TGF-β1 or CCN2 itself (Rachfal and Brigstock, 2003) while HSC from fa/fa rats incubated with glucose or insulin showed a TGF-β-independent increase in CCN2 mRNA and protein (Paradis *et al.*, 2001). Although these data show that CCN2 is produced as a function of HSC activation or under circumstances that support activation, the relative contribution to the activation process of CCN2 derived from HSC as compared to the other cells in the liver remains uncertain.

In addition to their production of CCN2, HSC also respond to exogenous stimulation by CCN2 suggesting that CCN2 is involved in autocrine and paracrine regulation of HSC function. Treatment of rat HSC with CCN2 induced migration and proliferation, the latter of which was associated with transient induction of c-fos activation and activation of the ERK1/2 signal pathway (Gao *et al.*, 2004; Paradis *et al.*, 2002). In addition, CCN2 induces expression of α-SMA and type I collagen by HSC, consistent with a role in activation and fibrogenesis (Paradis *et al.*, 2002). When targeted *in vivo* with antisense CCN2 oligonucleotides, HSC showed reduced type I collagen expression although fibrosis in general did not seem to be affected (Uchio *et al.*, 2004). The authors speculated this was due to the continued high levels of tissue inhibitor of matrix metalloproteinase-1 expression (Uchio *et al.*, 2004).

Several studies have shown that CCN2 is an adhesive substrate for HSC and that adhesion to CCN2 is sufficient to trigger specific changes in HSC gene expression. Both the HSC-T6 cell line and activated primary HSC demonstrate adhesion to CCN2, whereas freshly isolated HSC do not (Ball *et al.*, 2003; Gao and Brigstock, 2004; Rachfal and Brigstock, 2003). The principal cell surface receptors involved in HSC adhesion are integrin αvβ3, which binds to a unique non-RGD site in module 4 of CCN2 (Gao and Brigstock, 2004), and LRP, which binds to a site in module 3 (Gao and Brigstock, 2003). LRP is a multifunctional scavenger, signaling, and adhesion receptor (Herz and Strickland, 2001) that may be responsible for the binding by HSC of ^{125}I-labeled CCN2 and its subsequent internalization and degradation in the endosome (Chen *et al.*, 2001; Segarini *et al.*, 2001). The partnering of CCN2 with either LRP or integrin $\alpha v\beta 3$ is modulated by cell surface HSPG since binding to either site can be blocked by soluble heparin or by treating the cells with heparinase to sodium chlorate which prevents HSPG sulfation (Gao and Brigstock, 2003; Gao and Brigstock, 2004). The integrin $\alpha v\beta 3$- and heparin-binding properties of CCN2 are mimicked by a fragment of CCN2 comprising essentially module 4 alone (Ball *et al.*, 2003; Gao and Brigstock, 2004). This fragment arises naturally *in vivo* via limited proteolysis and stimulates fibroblast cell proliferation, α-SMA production in epithelial cells and HSC, and subcutaneous fibrosis *in vivo* (Ball *et al.*, 2003; Brigstock *et al.*, 1997; Rachfal and Brigstock, 2003). Limited proteolysis of CCN2 also occurs in HSC cultures (Gao *et al.*, 2004; Williams *et al.*, 2000) and yields a 20 kDa C-terminal fragment that comprises modules 3 and 4 that has similar biological activities (Ball *et al.*, 1998; Ball and Brigstock, 2001).

In assessing the potential biological significance of the effect of CCN2 on HSC adhesion and down-stream signaling, gene arrays were performed on HSC that has been allowed to bind either to CCN2 or poly-l-lysine. As shown in Fig. 2, CCN2 induced an 11-fold increase in collagen pro-$\alpha 1$ type mRNA, a 2-fold increase in FN, and a 6-fold increase in TIMP-1, while eliciting a 4-fold decrease in caspase 8 expression. As assessed by RT-PCR, expression of hepatocyte growth factor (HGF) by primary HSC was reduced 30% following adhesion of the cells to CCN2 for 8 hours. Further, HGF expression was reduced 80% following adhesion of the HSC-T6 cell line to CCN2 for 8 hours (D.R.B. and A.W.R., unpublished data). Thus adhesive signaling via CCN2 in HSC causes an up-regulation in the expression of pro-fibrogenic molecules but a down-regulation in the expression of pro-apoptotic molecules. Collectively, CCN2 likely drives or maintains HSC fibrogenic and survival pathways in activated HSC. That being the case, reversion or apoptosis of activated HSC may be achieved by the development of anti-fibrotic strategies that target CCN2 production or action.

SUMMARY AND PERSPECTIVES

CCN proteins appear to play a role in many aspects of liver dysfunction. Their expression is triggered in various cell types during injury or disease, although

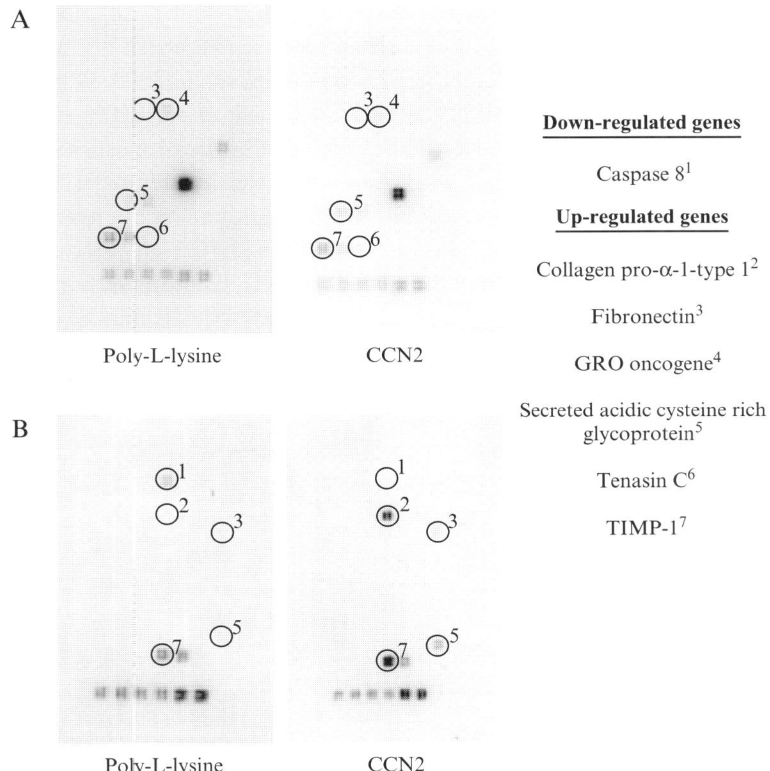

Fig. 2. Genes up- or down-regulated following adhesion of primary activated HSC to CCN2. Primary activated HSC were plated on 100 mm non-tissue culture plates that had been coated with 0.01% poly-L-lys (control) or 3 μg/ml 38 kDa full length CCN2. 8 hours later, total RNA was collected and labeled with α-[^{32}P]-dCTP and hybridized to a GEArray Q Series Mouse Angiogenesis Gene Array (**A**) or a GEArray Q Series Mouse Extracellular Matrix & Adhesion Molecules Gene Array (**B**). Developed blots were analyzed with the ScanAlyze program. This figure represents one of two experiments and is typical of the data obtained. Numbered circles refer to genes up- or down-regulated more than 0.5-fold from the control.

their precise function is not yet well defined. Additional studies need to be undertaken to more carefully establish the role of CCN proteins in hepatic development, differentiation, and pathology and to determine to what extent the various family members exhibit distinct, shared, or antagonistic functions. CCN signaling via integrins likely represents a highly adaptable mechanism, whereby cell function is regulated in a complex fashion according to the stimuli generated within normal, injured or diseased liver tissue.

Studies of CCN2 in liver, coupled with similar studies in other organ systems such as the skin and kidney, have suggested that CCN2 may offer a new lead as a therapeutic target in fibrosis (Blom *et al.*, 2002; Denton and Abraham, 2001; Goldschmeding *et al.*, 2000; Leask *et al.*, 2002; Riser and Cortes, 2001). It is not yet known whether the pro-fibrotic properties of CCN2 are unique or if they are shared with other CCN family members. Nonetheless, a strong case is emerging for an important role of CCN2 in liver fibrosis, especially as a downstream mediator of the actions of TGF-β. It is fairly clear that HSC are one of the principal target cells for CCN2 and there is optimism that the deleterious functions of HSC in fibrosis or cirrhosis may be targeted, at least in part, using anti-CCN2 strategies. The challenge ahead will be to determine which aspects of the CCN2 biosynthetic or signaling pathways are amenable to therapeutic intervention.

ACKNOWLEDGMENTS

This work was supported by NIH grant AA12817 awarded to D.R.B.

REFERENCES

Abou–Shady M, Friess H, Zimmermann A, di Mola FF, Guo XZ, Baer HU and Buchler MW (2000) Connective tissue growth factor in human liver cirrhosis. *Liver* **20**:296–304.

Ball DK, Surveyor GA, Diehl JR, Steffen CL, Uzumcu M, Mirando MA and Brigstock DR (1998) Characterization of 16- to 20-kilodalton (kDa) connective tissue growth factors (CTGFs) and demonstration of proteolytic activity for 38-kDa CTGF in pig uterine luminal flushings. *Biol Reprod* **59**:828–835.

Ball DK and Brigstock D (2001) Establishment of a recombinant CTGF expression system *in vitro* that models CTGF processing *in vivo*: structural and functional characterization of multiple mass CTGF proteins. *J Clin Pathol: Mol Pathol* **54**:P6.

Ball DK, Moussad EE, Rageh MA, Kemper SA and Brigstock DR (2003) Establishment of a recombinant expression system for connective tissue growth factor (CTGF) that models CTGF processing *in utero*. *Reproduction* **125**:271–284.

Ball DK, Rachfal AW, Kemper SA and Brigstock DR (2003) The heparin-binding 10 kDa fragment of connective tissue growth factor (CTGF) containing module 4 alone stimulates cell adhesion. *J Endocrinol* **176**:R1–7.

Blom IE, Goldschmeding R and Leask A (2002) Gene regulation of connective tissue growth factor: new targets for antifibrotic therapy? *Matrix Biol* **21**:473.

Brigstock DR, Steffen CL, Kim GY, Vegunta RK, Diehl JR and Harding PA (1997) Purification and characterization of novel heparin-binding growth factors in uterine secretory fluids. Identification as heparin-regulated Mr 10,000 forms of connective tissue growth factor. *J Biol Chem* **272**:20275–20282.

Brigstock DR (1999) The connective tissue growth factor/cysteine-rich 61/nephroblastoma overexpressed (CCN) family. *Endocr Rev* **20**:189–206.

Britton RS and Bacon BR (1999) Intracellular signaling pathways in stellate cell activation. *Alcohol Clin Exp Res* **23**:922–925.

Burroughs A, Hochhauser D and Meyer T (2004) Systemic treatment and liver transplantation for hepatocellular carcinoma: two ends of the therapeutic spectrum. *Lancet Oncol* **5**:409–418.

Burt AD (1999) Pathobiology of hepatic stellate cells. *J Gastroenterol* **34**:299–304.

Chen Y, Segarini P, Raoufi F, Bradham D and Leask A (2001) Connective tissue growth factor is secreted through the Golgi and is degraded in the endosome. *Exp Cell Res* **271**:109–117.

Coyle B, Freathy C, Gant TW, Roberts RA and Cain K (2003) Characterization of the transforming growth factor-beta 1-induced apoptotic transcriptome in FaO hepatoma cells. *J Biol Chem* **278**:5920–5928.

Crawford JM (2002) Liver Cirrhosis. In MacSween RNM, Burt AD, Portmann BC, Ishak KG, Scheuer PJ and Anthony PP, Pathology of the Liver, 4th ed. (Harcourt Publishers Limited, London), pp. 575–620.

Decker K (1990) Biologically active products of stimulated liver macrophages (Kupffer cells). *Eur J Biochem* **192**:245–261.

Denton CP and Abraham DJ (2001) Transforming growth factor-beta and connective tissue growth factor: key cytokines in scleroderma pathogenesis. *Curr Opin Rheumatol* **13**: 505–511.

Diehl AM (2002) Liver regeneration. *Front Biosci* **7**:e301–314.

Dufour JF, DeLellis R and Kaplan MM (1997) Reversibility of hepatic fibrosis in autoimmune hepatitis. *Ann Intern Med* **127**:981–985.

Eng FJ and Friedman SL (2000) Fibrogenesis I. New insights into hepatic stellate cell activation: the simple becomes complex. *Am J Physiol Gastrointest Liver Physiol* **279**:G7–G11.

Friedman L and Arthur MJP (2002) Reversing Hepatic Fibrosis. *Science and Medicine* **8**:194–205.

Friedman SL (1999) Stellate cell activation in alcoholic fibrosis — an overview. *Alcohol Clin Exp Res* **23**:904–910.

Friedman SL (2000) Molecular regulation of hepatic fibrosis, an integrated cellular response to tissue injury. *J Biol Chem* **275**:2247–2250.

Gao R and Brigstock DR (2003) Low density lipoprotein receptor-related protein (LRP) is a heparin-dependent adhesion receptor for connective tissue growth factor (CTGF) in rat activated hepatic stellate cells. *Hepatol Res* **27**:214–220.

Gao R, Ball DK, Perbal B and Brigstock DR (2004) Connective tissue growth factor induces c-fos gene activation and cell proliferation through p44/42 MAP kinase in primary rat hepatic stellate cells. *J Hepatol* **40**:431–438.

Gao R and Brigstock DR (2004) Connective tissue growth factor (CCN2) induces adhesion of rat activated hepatic stellate cells by binding of its C-terminal domain to integrin {alpha}v{beta}3 and heparan sulfate proteoglycan. *J Biol Chem* **279**:8848–8855.

Gardner CR, Laskin JD, Dambach DM, Sacco M, Durham SK, Bruno MK, Cohen SD, Gordon MK, Gerecke DR, Zhou P and Laskin DL (2002) Reduced hepatotoxicity of acetaminophen in mice lacking inducible nitric oxide synthase: potential role of tumor necrosis factor-alpha and interleukin-10. *Toxicol Appl Pharmacol* **184**:27–36.

Gardner CR, Laskin JD, Dambach DM, Chiu H, Durham SK, Zhou P, Bruno M, Gerecke DR, Gordon MK and Laskin DL (2003) Exaggerated hepatotoxicity of acetaminophen in mice lacking tumor necrosis factor receptor-1. Potential role of inflammatory mediators. *Toxicol Appl Pharmacol* **192**:119–130.

Goldschmeding R, Aten J, Ito Y, Blom I, Rabelink T and Weening JJ (2000) Connective tissue growth factor: just another factor in renal fibrosis? *Nephrol Dial Transplant* **15**:296–299.

Goodman ZD (2002) Drug hepatotoxicity. *Clin Liver Dis* **6**:381–397.

Hammel P, Couvelard A, O'Toole D, Ratouis A, Sauvanet A, Flejou JF, Degott C, Belghiti J, Bernades P, Valla D, Ruszniewski P and Levy P (2001) Regression of liver fibrosis after biliary drainage in patients with chronic pancreatitis and stenosis of the common bile duct. *N Engl J Med* **344**:418–423.

Hayashi N, Kakimuma T, Soma Y, Grotendorst GR, Tamaki K, Harada M and Igarashi A (2002) Connective tissue growth factor is directly related to liver fibrosis. *Hepatogastroenterology* **49**:133–135.

Herz J and Strickland DK (2001) LRP: a multifunctional scavenger and signaling receptor. *J Clin Invest* **108**:779–784.

Hirasaki S, Koide N, Ujike K, Shinji T and Tsuji T (2001) Expression of Nov, CYR61 and CTGF genes in human hepatocellular carcinoma. *Hepatol Res* **19**:294–305.

Kaplan MM, DeLellis RA and Wolfe HJ (1997) Sustained biochemical and histologic remission of primary biliary cirrhosis in response to medical treatment. *Ann Intern Med* **126**:682–688.

Konishi M, Kajihara M, Kato S, Horie Y, Aiso S and Ishii H (2001) Ethanol enhances expression of connective tissue growth factor (CTGF) in HEPG2 cell line stably expressing cytochrome P-4502E1. *Hepatology* **34**:377A.

Konishi M, Kato S, Kajihara M, Yamagishi Y, Horie Y, Aiso S and Ishii H (2002) Expression of connective tissue growth factor (CTGF) in alcoholic liver fibrosis and in HEPG2 cell line stably expressiong cytochrome P-4502E1. *Hepatology* **36**:248A.

Kweon YO, Goodman ZD, Dienstag JL, Schiff ER, Brown NA, Burchardt E, Schoonhoven R, Brenner DA, Fried MW and Burkhardt E (2001) Decreasing fibrogenesis: an immunohistochemical study of paired liver biopsies following lamivudine therapy for chronic hepatitis B. *J Hepatol* **35**:749–755.

Lau DT, Kleiner DE, Park Y, Di Bisceglie AM and Hoofnagle JH (1999) Resolution of chronic delta hepatitis after 12 years of interferon alfa therapy. *Gastroenterology* **117**:1229–1233.

Lau LF and Lam SC (1999) The CCN family of angiogenic regulators: the integrin connection. *Exp Cell Res* **248**:44–57.

Leask A, Holmes A and Abraham DJ (2002) Connective tissue growth factor: a new and important player in the pathogenesis of fibrosis. *Curr Rheumatol Rep* **4**:136–142.

Leask A and Abraham DJ (2003) The role of connective tissue growth factor, a multifunctional matricellular protein, in fibroblast biology. *Biochem Cell Biol* **81**:355–363.

Lee WM (2004) Acetaminophen and the U.S. Acute Liver Failure Study Group: lowering the risks of hepatic failure. *Hepatology* **40**:6–9.

Lin J, Liliensiek B, Kanitz M, Schimanski U, Bohrer H, Waldherr R, Martin E, Kauffmann G, Ziegler R and Nawroth PP (1998) Molecular cloning of genes differentially regulated by TNF-alpha in bovine aortic endothelial cells, fibroblasts and smooth muscle cells. *Cardiovasc Res* **38**:802–813.

Liu X, Wu H, Liu F, Huang M, Qiang O and Huang S (2001) Effects of tumor necrosis factor alpha on the expression of connective tissue growth factor in hepatic stellate cells. *Zhonghua Gan Zang Bing Za Zhi* **9** Suppl:15–17.

Minuk GY (2003) Hepatic regeneration: If it ain't broke, don't fix it. *Can J Gastroenterol* **17**:418–424.

Moussad EE and Brigstock DR (2000) Connective tissue growth factor: what's in a name? *Mol Genet Metab* **71**:276–292.

Paradis V, Dargere D, Vidaud M, De Gouville AC, Huet S, Martinez V, Gauthier JM, Ba N, Sobesky R, Ratziu V and Bedossa P (1999) Expression of connective tissue growth factor in experimental rat and human liver fibrosis. *Hepatology* **30**:968–976.

Paradis V, Perlemuter G, Bonvoust F, Dargere D, Parfait B, Vidaud M, Conti M, Huet S, Ba N, Buffet C and Bedossa P (2001) High glucose and hyperinsulinemia stimulate connective tissue growth factor expression: a potential mechanism involved in progression to fibrosis in nonalcoholic steatohepatitis. *Hepatology* **34**:738–744.

Paradis V, Dargere D, Bonvoust F, Vidaud M, Segarini P and Bedossa P (2002) Effects and regulation of connective tissue growth factor on hepatic stellate cells. *Lab Invest* **82**:767–774.

Perbal B (2001) NOV (nephroblastoma overexpressed) and the CCN family of genes: structural and functional issues. *Mol Pathol* **54**:57–79.

Pineiro–Carrero VM and Pineiro EO (2004) Liver. *Pediatrics* **113**:1097–1106.

Powell LW and Kerr JF (1970) Reversal of "cirrhosis" in idiopathic haemochromatosis following long-term intensive venesection therapy. *Australas Ann Med* **19**:54–57.

Poynard T, McHutchison J, Manns M, Trepo C, Lindsay K, Goodman Z, Ling MH and Albrecht J (2002) Impact of pegylated interferon alfa-2b and ribavirin on liver fibrosis in patients with chronic hepatitis C. *Gastroenterology* **122**:1303–1313.

Rachfal AW and Brigstock DR (2003) Connective tissue growth factor (CTGF/CCN2) in hepatic fibrosis. *Hepatol Res* **26**:1–9.

Riser BL and Cortes P (2001) Connective tissue growth factor and its regulation: a new element in diabetic glomerulosclerosis. *Ren Fail* **23**:459–470.

Sedlaczek N, Jia JD, Bauer M, Herbst H, Ruehl M, Hahn EG and Schuppan D (2001) Proliferating bile duct epithelial cells are a major source of connective tissue growth factor in rat biliary fibrosis. *Am J Pathol* **158**:1239–1244.

Segarini PR, Nesbitt JE, Li D, Hays LG, Yates JR, and 3rd Carmichael DF (2001) The low density lipoprotein receptor-related protein/alpha2-macroglobulin receptor is a receptor for connective tissue growth factor. *J Biol Chem* **276**:40659–40667.

Tamatani T, Kobayashi H, Tezuka K, Sakamoto S, Suzuki K, Nakanishi T, Takigawa M and Miyano T (1998) Establishment of the enzyme-linked immunosorbent assay for connective tissue growth factor (CTGF) and its detection in the sera of biliary atresia. *Biochem Biophys Res Commun* **251**:748–752.

Tanaka S, Sugimachi K, Kameyama T, Maehara S, Shirabe K, Shimada M, Wands JR and Maehara Y (2003) Human WISP1v, a member of the CCN family, is associated with invasive cholangiocarcinoma. *Hepatology* **37**:1122–1129.

Tsuneyama K, Kouda W and Nakanuma Y (2002) Portal and parenchymal alterations of the liver in idiopathic portal hypertension: a histological and immunochemical study. *Pathol Res Pract* **198**:597–603.

Uchio K, Graham M, Dean NM, Rosenbaum J and Desmouliere A (2004) Down-regulation of connective tissue growth factor and type I collagen mRNA expression by connective tissue growth factor antisense oligonucleotide during experimental liver fibrosis. *Wound Repair Regen* **12**:60–66.

Ujike K, Shinji T, Hirasaki S, Shiraha H, Nakamura M, Tsuji T and Koide N (2000) Kinetics of expression of connective tissue growth factor gene during liver regeneration after partial hepatectomy and D-galactosamine-induced liver injury in rats. *Biochem Biophys Res Commun* **277**:448–454.

Wanless IR (2001) Use of corticosteroid therapy in autoimmune hepatitis resulting in the resolution of cirrhosis. *J Clin Gastroenterol* **32**:371–372.

Williams EJ, Gaca MD, Brigstock DR, Arthur MJ and Benyon RC (2000) Increased expression of connective tissue growth factor in fibrotic human liver and in activated hepatic stellate cells. *J Hepatol* **32**:754–761.

Zimmermann A (2004) Regulation of liver regeneration. *Nephrol Dial Transplant* **19**(4):6–10.

CHAPTER 7

GENETIC ANALYSIS OF CCN GENE FUNCTION IN MAMMALIAN DEVELOPMENT

Lisa M. Dornbach* and Karen M. Lyons*,†,‡

*Departments of *Biological Chemistry, †Molecular, Cellular, and Developmental Biology, and ‡Orthopaedic Surgery, University of California at Los Angeles, Los Angeles, CA, USA*

Gene expression patterns and numerous *in vitro* experiments suggest that CCNs may be required for multiple aspects of mammalian development. Roles for several CCN genes during mammalian development have been elucidated through genetic analysis in mice and humans. Mice lacking *Ccn1* are characterized by insufficient placental vascularization and defective embryonic vasculature, confirming an essential and novel role for CCN1 in angiogenesis. Loss of *Ccn2* in mice results in multiple defects in skeletal development, including impaired chondrocyte proliferation, decreased production of extracellular matrix components, and defective growth plate angiogenesis. Mutations in *Ccn6* in humans are associated with defects in post-natal maintenance of articular cartilage. These genetic analyses have also shed light on signaling pathways utilized by CCNs and have revealed key molecular targets of CCN gene action.

INTRODUCTION

Ccn genes encode secreted proteins that are associated with the extracellular matrix. Gain-of-function studies have demonstrated that they are capable of mediating diverse cellular processes such as adhesion, proliferation, migration, and differentiation. With respect to *in vivo* activities, CCNs have been studied most extensively in the context of disease. CCN1, CCN2, CCN3, and CCN4 have all been implicated in tumor progression and metastasis. Studies employing gain-of-function and anti-sense approaches, in conjunction with correlations between CCN expression levels and tumor progression, indicate that CCNs promote or inhibit tumor progression, depending on the tumor type (reviewed in Planque and Perbal, 2003). Aberrant expression of CCN1 is associated with vascular disease, but its role in disease progression is unknown (reviewed in Mo *et al.*, 2002). The most extensively studied member of the CCN family, CCN2, plays a vital role in all fibrotic diseases studied to date. CCN2 is induced by TGFβ and potentiates the ability of TGFβ to induce excess extracellular matrix (ECM) synthesis (reviewed in Leask *et al.*, 2004).

The roles of CCNs in normal physiological events are less understood. CCN genes are expressed broadly throughout the developing mammalian embryo starting from early stages of development. Regions of CCN expression include the cardiovascular, nervous, pulmonary, reproductive, secretory, and skeletal systems (reviewed in Brigstock, 1999). Hints about the functions of CCNs in development have come from *in vitro* assays, but in order to determine which activities are relevant for mammalian development, the CCN gene must be rendered non-functional *in vivo*. This review will discuss how the inactivation of three CCN genes — *Ccn1* and *Ccn2* in mice and *Ccn6* in humans — reveals that CCNs are essential regulators of angiogenesis, chondrogenesis, and skeletal maintenance in mammals.

CCN1

CCN1 (Cyr61) is expressed throughout the developing mouse embryo and placenta, with highest levels in the cardiovascular and skeletal systems. Additional sites of expression include portions of the developing central nervous system, including the forebrain and floorplate, somites, and notochord (Latinkic *et al.*, 2001). Within the placenta, CCN1 is expressed in trophoblast giant cells and in endothelial cells of placental vessels (O'Brien and Lau, 1992; Latinkic *et al.*, 2001). In the cardiovascular system, CCN1 expression is detected in major blood vessels and the heart (O'Brien and Lau, 1992; Latinkic *et al.*, 2001). CCN1 is also expressed in the allantois and chorion prior to chorioallantoic fusion (Mo *et al.*, 2002). In the developing skeletal system, CCN1 is expressed in the cartilage of many skeletal elements. In these elements, CCN1 is first expressed in condensing mesenchyme, with expression persisting throughout the stages of chondrogenesis (O'Brien and Lau, 1992). The strong expression of CCN1 in the developing cardiovascular and skeletal systems suggests that it might play important roles in angiogenesis and/or chondrogenesis.

A potential role for CCN1 as a regulator of cardiovascular development is suggested by its potent pro-angiogenic properties. These have been studied extensively through *in vitro* assays which show that CCN1 positively regulates endothelial cell adhesion, migration, survival, tubule formation, and neovascularization at least in part via binding to integrin receptors (Kireeva *et al.*, 1996; Babic *et al.*, 1998; Leu *et al.*, 2002; Leu *et al.*, 2003). CCN1 can also promote endothelial cell proliferation by cooperating with growth factors (Kireeva *et al.*, 1996). In accordance with the expression of *Ccn1* in developing cartilage, *in vitro* assays also suggest a role in chondrogenesis. CCN1 promotes aggregation and differentiation, demonstrated by enhanced nodule formation and expression of type II collagen (Wong *et al.*, 1997).

The generation and analysis of *Ccn1* null mice confirms an essential and novel role for CCN1 in angiogenesis. The majority of mutants die in midgestation. Approximately one-third of all mutants die by E10.5 due to failure of chorioallantoic fusion [Fig. 1(A)] (Mo *et al.*, 2002). A second class of mutants survives past the chorioallantoic fusion event, but these mice die during midgestation of embryonic and placental vascularization defects [Figs. 1(E), (F)]. Placental vascularization is accomplished by sequential processes of non-sprouting angiogenesis (in which one set of parental vessels elongates and bifurcates extensively into a vessel network), and sprouting angiogenesis (in which new vessels sprout from the existing vessel network). *Ccn1* mutant placentas have impaired vessel bifurcation and fewer vessels of embryonic origin. Embryonic vasculature is also defective. Major blood vessels such as the dorsal aorta and umbilical artery are weak, dilated, and susceptible to hemorrhage as a consequence of disorganized vessel structure (Fig. 2). It is hypothesized that these defects arise from impaired interactions between smooth muscle cells, endothelial cells, and the ECM (Mo *et al.*, 2002). In addition to promoting the growth of the initial vessel network, the mutant shows that CCN1 is an essential regulator of later vascular remodeling. While recruitment of smooth muscle cells and pericytes to the blood vessel still occurs in *Ccn1* mutants, the highly disorganized nature of the vessels suggests that CCN1 is required for the proper interaction of the cells and ECM that comprise the vessels. Thus, the mutant phenotype confirms essential and novel roles for CCN1 in multiple aspects of angiogenesis. In accordance with the phenotype, in which non-sprouting angiogenesis is affected but sprouting angiogenesis is not, the non-sprouting angiogenesis marker *Vegf-C* is decreased in mutant vasculature, while other angiogenesis markers are not affected. *In vitro* studies with fibroblasts, in which CCN1 treatment upregulates *Vegf-C* mRNA expression, further support the idea that CCN1 is an endogenous regulator of *Vegf-C* expression (Mo *et al.*, 2002). Interestingly, not all VEGF isoforms are downregulated in mutants; *Vegf-A*, the predominant VEGF isoform and an essential regulator of sprouting angiogenesis, is detected at wild-type levels.

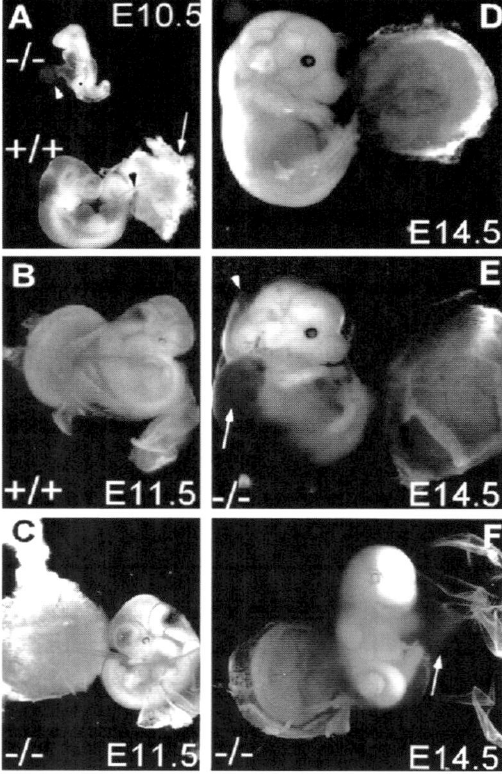

Fig. 1. (A) Chorioallantoic fusion in WT embryos, with the allantois attached to the placenta (white arrow). In some *Ccn1*$^{-/-}$ embryos, the allantois (white arrowhead) fails to fuse with the placenta, resulting in atrophy. (B) WT E11.5 embryos. (C) *Ccn1*$^{-/-}$ embryos at E11.5 show successful chorioallantoic fusion but display vascular deficiency in the chorionic plate. (D) WT E14.5 embryo. (E) *Ccn1*$^{-/-}$ embryo that developed to E14.5, exhibiting edema (arrowhead) and hemorrhage (arrow). (F) Other embryos obtained from E14.5 litters. These embryos were moribund, showing evidence of hemorrhage from the umbilical artery, filling the amnion. (Adapted from Mo *et al.*, 2002. Reprinted with permission.)

The apparent requirement for *Ccn1* in some angiogenic processes, such as nonsprouting angiogenesis, but not in others (sprouting angiogenesis), is unexpected given the potent pro-angiogenic activities of CCN1 *in vitro*. In contrast to the situation *in vivo*, in microarrays and *in vitro* assays CCN1 upregulates both *Vegf-A* and *Vegf-C* levels in fibroblasts (Chen *et al.*, 2001). The basis for this difference warrants further investigation, as does the role of CCN1 in promoting chorioallantoic fusion.

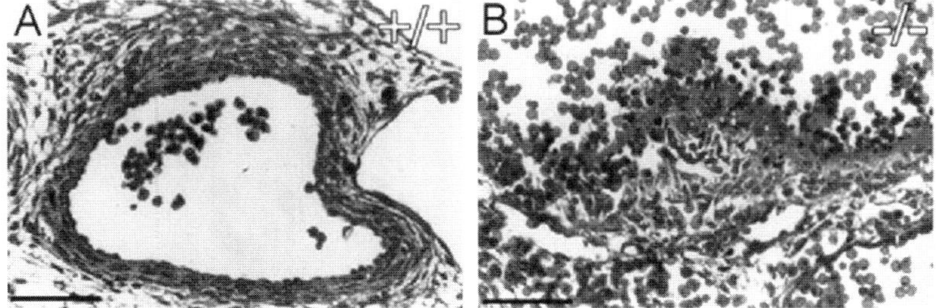

Fig. 2.   Hematoxylin and eosin staining shows disorganized vascular cells (B) in the *Ccn1−/−* dorsal aorta compared to that for the wild-type (A). (Adapted from Mo *et al.*, 2002. Reprinted with permission.)

CCN2

CCN2 (Connective tissue growth factor, or CTGF) is expressed in many tissues during development, including the vascular, skeletal, and nervous systems. Thus, CCN1 and CCN2 are extensively co-expressed. *Ccn2* mRNA is found in the heart and endothelial linings of major arterial blood vessels (Friedrichsen *et al.*, 2003). In neural tissues, *Ccn2* is expressed in the brain and neural tube, where it marks the cerebral cortex and olfactory bulb (Friedrichsen *et al.*, 2003; Ivkovic *et al.*, 2003). *Ccn2* expression in cartilage is first detected at E12.5 in the perichondrium. This domain of expression expands to include proliferating and maturing cells, with highest levels in hypertrophic chondrocytes (Nakanishi *et al.*, 1997; Ivkovic *et al.*, 2003). CCN2 expression persists in adult cartilage in hypertrophic and articular chondrocytes (Fukunaga *et al.*, 2003). The strong expression of *Ccn2* in vascular and skeletal tissues suggests that it may play an important role in their development. CCN2 promotes multiple processes related to angiogenesis, including endothelial cell migration, adhesion, survival, and proliferation *in vitro*, as well as neovascularization *in vivo* (Babic *et al.*, 1999; Brigstock, 2002). With respect to chondrogenesis, CCN2 stimulates proliferation of chondrocytes and promotes type II collagen and proteoglycan synthesis, a hallmark of differentiation. CCN2 also increases alkaline phosphatase activity, a marker for terminally differentiated hypertrophic chondrocytes (Nakanishi *et al.*, 2000; Nishida *et al.*, 2002).

Loss of CCN2 has significant consequences for skeletal development. *Ccn2* null mutants die within minutes of birth due to skeletal defects; many bones are kinked as a consequence of deformities in the cartilaginous template prior to the onset of ossification [Figs. 3(A), (C)]. Mutants also have defects in the

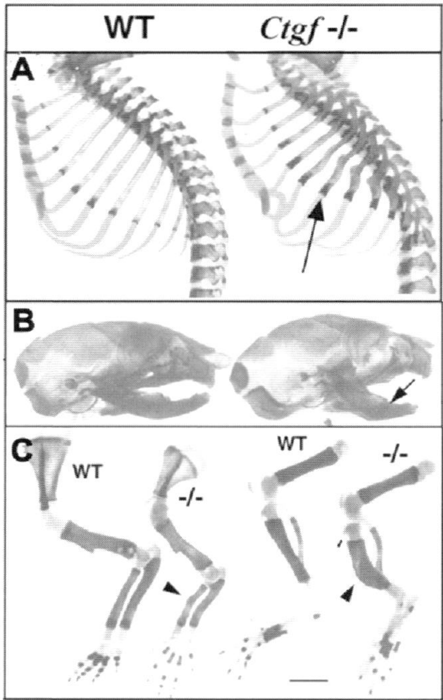

Fig. 3.   (A) Sagittal views of neonatal rib cages showing deformation of cartilage and kinks in bone in *Ccn2* mutant. (B) Side views of neonatal skulls showing domed skull and shortened mandibles (arrow). (C) Cleared skeletal preparations of neonatal limbs showing deformations (arrowheads) in the radius and ulna, and tibia and fibula. (Adapted from Ivkovic *et al.*, 2003.)

craniofacial skeleton, including a domed skull and deformed Meckels and nasal cartilages [Fig. 3(B) and Ivkovic *et al.*, 2003]. The earliest stages of chondrogenesis are not impaired in the *Ccn2* mutant. Differences between wild type and mutant littermates are first apparent at E14.5, when mutant growth plates have decreased numbers of proliferating cells and enlarged hypertrophic zones, suggesting that CCN2 functions in later stages of chondrocyte proliferation and differentiation (Ivkovic *et al.*, 2003). There are also differences in the cartilage ECM. Matrix components such as aggrecan and link protein are decreased, demonstrating that CCN2 is an important regulator of ECM synthesis *in vivo* and confirming previous *in vitro* studies (Nakanishi *et al.*, 2000; Ivkovic *et al.*, 2003). As cartilage derives its elasticity and tensile strength from the ECM, defective ECM synthesis is probably responsible for the defects in the shapes of bones.

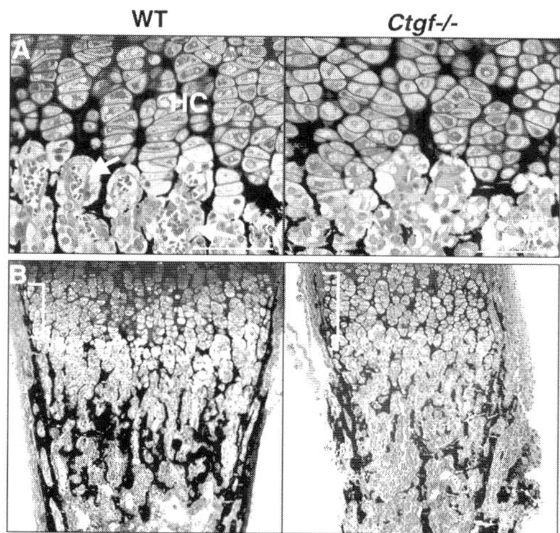

Fig. 4.   (A) Plastic sections through the growth plate of P0 femora stained by the method of von Kossa. Hypertrophic chondrocyte columns are disorganized in *Ccn2* mutants, with fewer capillaries invading the cartilage matrix. (B) Sections through P0 femora demonstrate that mutants are osteopenic. (Adapted from Ivkovic *et al.*, 2003.)

In the growth plate of *Ccn2* mutants, the hypertrophic zone is disorganized and contains few intact blood vessels (Fig. 4). Within the growth plate, VEGF-A promotes angiogenesis (Gerber *et al.*, 1999). *Vegf-A* levels are reduced in hypertrophic cartilage, suggesting that CCN2 regulates growth plate angiogenesis by regulating *Vegf-A* expression (Ivkovic *et al.*, 2003). In addition to decreased *Vegf-A* expression, *Ccn2* mutants have decreased numbers of matrix metalloproteinase 9 (MMP9)-producing osteoclasts/chondroclasts in the region of the hypertrophic zone. MMPs are essential regulators of growth plate angiogenesis that degrade ECM, permitting the invasion of blood vessels and promoting the release of sequestered angiogenic factors (Sternlicht and Werb, 2001). Decreased numbers of MMP9-producing cells in growth plates of *Ccn2* mutants, in conjunction with decreased expression of specific ECM components, suggests that CCN2 regulates growth plate angiogenesis in part by modifying ECM turnover (Ivkovic *et al.*, 2003).

The phenotype of the *Ccn2* mutant highlights the importance of loss-of-function animal models in discerning gene action. Some aspects of the *Ccn2* mutant phenotype are consistent with *in vitro* studies, such as the role of CCN2 in promoting chondrocyte proliferation and expression of several ECM components. However, other aspects of the phenotype are unexpected. For example, while CCN2 upregulates the expression of types II and X collagen *in vitro*, *Ccn2* mutants do not have lower levels of collagens II and X, demonstrating that not every *in vitro* function is relevant under physiological conditions. The growth plate angiogenesis phenotype is also somewhat unexpected; while CCN2 has been implicated as a pro-angiogenic factor *in vitro* (Babic *et al.*, 1999; Shimo *et al.*, 1999), the specific requirement for CCN2 in growth plate angiogenesis was not expected. Thus, the angiogenic activity of CCN2 during development is more restricted than its expression pattern would suggest. Whether this is due to functional redundancy with other CCN family members is an important, and unanswered, question. Finally, the *Ccn2* mutant phenotype reveals that VEGF expression is dependent on CCN2 activity in hypertrophic chondrocytes, a regulation that was not previously suspected from *in vitro* data.

CCN6

Compared to CCN1 and CCN2, far less is known about the *in vivo* functions of other CCN family members. However, an essential role for CCN6 (WISP-3) has been demonstrated in humans. CCN6 was first identified as a target of the canonical Wnt signaling pathway (Pennica *et al.*, 1998). While CCN6 is expressed in several tumor types, little is known about its expression in the embryo.

In humans, *Ccn6* mRNA is detected in fetal kidney (Pennica *et al.*, 1998), as well as in primary synoviocytes and chondrocytes (Hurvitz *et al.*, 1999). However, *in situ* hybridization studies have failed to localize *Ccn6* within developing skeletal elements, perhaps due to low levels of expression (Hurvitz *et al.*, 1999). Overexpression of CCN6 in human chondrocytic cell lines increases expression of type II collagen and aggrecan (Sen *et al.*, 2004), a finding reminiscent of the ECM-inducing properties of CCN1 and CCN2, but the potency of CCN6 compared to CCNs 1 and 2 is unknown. Loss-of-function mutations in *Ccn6* cause the recessive disorder progressive pseudorheumatoid dysplasia (PPD), a condition involving joint degeneration manifesting in children as stiffness and swelling of joints (Fig. 5). Articular cartilage is progressively destroyed (Hurvitz *et al.*, 1999). Examination of the cartilage reveals that the columnar organization of the growth plate is disturbed, and abnormal clusters of chondrocytes appear, as seen in arthritic cartilage attempting repair. The apparent lack of an embryonic phenotype indicates that CCN6 is required for post-natal skeletal growth and cartilage homeostasis. The phenotype is thus distinct from that of *Ccn2* mutants.

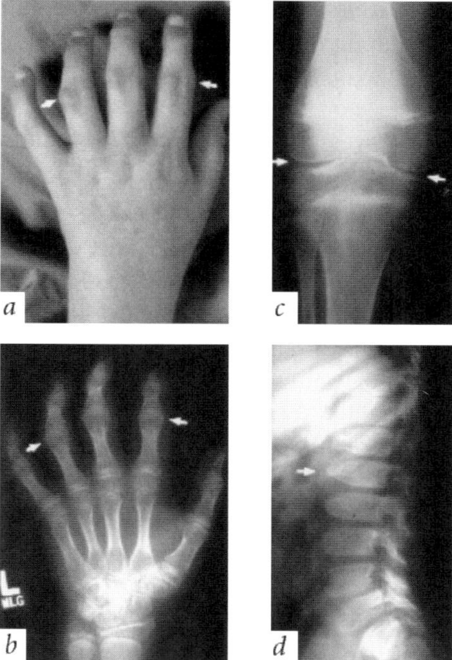

Fig. 5.   (A) Enlargement of proximal interphalangeal joints (arrows) of a PPD patient. (B) Hand radiograph demonstrating enlargement of epiphyseal and metaphyseal portions of metacarpals and phalanges. (C) Knee radiograph demonstrating enlargement of femoral and tibial epiphyses and narrowing of the joint space (arrows). (D) Lateral spine radiograph demonstrating flattening and interior beaking (arrow) of the thoraco-lumbar spine. (Adapted from Hurvitz *et al.*, 1999. Reprinted with permission from Nature Publishing Group.)

CCNS, INTEGRINS, AND THE ECM

While the *Ccn1* and *Ccn2* knockouts have provided novel insights into the functions of these genes, there are many unanswered questions. One of the most important of these is the mechanism of CCN action. Both CCN1 and CCN2 bind to multiple integrins (reviewed in Lau and Lam, 1999), triggering intracellular signaling cascades that regulate transcription, alter cytoskeletal organization, and promote cell proliferation, differentiation, survival, and shape changes (reviewed in Bökel and Brown, 2002). While these integrin-mediated activities are sufficient to account for the mutant phenotypes, it not known whether integrins are required to mediate CCN signaling *in vivo*. The phenotypes of integrin mutants (e.g., Bader *et al.*, 1998; Terpstra *et al.*, 2003) are consistent with the possibility that most, if not all, of the actions of CCN1 and CCN2 in angiogenesis and chondrogenesis are mediated by integrins, but combinatorial and overlapping functions of integrins and their ligands do not permit the unambiguous assignment of any specific activity of CCN1 or CCN2 to engagement of a specific integrin receptor. Moreover, the fact that CCN1 and CCN2 bind to multiple integrin receptors further complicates genetic analysis of these interactions.

CCN1 acts as a ligand for the αv integrins $\alpha_v\beta_3$ and $\alpha_v\beta_5$ to promote cell adhesion and migration (Grzeszkiewicz *et al.*, 2001), and mice lacking αv integrins display a phenotype similar to that of *Ccn1* mutants. As in *Ccn1* mutants, mice lacking αv integrins have normal yolk sac vasculature, but the placenta is frequently undervascularized with fewer embryonic blood vessels (Bader *et al.*, 1998). Some αv mutants have normal placental development, but like *Ccn1* mutants, these suffer from hemorrhage (Bader *et al.*, 1998). In *Ccn1* mutants, the VEGF-R3 signaling pathway may be a downstream target of αv integrin activation. Moreover, the ability of VEGF to promote endothelial cell survival requires engagement of αvβ3 (Hutchings *et al.*, 2003), and CCN1 may somehow enhance this interaction. The αv integrin knockout is not entirely similar to the *Ccn1* knockout, however; embryonic vascularization outside of the brain is largely normal (Bader *et al.*, 1998). Similarly, the ability of CCN1 to promote chorioallantoic fusion is most likely not mediated through αv integrins, since this process is normal in αv integrin mutants (Bader *et al.*, 1998). Thus αv integrins most likely mediate some, but not all, aspects of CCN1 activity *in vivo*.

Similarities can also be found between the CCN2 knockout and mutants affecting integrin pathways. CCN2 binds to integrin $\alpha_6\beta_1$ (Leu *et al.*, 2003), and this integrin is expressed in cartilage (Aszodi *et al.*, 2003). A requirement for β1 integrin-mediated signaling in chondrogenesis is revealed by the conditional loss of β1 integrin in chondrocytes. β1 mutants exhibit shortened bones, disrupted growth plate organization, and delayed ossification and growth plate angiogenesis. Mutant chondrocytes exhibit defects in cell shape, cytokinesis, adhesion and motility (Aszodi *et al.*, 2003). *Ccn2* mutant chondrocytes exhibit similar disruptions in columnar organization and in cell shape in the hypertrophic zone (Fig. 4). It is thus possible that these phenotypic similarities arise because CCN2 is a ligand for β1 integrins in chondrocytes. However, β1 mutants do not exhibit the deformations in long bones seen in *Ccn2* mutants, indicating that the ability of *Ccn2* to regulate the synthesis of multiple ECM components does not exclusively require β1 integrins *in vivo*. Another mutant affecting integrin signaling in cartilage, a conditional knockout of integrin-linked kinase (ILK), displays moderate chondrodysplasia. As in *Ccn2* mutants, chondrocyte proliferation is impaired (Terpstra *et al.*, 2003). Unlike *Ccn2* mutants, however, *Ilk* mutants have limbs that are shorter but not deformed. Thus, the $\beta1$ and *Ilk* phenotypes are consistent with the possibility that CCN2 promotes chondrocyte proliferation through integrins, but do not shed light on the mechanisms by which CCN2 regulates ECM production and VEGF expression in hypertrophic chondrocytes.

The *Ccn2* knockout also demonstrates that CCN2 is an essential regulator of the production of multiple ECM components. Among existing mutant strains, the *Ccn2* mutant phenotype most closely resembles that of *Sox9* $+/-$ mice (Bi *et al.*, 2001). Sox9 is a transcription factor that is required not only for commitment to the chondrogenic lineage, but also for the production of many ECM components, and the regulatory regions of genes encoding these components contain binding sites for Sox9. Other mutant phenotypes also support the hypothesis that a major consequence of loss of *Ccn2* is disruption of ECM synthesis. For example, link protein is diminished in *Ccn2* mutants, and mice lacking the gene that encodes link protein display some of the same skeletal defects seen in *Ccn2* mutants (Watanabe and Yamada, 1999; Ivkovic *et al.*, 2003).

INTERACTIONS OF CCNS WITH GROWTH FACTORS AND THEIR RECEPTORS

While it is possible that CCNs mediate some of their functions through unique, non-integrin receptors, specific receptors have not been identified despite intensive searches for them (reviewed in Brigstock, 2003). However, there is evidence to support an interaction with Wnt receptors. In *Xenopus*, overexpression and inhibition of CCN1 have the same effect of disrupting gastrulation, possibly due to defective cell adhesion (Latinkic *et al.*, 2003). Overexpression of *Ccn1* mRNA in *Xenopus* weakly activates Wnt signaling on its own, but CCN1 inhibits Wnt signaling in the presence of Wnt ligand. These data suggest that CCN1 might be important to maintain Wnt signaling at specific levels (Latinkic *et al.*, 2003). *Ccn2* mRNA overexpression in *Xenopus* results in an expanded neural plate, a phenotype that resembles anti-Wnt phenotypes, and CCN2 can interfere with the canonical Wnt pathway by binding to the Wnt receptor LRP-6 (Mercurio *et al.*, 2004). The relevance of these findings to the *Ccn1* and *Ccn2* mutant mouse phenotypes is unclear, in part because the roles of LRP-6 in angiogenesis and chondrogenesis are as yet unknown. In *Xenopus*, antisense inhibition of *Ccn1* leads to gastrulation phenotypes, whereas these are not seen in the mouse. This may reflect species-specific differences in gene function. However, the possibility that CCN1 and CCN2 mediate some of their functions in mammalian development via binding to the Wnt receptors LRP5 and LRP-6 warrants further investigation. The possibility of physiologically significant effects via LRPs in mammals is strengthened by independent reports showing that can CCN2 bind to LRP (Segarini *et al.*, 2001) and mediate adhesion (Gao and Brigstock, 2004) in mammalian cells.

CCNs bind to and modify the activity of growth factors *in vitro*, as is the case for other matricellular proteins. For example, CCN2 can bind to TGFβ and BMP *in vitro*, and activate or inhibit, respectively, the abilities of these growth factors to bind to their receptors (Abreu *et al.*, 2002). At present, the physiological relevance of these associations is unclear. Loss of *Ccn2* does not yield phenotypes consistent with increased BMP activity, as would be predicted if a physiological role for CCN2 is to inhibit BMP activity. On the other hand, the *Ccn2* phenotype does bear some similarity to mutants exhibiting reduced TGFβ activity. Mice lacking *Tgfβ2* exhibit shortened mandibles and irregular ribs, defects similar to but less severe than those seen in *Ccn2* mutants (Sanford *et al.*, 1997). Mice lacking *Smad3* signals display reduced chondrocyte proliferation (Yang *et al.*, 2001), as is seen in *Ccn2* mutants. However, interpretation of these phenotypic data in terms of physical associations between CCN2 and TGFβ is complicated by the fact that CCN2 is an immediate downstream target of TGFβ, and is known to mediate many of the effects of TGFβ on cell proliferation and apoptosis (reviewed in Brigstock, 1999). Thus, these phenotypic similarities may reflect the presence of a pathway in which TGFβ signaling induces CCN2, which then acts as a downstream mediator of TGFβ action.

VEGFS ARE DOWNSTREAM TARGETS OF CCN1 AND CCN2 *IN VIVO*

Both the *Ccn1* and *Ccn2* mutant phenotypes demonstrate that VEGFs are major, physiologically relevant targets of CCN action during development. In the *Ccn1* knockout, the bifurcation and remodeling of initial vessels by non-sprouting angiogenesis to form a vascular network is severely impaired (Mo *et al.*, 2002). This unique phenotype is accompanied by decreased expression of the VEGF-R3 ligand VEGF-C in allantoic mesoderm. Mice lacking *Vegf-R3* are defective in non-sprouting angiogenesis. *Vegf-R3* mutants form an immature yolk sac vascular plexus consisting of uniformly sized vessels, but this immature plexus does not become remodeled into a complex network of small and large vessels (Dumont *et al.*, 1998). The requirement for VEGF-R3 signaling in vascular remodeling, coupled with defects in vessel bifurcation and VEGF-C regulation in *Ccn1* mutants, suggests that CCN1 exerts some of its effects on angiogenesis by activating a VEGF-R3 pathway to promote vascular remodeling in the placenta. However, defects in expression of VEGF-C alone are not sufficient to account for the *Ccn1* mutant phenotype, as loss of VEGF-C does not cause vascular remodeling defects *in vivo* (Karkkainen *et al.*, 2004). It is likely that CCN1 regulates the expression and/or activity of other VEGF isoforms and receptors.

In *Ccn2* knockouts, VEGF-A mRNA and protein are downregulated in hypertrophic chondrocytes where angiogenesis occurs. The importance of VEGF-A activity in the growth plate has been established *in vivo*. Inactivation of *Vegf-A* in chondrocytes leads to a loss of blood vessel invasion and concomitant expansion of the hypertrophic zone (Gerber *et al.*, 1999; Zelzer *et al.*, 2004). The mechanism by which CCN2 regulates *Vegf-A* in the growth plate is not known, and the relationship between CCN2 and VEGF-A is likely to be complex. The relationship may be indirect, with CCN2 activating integrin pathways that in turn induce *Vegf-A*. CCN2 *in vitro* activates a p42/44 MAPK pathway via integrins (Chen *et al.*, 2001), and the p42/44 MAPK pathway can induce *Vegf-A* expression (Milanini *et al.*, 1998), although these activities have not yet been shown to be operative in chondrocytes. Alternatively, CCN2 could regulate *Vegf-A* via hypoxia-inducible factor (HIF1α) and/or TGFβ. HIF1α promotes *Vegf-A* expression in hypertrophic cartilage (Schipani *et al.*, 2001), and HIF1α and TGFβ can synergize to activate VEGF expression *in vitro* (Sanchez–Elsner *et al.*, 2001). CCN2 might also participate in this process by activating pathways such as p42/44 MAPK kinase that lead to stabilization and/or increased transactivational activity of HIF1α. It is also possible that CCN2 regulates VEGF activity in the growth plate through physical association. CCN2 can bind to and inhibit the ability of VEGF-A to induce angiogenesis (Inoki *et al.*, 2001). Thus, it is possible that CCN2 modulates growth plate angiogenesis by positively regulating *Vegf-A* expression and negatively regulating VEGF-A activity. This latter regulation of activity may involve MMP9, which can release VEGF from ECM, thereby promoting angiogenesis (Hashimoto *et al.*, 2002). Exploration of the mechanisms by which CCN1 and CCN2 regulate expression and activity of VEGF-C and VEGF-A, respectively, should shed considerable light on physiologically relevant signaling pathways activated by CCNs.

FUTURE DIRECTIONS FOR *IN VIVO* ANALYSIS OF CCN FUNCTION

Though CCN1 and CCN2 have broad expression patterns, suggesting that they participate in the development of multiple tissues, the knockouts exhibit restricted phenotypes. For example, CCN1 and CCN2 are co-expressed in a number of sites, such as the developing nervous system, in which no phenotype has been reported. One explanation for this would be functional redundancy; hence the construction of *Ccn1/Ccn2* double mutants is required. However, the early lethality of *Ccn1* mutants will restrict this analysis to examination of early developmental stages.

To examine potential functional overlap in other tissues, the generation of conditional alleles of *Ccn1* and *Ccn2* is required. A major unanswered question is whether CCN1 collaborates with CCN2 during angiogenesis and/or chondrogenesis, or whether these two CCN family members serve unique functions in these processes. This can be investigated by examination of the phenotypes of cartilage-specific *Ccn1* mutants and *Ccn1/Ccn2* double mutants. These experiments will test the possibility that, in spite of nearly identical *in vitro* activities, CCN1 and CCN2 fulfill unique functions in tissues in which they are co-expressed. Conditional alleles will also be a valuable tool for studying the roles of CCN1 and CCN2 in normal adult tissue homeostasis and in disease processes where these proteins have been implicated, such as vascular disease, fibrotic disease, and cancer. Additional genes misregulated in *Ccn* mutants need to be identified, as these targets will permit investigations of the signaling pathways by which CCNs control their expression, providing essential clues about which of the many proposed modes of CCN action revealed by *in vitro* and overexpression studies are physiologically relevant. Finally, generation of mutant alleles for other CCN family members is required to understand their individual functions, whether they exhibit functional redundancy with other CCN family members, and the full spectrum of cellular events regulated by CCN genes *in vivo*.

ACKNOWLEDGMENTS

The authors thank Sean Brugger and Bernadette So for helpful comments on the manuscript.

REFERENCES

Abreu JG, Ketpura NI, Reversade B and DeRobertis EM (2002) Connective-tissue growth factor (CTGF) modulates cell signaling by BMP and TGF-β. *Nature Cell Biol* **4**:599–603.

Aszodi A, Hunziker EB, Brakebusch C and Fässler R (2003) β1 integrins regulate chondrocyte rotation, G1 progression, and cytokinesis. *Genes Dev* **17**:2465–2479.

Babic AM, Kireeva ML, Kolesnikova TV and Lau LF (1998) CYR61, a product of a growth-factor-inducible immediate early gene, promotes angiogenesis and tumor growth. *Proc Natl Acad Sci USA* **95**:6355–6360.

Babic AM, Chen C-C and Lau LF (1999) Fisp12/mouse connective tissue growth factor mediates endothelial cell adhesion and migration through integrin $\alpha_v\beta_3$, promotes endothelial cell survival, and induces angiogenesis *in vivo*. *Mol Cell Biol* **19**:2958–2966.

Bader BL, Rayburn H, Crowley D and Hynes RO (1998) Extensive vasculogenesis, angiogenesis, and organogenesis precede lethality in mice lacking all αv integins. *Cell* **95**:507–519.

Bi W, *et al.* (2001) Haploinsufficiency of *Sox9* results in defective cartilage primordial and premature skeletal mineralization. *Proc Natl Acad Sci USA* **98**:6698–6703.

Bökel C and Brown NH (2002) Integrins in development: moving on, responding to, and sticking to the extracellular matrix. *Dev Cell* **3**:311–321.

Brigstock DR (1999) The connective tissue growth factor/cysteine-rich 61/nephroblastoma overexpressed (CCN) family. *Endocr Rev* **20**:189–206.

Brigstock DR (2002) Regulation of angiogenesis and endothelial cell function by connective tissue growth factor (CTGF) and cysteine-rich 61 (CYR61). *Angiogenesis* **5**:153–165.

Brigstock DR (2003) The CCN family: a new stimulus package. *J Endocrinol* **178**: 169–175.

Chen C-C, Chen N and Lau LF (2001) The angiogenic factor Cyr61 activates a genetic program for wound healing in human skin fibroblasts. *J Biol Chem* **276**: 10443–10452.

Dumont DJ, *et al.* (1998) Cardiovascular failure in mouse embryos deficient in VEGF receptor-3. *Science* **282**:946–949.

Friedrichsen S, *et al.* (2003) CTGF expression during mouse embryonic development. *Cell Tissue Res* **312**:175–188.

Fukunaga T, Yamashiro, T, Oya S, Takeshita N, Takigawa N, Takano–Yamamoto T (2003) Connective tissue growth factor mRNA expression pattern in cartilages is associated with their type I collagen expression. *Bone* **33**:911–918.

Gao R and Brigstock DR (2003) Low density lipoprotein receptor-related protein (LRP) is a heparin-dependent adhesion receptor for connective tissue growth factor (CTGF) in rat activated hepatic stellate cells. *Hepatol Res* **27**:214–220.

Gerber H-P, *et al.* (1999) VEGF couples hypertrophic cartilage remodeling, ossification and angiogenesis during endochondral bone formation. *Nature Med* **5**: 623–628.

Grzeszkiewicz TM, Kirschling DJ, Chen N and Lau LF (2001) CYR61 stimulates human skin fibroblast migration through integrin $\alpha_v\beta_5$ and enhances mitogenesis through integin $\alpha_v\beta_3$, independent of its carboxyl-terminal domain. *J Biol Chem* **276**:21943–21950.

Hashimoto G, *et al.* (2002) Matrix metalloproteinases cleave connective tissue growth factor and reactivate angiogenic activity of vascular endothelial growth factor 165. *J Biol Chem* **277**:36288–36295.

Hurvitz JR, *et al.* (1999) Mutations in the CCN gene family member WISP3 cause progressive pseudorheumatoid dysplasia. *Nature Genet* **23**:94–98.

Hutchings H, Ortega N and Plouet J (2003) Extracellular matrix-bound vascular endothelial growth factor promotes endothelial cell adhesion, migration, and survival through integrin ligation. *FASEB J* **17**:1520–1522.

Inoki I, *et al.* (2002) Connective tissue growth factor binds vascular endothelial growth factor (VEGF) and inhibits VEGF-induced angiogenesis. *FASEB J* **16**:219–221.

Ivkovic S, *et al.* (2003) Connective tissue growth factor coordinates chondrogenesis and angiogenesis during skeletal development. *Development* **130**:2779–2791.

Karkkainen MJ, *et al.* (2004) Vascular endothelial growth factor C is required for sprouting of the first lymphatic vessels from embryonic veins. *Nat Immunol* **5**:74–80.

Kireeva ML, Mo F-E, Yang GP and Lau LF (1996) Cyr61, a product of a growth factor-inducible immediate-early gene, promotes cell proliferation, migration, and adhesion. *Mol Cell Biol* **16**:1326–1334.

Latinkic BV, *et al.* (2001) Promoter function of the angiogenic inducer *Cyr61* gene in transgenic mice: tissue specificity, inducibility during wound healing, and role of the serum response element. *Endocrinology* **142**:2549–2557.

Latinkic BV, *et al.* (2003) *Xenopus Cyr61* regulates gastrulation movements and modulates Wnt signaling. *Development* **130**:2429–2441.

Lau LF and Lam S-J (1999) The CCN family of angiogenic regulators: the integrin connection. *Exp Cell Res* **248**:44–57.

Leask A, Denton CP and Abraham DJ (2004) Insights into the molecular mechanism of chronic fibrosis: the role of connective tissue growth factor in scleroderma. *J Invest Dermatol* **122**:1–6.

Leu S-J, Lam SC-T and Lau LF (2002) Pro-angiogenic activities of CYR61 (CCN1) mediated through integrins $\alpha_v\beta_3$ and $\alpha_6\beta_1$ in human umbilical vein endothelial cells. *J Biol Chem* **277**:46248–46255.

Leu SJ, Liu Y, Chen N, Chen C-C, Lam SC and Lau LF (2003) Identification of a novel integrin $\alpha_6\beta_1$ in the angiogenic inducer CCN1 (Cyr61). *J Biol Chem* **278**: 33801–33806.

Mercurio S, Latinkic BV, Itasaki N, Krumlauf R and Smith JC (2004) Connective-tissue growth factor modulates WNT signaling and interacts with the WNT receptor complex. *Development* **131**:2137–2147.

Milanini J, Vinals F, Pouyssegur J and Pages G (1998) p42/p44 MAP kinase module plays a key role in the transcriptional regulation of the vascular endothelial growth factor gene in fibroblasts. *J Biol Chem* **273**:18165–18172.

Mo F-E, Muntean AG, Chen C-C, Stolz DB, Watkins SC and Lau LF (2002) CYR61 (CCN1) is essential for placental development and vascular integrity. *Mol Cell Biol* **22**:8709–8720.

Nakanishi T, *et al.* (1997) Overexpression of connective tissue growth factor/hypertrophic chondrocyte-specific gene product 24 decreases bone density in adult mice and induces dwarfism. *Biochem Biophys Res Commun* **234**:206–210.

Nakanishi T, *et al.* (2000) Effects of CTGF/Hcs24, a product of a hypertrophic chondrocyte-specific gene, on the proliferation and differentiation of chondrocytes in culture. *Endocrinology* **141**:264–273.

Nishida T, *et al.* (2002) Effects of CTGF/Hcs24, a hypertrophic chondrocyte-specific gene product, on the proliferation and differentiation of osteoblastic cells *in vitro*. *J Cell Biochem* **192**:55–63.

O'Brien TP and Lau LF (1992) Expression of the growth factor-inducible immediate early gene *cyr61* correlates with chondrogenesis during mouse embryonic development. *Cell Growth Differ* **3**:645–654.

Pennica D, *et al.* (1998) *WISP* genes are members of the connective tissue growth factor family that are up-regulated in Wnt1-transformed cells and aberrantly expressed in human colon tumors. *Proc Natl Acad Sci USA* **95**:14717–14722.

Planque N and Perbal B (2003) A structural approach to the role of CCN (CYR61/CTGF/NOV) proteins in tumorigenesis. *Cancer Cell Int* **3**:15–29.

Sánchez–Elsner T, Botella LM, Velasco B, Corbi A, Attisano L and Bernabéu C (2001) Synergistic cooperation between hypoxia and transforming growth factor-β pathways on human vascular endothelial growth factor gene expression. *J Biol Chem* **276**:38527–38535.

Sanford LP, *et al.* (1997) TGFβ2 knockout mice have multiple developmental defects that are non-overlapping with other TGFβ knockout phenotypes. *Development* **124**:2659–2670.

Schipani E, Ryan HE, Didrickson S, Kobayashi T, Knight M and Johnson RS (2001) Hypoxia in cartilage: HIF-1α is essential for chondrocyte growth arrest and survival. *Genes Dev* **15**:2865–2876.

Segarini PR, Nesbitt JE, Li D, Hays LG, Yates JR 3rd, and Carmichael DF (2001) The low density lipoprotein receptor-related protein/α_2-macroglobulin receptor is a receptor for connective tissue growth factor. *J Biol Chem* **276**:40659–40667.

Sen M, *et al.* (2004) WISP3-dependent regulation of type II collagen and aggrecan production in chondrocytes. *Arthritis Rheum* **50**:488–497.

Shimo T, *et al.* (1999) Connective tissue growth factor induces the proliferation, migration, and tube formation of vascular endothelial cells *in vitro*, and angiogenesis *in vivo*. *J Biochem (Tokyo)* **126**:137–145.

Sternlicht MD and Werb Z (2001) How matrix metalloproteinases regulate cell behavior. *Annu Rev Cell Dev Biol* **17**:463–516.

Terpstra L, *et al.* (2003) Reduced chondrocyte proliferation and chondrodysplasia in mice lacking the integrin-linked kinase in chondrocytes. *J Cell Biol* **162**:139–148.

Watanabe H and Yamada Y (1999) Mice lacking link protein develop dwarfism and craniofacial abnormalities. *Nature Genet* **21**:225–229.

Wong M, Kireeva ML, Kolesnikova TV and Lau LF (1997) Cyr61, product of a growth factor-inducible immediate-early gene, regulates chondrogenesis in mouse limb bud mesenchymal cells. *Dev Biol* **192**:492–508.

Yang X, Chen L, Xu X, Li C, Huang C and Deng CX (2001) TGF-β/Smad3 signals repress chondrocyte hypertrophic differentiation and are required for maintaining articular cartilage. *J Cell Biol* **153**:35–46.

Zelzer E, Mamluk R, Ferrara N, Johnson RS, Schipani E and Olsen BR (2004) VEGFA is necessary for chondrocyte survival during bone development. *Development* **131**:2161–2171.

CHAPTER 8

CCN FAMILY IN EMBRYONIC DEVELOPMENT (NON-MAMMALIAN MODELS)

Branko V. Latinkic

School of Biosciences, Cardiff University, Cardiff CF10 3US, Wales, UK

The study of CCN genes in the development of the zebrafish, the frog *Xenopus* and the chick has only just begun. The main findings thus far complement the results obtained in better studied mammalian models. CCN1 in *Xenopus* embryos regulates gastrulation movements and can either antagonize or stimulate Wnt signaling, in a context dependent manner. In the same model system, CCN2 can also antagonize Wnt signaling, through interaction of its CT domain with the Wnt receptor complex. In *Xenopus* embryos CCN2 additionally antagonizes BMP pathway. The basis of this activity is binding of the VWC domain to BMP4. These activities likely contribute to the anteriorization of the nervous system and to the delay in neural differentiation seen in gain-of-function experiments.

During embryonic development, a single cell is gradually transformed into a functional organism. This transformation involves numerous regulated processes, including cell division, differentiation and migration. These aspects of development have been studied in great detail, but usually in isolation. However, it is obvious that all processes occurring in the embryo are carefully orchestrated. In the developing nervous system, for example, asymmetrical cell divisions are important not only to generate the bulk of the tissue whilst maintaining the stem cell population, but are also used to specify different cell fates through unequal distribution of cellular determinants (Betschinger and Knoblich, 2004). The formation of the extracellular matrix during development does not merely provide the cells with an adhesive surface: for example, the structural integrity of the notochord, provided by the basement membrane laminins, is not only important for the maintenance of the shape of the organ, but equally for its correct differentiation (Parsons *et al.*, 2002). It is clear that providing the missing links that integrate numerous embryonic cellular and molecular processes is the key to a more complete understanding of how an embryo develops.

As reviewed elsewhere in this book, the versatile CCN proteins can promote cell adhesion and migration and can modulate cellular proliferation and signaling, and may therefore have the potential to co-ordinate those activities during development. Thus, a study of CCN proteins in embryonic development offers an opportunity to provide an insight into the molecular mechanisms of integration of multitude of cellular processes during embryonic development. However, thus far, most of the work on the CCN family has focused on their biochemical activities and the roles they play in various pathological conditions; comparatively little effort has been devoted to investigation of their function in embryonic development.

More recently, this has begun to change. For example, very informative null mutations in CCN1 and CCN2 in mouse embryos have been generated (Mo *et al.*, 2002; Ivkovic *et al.*, 2003), and are discussed in detail by Dornbach and Lyons in Chapter 7.

In addition to the work on mammalian models, the initial work on CCN genes in other model systems, such as *Xenopus*, zebrafish and chick, has been reported. Embryos of these organisms develop externally and are amenable to experimental manipulation. The frog and the fish embryo additionally do not have an early requirement for functional cardiovascular system, unlike mouse embryos. Therefore, non-mammalian models are well suited to provide complementary insight into the role of the CCN family in development. Here I will provide an overview of these findings, and place them in the context of CCN biology.

CCN FAMILY ONLY EXISTS IN VERTEBRATES

A hint about the nature of the roles that CCN family might play in development comes from comparative genomics. It appears that the CCN family is vertebrate-specific, as it is not found in the *Drosophila* genome. Even though the unique arrangement of the four domains that characterize CCN proteins seems to be specific for the vertebrates, each of the building blocks (modules) is similar to the domains present in diverse Metazoan classes.

This is not a unique feature of the CCN family: other modular extracellular proteins that appear to be present only in vertebrates include fibronectin, fibrinogen, vitronectin, VEGF and the collagens. It has been suggested that vertebrate genes not represented by the fly orthologues may be involved in the development and homeostasis of the vascular system (Hynes and Zhao, 2000). Elaborate circulatory network and tightly controlled blood homeostasis are vertebrate novelty, and it is clear that vertebrate-specific extracellular proteins mentioned above are involved in normal development and homeostasis of blood vessels and blood (Hynes and Zhao, 2000).

For the CCN family, the available genetic information on the consequences of inactivation of CCN1 and CCN2 in the mouse embryo supports this view (Mo *et al.*, 2002; Ivkovic *et al.*, 2003). Both mutants are characterized by the defects in blood vessels formation, and are the focus of the Chapter 7 by Dornbach and Lyons.

CCN FAMILY IN ANAMNIOTES AND BIRDS

Multifunctional CCN proteins are likely to play the roles in development which are in addition to those in blood vessel formation. This is suggested by their complex and dynamic expression patterns outside the cardiovascular system and also by the multitude of their activities. Some of the roles and activities of CCN genes in early development are beginning to emerge from the work using *Xenopus* as a model.

CCN1

CCN1 cDNAs from the South African clawed frog *Xenopus laevis* (xCCN1) were isolated by degenerate PCR approach (Latinkic *et al.*, 2003) and independently in EST projects. *Xenopus* CCN1 is expressed maternally and zygotic activation occurs at late neurula stages (Latinkic *et al.*, 2003). The spatial pattern of expression of xCCN1 is similar to the well-described mouse expression pattern (O'Brien and Lau, 1992; Kireeva *et al.*, 1997; Latinkic *et al.*, 2001; Mo *et al.*, 2002), and includes the somites, notochord and the heart. The only current study on the roles of xCCN1 in development, reviewed below, most likely relates to the early, maternal phase of expression.

xCCN1 regulates gastrulation movements

The roles of CCN1 in early development of the frog embryo were investigated by a combination of gain-of-function and loss-of-function approaches (Latinkic *et al.*, 2003). Interference with the early function of *Xenopus* CCN1 by antisense morpholino oligonucleotides (stable and non-toxic modified oligonucleotides that block translation of their targets) disrupts gastrulation, as does overexpression of CCN1 mRNA. In addition, intrablastocoelic (blastocoel is a primary embryonic cavity that is defined on its top or animal pole by the "roof" of extracellular matrix that serves as a substrate for migration of the mesoderm) injection of purified CCN1 protein has the same effect, arguing that CCN1 regulates gastrulation movements relatively directly.

Gastrulation is a complex and tightly regulated process that involves polarized changes in cell shape, directed cell migration, and modulation of the cell cycle. Simultaneously, the fate of gastrulating cells is specified by inductive signaling. In the past, very few studies have tackled this process of intimidating complexity, but recently, our understanding of vertebrate gastrulation has improved significantly. We now know that Wnt signaling, acting through the non-canonical planar polarity pathway, regulates cell movements during gastrulation in *Xenopus* and zebrafish embryos (Wallingford *et al.*, 2002). For example, Dishevelled constructs, which specifically disrupt planar cell polarity signalling, interfere with convergent extension (cellular movements that extend the main body axis through migration towards the midline and through intercalation) in both *Xenopus* (Tada and Smith, 2000; Wallingford *et al.*, 2000) and zebrafish (Heisenberg *et al.*, 2000). Other components of the Wnt planar cell polarity pathway, such as RhoA, Rac and JNK, regulate cytoskeletal function, and thereby cell shape and polarity (Habas *et al.*, 2001; Habas *et al.*, 2003). In addition, fibronectin and its integrin receptors are now known to be required for proper execution of convergent extension movements in *Xenopus* (Marsden and DeSimone, 2001; Marsden and DeSimone, 2003).

As CCN1 associates with the extracellular matrix, mediates cell adhesion and cell migration and chemotaxis and activates intracellular signaling, it is perhaps not surprising that it might regulate gastrulation through some or all of those activities. The initial report has suggested that the correct amount of CCN1 is critical for normal gastrulation to proceed, as both overexpression of CCN1 and its downregulation cause similar defects in morphogenesis (Latinkic *et al.*, 2003). In this respect, CCN1 resembles other genes involved in gastrulation movements, where overexpression and inhibition can both cause disruption of gastrulation (Tada and Smith, 2000).

Some of the effects of CCN1 on gastrulation may derive from its ability to support assembly of a fibronectin-rich extracellular matrix in blastocoel roof, which is the main substrate for migrating mesoderm; it was found that downregulation of CCN1 disrupts the formation of fibronectin fibrils, without significantly affecting the total levels of fibronectin. Other effects of CCN1 on *Xenopus* gastrulation may derive from its ability to support cell adhesion of the early gastrula cells, and to regulate cell-cell interactions as well.

The morphology of embryonic cells plated on CCN1, with large lamellipodia and few or no filopodia, differs from that of cells adherent to fibronectin (Latinkic *et al.*, 2003) and it seems likely that the behavior of embryonic cells adherent to the two substrates would differ, with those attached to fibronectin being more motile than those attached to CCN1. The attachment of embryonic cells to CCN1 is mediated through its CT domain and it likely requires HSPGs, since it can be blocked by exogenous heparin (Latinkic *et al.*, 2003). The cellular receptor that mediates adhesion to CCN1 in this system was not determined, but one likely candidate is integrin $\alpha 5\beta 1$ (Ransom *et al.*, 1993; Joos *et al.*, 1995).

Interference with cell-matrix adhesion is one way in which manipulation of CCN1 levels might disrupt gastrulation. Another is through interference with cell-cell adhesion. The depletion of CCN1 from the embryo compromises Ca^{2+}-induced cell adhesion, arguing that CCN1 may be important for this process during gastrulation as well. During gastrulation cells migrate in groups, or sheets. In addition, radial intercalation and cell sorting during gastrulation require cell-cell interactions. Thus, interference with this cellular behavior may contribute to defective convergent extension movements.

The link between cell-extracellular matrix and cell-cell interactions may be direct: recently it was shown that integrin-fibronectin complex activates cadherin-mediated cell-cell interactions in *Xenopus* embryo (Marsden and DeSimone, 2003).

The results presented above implicate CCN1 as an important regulator of gastrulation movements through its ability to support cell-substrate adhesion, through its role in the assembly of the extracellular matrix and through its influence on cell-cell interactions. Given that CCN1 can cause adhesive signaling and cytoskeletal rearrangements (Chen *et al.*, 2001) it is possible that these activities may be involved in regulation of gastrulation movements as well.

xCCN1 can modulate Wnt signaling

In addition to regulating gastrulation movements, xCCN1 was shown to modulate Wnt signaling (Latinkic *et al.*, 2003). Thus, CCN1 was found to cause the formation of complete secondary body axes in *Xenopus* embryos, a hallmark of activation of the canonical Wnt/β-catenin signaling pathway. This effect is likely to be direct, since CCN1 also induces the expression of the direct Wnt target gene *Siamois*. The ability of CCN1 to activate the Wnt/β-catenin pathway, which is weak compared with that of Wnt pathway components Wnt8 or Dishevelled, is likely to be mediated by the IGFBP domain (domain 1; Fig. 1) (Latinkic *et al.*, 2003).

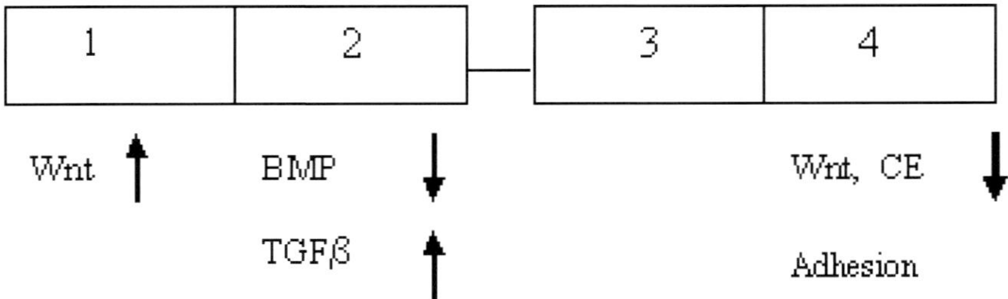

Fig. 1.   The activities of CCN modules in development, based on the work on xCCN1 and xCCN2 (Abreu *et al.*, 2002; Latinkic *et al.*, 2003; Mercurio *et al.*, 2004).

1. (IGFBP domain). Positive Wnt/β-catenin modulation (IGF antagonism?) (CCN1).
2. (VWC domain). Weak BMP antagonism; stimulation of TGFβ signaling in cell culture (CCN2).
3. (TSP domain). Activities *in vivo* have not been described.
4. (CT domain). Wnt/β-catenin antagonism; interference with convergence-extension movements (CE; CCN1 and CCN2). Interaction with the Wnt receptor complex (CCN2); HSPG-dependent adhesion (CCN1).

How CCN1 activates Wnt/β-catenin pathway is unclear at the present. One possibility is suggested by recent work, which has shown that the activation of the IGF receptor in *Xenopus* embryos inhibits the Wnt pathway (Pera *et al.*, 2001; Richard–Parpaillon *et al.*, 2002). We can speculate that the domain 1 of CCN1 binds endogenous members of the insulin-like growth factor family and relieves this inhibition.

A separate series of experiments has revealed that, surprisingly, xCCN1 can also inhibit Wnt/β-catenin signaling, in addition to activating it. Thus, CCN1 prevented the formation of secondary axes in response to Wnt8 and inhibited activation of a transcriptional target by Wnt8. This inhibition was found to be mediated by the CT domain (domain 4; see Fig. 1).

Whilst it is currently not known how xCCN1 domain 4 inhibits Wnt signaling, it is interesting to speculate that the abilities of the CT domain to regulate cell adhesion and to inhibit Wnt signaling may be related. This domain is required for the adhesion of fibroblasts (Grzeszkiewicz *et al.*, 2001) and of *Xenopus* embryonic cells to CCN1, and it also mediates the interaction of CCN1 with heparan sulphate proteoglycans (HSPGs) (Chen *et al.*, 2000). Exogenous heparin blocks the adhesion of *Xenopus* cells to CCN1 and this is likely to occur as a result of competition with cell-associated HSPGs for sites on CCN1. HSPGs have also been implicated in the regulation of Wnt signaling (Tsuda *et al.*, 1999; Topczewski *et al.*, 2001), and it is possible that the ability of CCN1 to bind HSPGs is related to its ability to inhibit Wnt signaling. Another potential link between cell adhesion and the modulation of Wnt signaling by CCN1 might be provided by the integrins, which are the only known receptors for CCN1 (Bokel and Brown, 2002); (Lau and Lam, 1999; Grzeszkiewicz *et al.*, 2001). Integrin-mediated adhesion of *Xenopus* embryonic cells to the extracellular matrix recruits Wnt pathway transducer Dishevelled to the plasma membrane (Marsden and DeSimone, 2001), and this may enhance the ability of cells to respond to a Wnt signal.

The non-canonical Wnt/planar polarity pathway regulates gastrulation movements (see above), and it is possible that CCN1 exerts some of its effects on the *Xenopus* embryo through this route. In support of this view, xCCN1 can block convergence-extension movements in a *Xenopus* embryo explant assay that is used for characterization of the Wnt/planar polarity pathway. This activity resides in the domain 4 (CT; Fig. 1); however, it has not yet been demonstrated that CCN1 directly acts on the Wnt/planar polarity pathway.

It is currently not known what, if any, might be the relevance of the Wnt-modulating activities of CCN1 for normal development. Downregulation of CCN1 by antisense oligonucleotides in *Xenopus* embryos leads to a reduction of *Brachyury*, an early mesodermal marker known to require Wnt signaling, and this might be a consequence of attenuation of the canonical Wnt pathway.

As the phenotype produced by downregulation or misexpression of xCCN1 is complex and severe, in the future it will be necessary to separate the different aspects of the phenotype by correlating them with particular domains and aminoacids within the protein. This work might also reveal how can xCCN1 both activate and inhibit Wnt signaling.

CCN1 in other models

Given the well known advantages of the chick and zebrafish systems, it is surprising that they have not been used to investigate the roles played by CCN1 (as well as other CCN genes) in early development. We anticipate that in the near future these excellent models will significantly contribute to the knowledge of CCN1 functions and of processes that it regulates.

CCN2

In what is currently the only report on the zebrafish CCN genes, CCN2 was identified in a study of regulated gene expression in the floor plate of the central nervous system (Dickmeis *et al.*, 2004). It was found that the promoter-proximal region in the CCN2 gene that is well conserved in another bony fish, Takifugu, can direct expression of a reporter gene in the floor plate, and also ectopically in the notochord (Dickmeis *et al.*, 2004). This study has suggested that a single compact regulatory element is sufficient to direct the expression of CCN2 in the floor plate, and also that additional element(s) are required to suppress inappropriate expression. The functional characterization of the zebrafish CCN2 protein has not been reported.

Two recent studies have focused on the *Xenopus laevis* CCN2 (xCCN2) (Abreu *et al.*, 2002; Mercurio *et al.*, 2004). In the *Xenopus* embryo, CCN2 expression resembles that of its mouse orthologue, and includes the somites, floor plate and the heart. The expression commences at the early neurula stage, and persists until at least the tadpole stages.

xCCN2 antagonizes BMP and Wnt signaling

Gain of function approach, combined with biochemical studies, has uncovered that xCCN2 can antagonize BMP signaling by binding to BMP4 and preventing it from interaction with its signal transducing receptor (Abreu *et al.*, 2002). This activity is mediated by the domain 2 (VWC domain) (Abreu *et al.*, 2002), which has significant homology with cysteine-rich BMP binding repeats of Chordin, a well-described BMP antagonist. Forced expression of xCCN2 on the ventral side of *Xenopus* embryos causes formation of partial secondary axes, which are characteristic of BMP antagonizing activity.

In the same study, xCCN2 was found to enhance the binding of TGFβ to its receptors and to lead to a corresponding increase in downstream signaling in cell culture (Abreu *et al.*, 2002). The importance of this activity of xCCN2 for embryonic development was not reported.

A separate line of investigation has found that xCCN2 can antagonize Wnt signaling in *Xenopus* embryos, in addition to its ability to antagonize BMP signaling (Mercurio *et al.*, 2004). Biochemical analyses have suggested that the basis of this antagonism is likely to be the interaction of xCCN2 with the Wnt receptor complex. Within this complex, xCCN2 can interact avidly with LRP-6, a co-receptor of the main Wnt receptor Frizzled, and can also weakly interact with Frizzled. The interaction of xCCN2 with the Wnt receptor complex may inhibit Wnt signaling in at least two ways: by competing with a Wnt ligand for binding to the receptor, and by mediating internalization of the receptor complex through binding to LRP-6, in a manner similar to the mechanism of action of another Wnt antagonist, Dkk (Mercurio *et al.*, 2004).

Can xCCN2 antagonize the Wnt pathway in the embryo? Gain-of-function experiments revealed that xCCN2 can block the formation of secondary axes by Wnt8. They have also shown that xCCN2 has effects on neural patterning that can be explained by interference with the Wnt/β-catenin pathway (Mercurio *et al.*, 2004). Ectopic expression of xCCN2 leads to the expansion of the anterior neural territory, and it inhibits or delays primary neurogenesis as well as neural crest formation (Mercurio *et al.*, 2004). Whether endogenous xCCN2 plays a role in patterning of the nervous system and neural crest is not known.

In addition to antagonizing the Wnt/β-catenin pathway, xCCN2 shares with xCCN1 its ability to block convergence-extension movements in an explant assay (Mercurio *et al.*, 2004). This activity resides in the domain 4 (CT; Fig. 1), but whether it reflects more direct modulation of the Wnt/planar polarity pathway, which regulates cell movements during gastrulation, or it may be a consequence of its independent effects on cell adhesion and migration, is not known.

In addition to the *Xenopus* CCN2, CCN2 was described in urodele amphibians. The newt is used as an excellent model of tissue and organ regeneration, and CCN2 was identified in a search for transcripts regulated during this process. It was found that CCN2 is expressed throughout the limb blastema mesenchyme (dedifferentiated, proliferative cells found at the site of amputation, from which the regenerated structure will form), and is down-regulated by retinoic acid, which is known to inhibit cell proliferation within the blastema. Because the expression of CCN2 was found in the limb blastema at more advanced stages of regeneration, it was suggested that its role may include regulation of cell proliferation and differentiation instead of (or in addition to) a potential role in early wound healing (Cash *et al.*, 1998).

CCN3

CCN3 was identified in *Xenopus laevis* embryos and oocytes (Ying and King, 1996). The mRNA was found in all stages examined (from gastrulation onward) but the spatial pattern of expression has not been reported to date (Ying and King, 1996). Preliminary gain of function studies with xCCN3 have suggested that it does not share the activities of xCCN1 and xCCN2 (BL, unpublished data).

In the chick embryo, CCN3 is expressed in a highly dynamic and intriguing pattern, starting with expression in the Hensen's node at stage HH4 (primitive streak formation) (Katsube *et al.*, 2001). This early domain of expression is asymmetric, resembling in this aspect the expression of signaling factor sonic hedgehog (Katsube *et al.*, 2001). Later on, several important signaling centres-the notochord, floor plate, and the floor of mesencephalon-all express CCN3 (Katsube *et al.*, 2001). Virtually nothing is known about the function of CCN3 in the chick embryo.

In one promising recent study it was reported that the CT domain of CCN3 interacts with the EGF repeats of the Notch receptor, and that this interaction may lead to attenuation of myogenesis in cell culture via activation of the Notch pathway (Sakamoto *et al.*, 2002). The chick CCN3, Notch1 and Delta all appear to be expressed in presomitic mesoderm, adding to the potential importance of these findings (Sakamoto *et al.*, 2002). It remains to be seen whether CCN3 will prove to have a role in myogenesis *in vivo*.

CONCLUDING REMARKS

Even though very few studies on the CCN family in non-mammalian models of vertebrate development have been reported, the current data indicate that they may play important roles in gastrulation and neural system development.

The scarcity of data clearly shows that we are at the beginning of the research into the roles of CCN proteins in embryonic development. The basic characterization of CCN4–6 in the zebrafish, frog and chick systems has not been reported, and, as remarked above, CCN1 and CCN2 have only been functionally analyzed in the *Xenopus*, and CCN3 in the chick. Amongst the multitude of questions that are worth addressing, it might be useful to highlight just a few.

The best-studied members of the family, CCN1 and CCN2, have largely overlapping activities *in vitro* and also have overlapping expression patterns *in vivo*. The gene knock-outs have demonstrated that CCN1 and CCN2 have unique non-overlapping functions, but the extent of overlap will only become clear after further genetic experiments such as the replacement of one gene with the other ("knock-in") and double knock-outs. The experiments with *Xenopus* embryos have indicated similarities and differences in activities of CCN1 and CCN2. For example, both can block Wnt signaling, but CCN1 can also stimulate it. On the other hand, CCN2 has effects on the patterning of the nervous system that it does not share with CCN1. In the future it should be possible to rapidly map the molecular basis for this difference using the *Xenopus* assay system, and this also applies to any similar differences between other CCN family members.

Is the early role of CCN1 revealed by experiments in *Xenopus* conserved? The mouse CCN1 is expressed in gastrulating embryos and earlier, during preimplantation stages (Mo *et al.*, 2002) (GenBank EST AA467423 isolated from blastocysts). Whether this early CCN1 has any role during these early stages of mammalian development has not yet been addressed. Even though the architecture of the frog, zebrafish, chick and mouse gastrula stage embryos differs greatly, the underlying molecular and cellular processes are remarkably similar, raising the possibility that CCN1 may be involved in mammalian preimplantation and gastrulation stages.

At the beginning of this chapter we commented that CCN proteins may be involved in integration of cell migration, adhesion, proliferation and signaling. One way of more directly addressing this issue of considerable interest, will be to clarify the relationship between the Wnt, BMP and integrin binding of CCN1/CCN2 in the context of the developing embryo.

ACKNOWLEDGEMENTS

I would like to thank Dr. Sara Mercurio for helpful comments on the manuscript.

REFERENCES

Abreu JG, Ketpura NI, Reversade B and De Robertis EM (2002) Connective-tissue growth factor (CTGF) modulates cell signalling by BMP and TGF-beta. *Nat Cell Biol* **4**(8):599–604.

Betschinger J and Knoblich JA (2004) Dare to be different: asymmetric cell division in Drosophila, *C. elegans* and vertebrates. *Curr Biol* **14**(16):R674–685.

Bokel C and Brown NH (2002) Integrins in development: moving on, responding to, and sticking to the extracellular matrix. *Dev Cell* **3**(3):311–321.

Cash DE, Gates PB, Imokawa Y and Brockes JP (1998) Identification of newt connective tissue growth factor as a target of retinoid regulation in limb blastemal cells. *Gene* **222**(1):119–124.

Chen CC, Mo FE and Lau LF (2001) The angiogenic factor Cyr61 activates a genetic program for wound healing in human skin fibroblasts. *J Biol Chem* **276**(50): 47329–47337.

Chen N, Chen CC and Lau LF (2000) Adhesion of human skin fibroblasts to Cyr61 is mediated through integrin alpha 6beta 1 and cell surface heparan sulfate proteoglycans. *J Biol Chem* **275**(32):24953–24961.

Dickmeis T, Plessy C, Rastegar S, Aanstad P, Herwig R, Chalmel F, Fischer N and Strahle U (2004) Expression profiling and comparative genomics identify a conserved regulatory region controlling midline expression in the zebrafish embryo. *Genome Res* **14**(2):228–238.

Grzeszkiewicz TM, Kirschling DJ, Chen N and Lau LF (2001) CYR61 stimulates human skin fibroblast migration through Integrin alpha vbeta 5 and enhances mitogenesis through integrin alpha vbeta 3, independent of its carboxyl-terminal domain. *J Biol Chem* **276**(24):21943–21950.

Habas R, Dawid IB and He X (2003) Coactivation of Rac and Rho by Wnt/Frizzled signaling is required for vertebrate gastrulation. *Genes Dev* **17**(2):295–309.

Habas R, Kato Y and He X (2001) Wnt/Frizzled activation of Rho regulates verte-brate gastrulation and requires a novel Formin homology protein Daam1. *Cell* **107**(7):843–854.

Heisenberg CP, Tada M, Rauch GJ, Saude L, Concha ML, Geisler R, Stemple DL, Smith JC and Wilson SW (2000) Silberblick/Wnt11 mediates convergent extension movements during zebrafish gastrulation. *Nature* **405**(6782):76–81.

Hynes RO and Zhao Q (2000) The evolution of cell adhesion. *J Cell Biol* **150**(2): F89–96.

Ivkovic S, Yoon BS, Popoff SN, Safadi FF, Libuda DE, Stephenson RC, Daluiski A and Lyons KM (2003) Connective tissue growth factor coordinates chondrogenesis and angiogenesis during skeletal development. *Development* **130**(12):2779–2791.

Joos TO, Whittaker CA, Meng F, DeSimone DW, Gnau V and Hausen P (1995) Integrin alpha 5 during early development of Xenopus laevis. *Mech Dev* **50**(2–3): 187–199.

Katsube K, Chuai ML, Liu YC, Kabasawa Y, Takagi M, Perbal B and Sakamoto K (2001) The expression of chicken NOV, a member of the CCN gene family, in early stage development. *Gene Expr Patterns* **1**(1):61–65.

Kireeva ML, Latinkic BV, Kolesnikova TV, Chen CC, Yang GP, Abler AS and Lau LF (1997) Cyr61 and Fisp12 are both ECM-associated signaling molecules: activities, metabolism, and localization during development. *Exp Cell Res* **233**(1): 63–77.

Latinkic BV, Mercurio S, Bennett B, Hirst EM, Xu Q, Lau LF, Mohun TJ and Smith JC (2003) Xenopus Cyr61 regulates gastrulation movements and modulates Wnt sig-nalling. *Development* **130**(11):2429–2441.

Latinkic BV, Mo FE, Greenspan JA, Copeland NG, Gilbert DJ, Jenkins NA, Ross SR and Lau LF (2001) Promoter function of the angiogenic inducer Cyr61gene in transgenic mice: tissue specificity, inducibility during wound healing, and role of the serum response element. *Endocrinology* **142**(6):2549–2557.

Lau LF and Lam SC (1999) The CCN family of angiogenic regulators: the integrin connection. *Exp Cell Res* **248**(1):44–57.

Marsden M and DeSimone DW (2001) Regulation of cell polarity, radial intercala-tion and epiboly in Xenopus: novel roles for integrin and fibronectin. *Development* **128**(18):3635–3647.

Marsden M and DeSimone DW (2003) Integrin-ECM interactions regulate cadherin-dependent cell adhesion and are required for convergent extension in Xenopus. *Curr Biol* **13**(14):1182–1191.

Mercurio S, Latinkic B, Itasaki N, Krumlauf R and Smith JC (2004) Connective-tissue growth factor modulates WNT signalling and interacts with the WNT receptor complex. *Development* **131**(9):2137–2147.

Mo FE, Muntean AG, Chen CC, Stolz DB, Watkins SC and Lau LF (2002) CYR61 (CCN1) is essential for placental development and vascular integrity. *Mol Cell Biol* **22**(24):8709–8720.

O'Brien TP and Lau LF (1992) Expression of the growth factor-inducible immediate early gene cyr61 correlates with chondrogenesis during mouse embryonic development. *Cell Growth Differ* **3**(9):645–654.

Parsons MJ, Pollard SM, Saude L, Feldman B, Coutinho P, Hirst EM and Stemple DL (2002) Zebrafish mutants identify an essential role for laminins in notochord formation. *Development* **129**(13):3137–3146.

Pera EM, Wessely O, Li SY and De Robertis EM (2001) Neural and head induction by insulin-like growth factor signals. *Dev Cell* **1**(5):655–665.

Ransom DG, Hens MD and DeSimone DW (1993) Integrin expression in early amphibian embryos: cDNA cloning and characterization of Xenopus beta 1, beta 2, beta 3, and beta 6 subunits. *Dev Biol* **160**(1):265–275.

Richard–Parpaillon L, Heligon C, Chesnel F, Boujard D and Philpott A (2002) The IGF pathway regulates head formation by inhibiting Wnt signaling in Xenopus. *Dev Biol* **244**(2):407–417.

Sakamoto K, Yamaguchi S, Ando R, Miyawaki A, Kabasawa Y, Takagi M, Li CL, Perbal B and Katsube K (2002) The nephroblastoma overexpressed gene (NOV/ccn3) protein associates with Notch1 extracellular domain and inhibits myoblast differentiation via Notch signaling pathway. *J Biol Chem* **277**(33): 29399–29405.

Tada M and Smith JC (2000) Xwnt11 is a target of Xenopus Brachyury: regulation of gastrulation movements via Dishevelled, but not through the canonical Wnt pathway. *Development* **127**(10):2227–2238.

Topczewski J, Sepich DS, Myers DC, Walker C, Amores A, Lele Z, Hammerschmidt M, Postlethwait J and Solnica–Krezel L (2001) The zebrafish glypican knypek controls cell polarity during gastrulation movements of convergent extension. *Dev Cell* **1**(2):251–264.

Tsuda M, Kamimura K, Nakato H, Archer M, Staatz W, Fox B, Humphrey M, Olson S, Futch T, Kaluza V, Siegfried E, Stam L and Selleck SB (1999) The cell-surface proteoglycan Dally regulates Wingless signalling in Drosophila. *Nature* **400**(6741): 276–280.

Wallingford JB, Fraser SE and Harland RM (2002) Convergent extension: the molecular control of polarized cell movement during embryonic development. *Dev Cell* **2**(6):695–706.

Wallingford JB, Rowning BA, Vogeli KM, Rothbacher U, Fraser SE and Harland RM (2000) Dishevelled controls cell polarity during Xenopus gastrulation. *Nature* **405**(6782):81–85.

Ying Z and King ML (1996) Isolation and characterization of xnov, a Xenopus laevis ortholog of the chicken nov gene. *Gene* **171**(2):243–248.

CHAPTER 9

CCN3 EXPRESSION AND ITS ROLE DURING DEVELOPMENT

Ken-ichi Katsube

Molecular Pathology Graduate School of Tokyo Medical and Dental University
katsube.mpa@tmd.ac.jp

Many types of Drosophila mutants have been systematically established and they have become a novel powerful tool for genetic study (Nusslein–Volhard, 1993). Phenotype of hybrids between mutants (especially in embryonic stage) indicated the genetic interaction of mutated genes themselves. Various genes were classified into different signal cascades by this analysis and the established cascades were often found again in vertebrate. Therefore, the study of Drosophila genetic interaction becomes essential for molecular biology in vertebrate. For the CCN gene family, there seem many unknown factors that are related to its function, but unfortunately this family does not have an equivalent in invertebrate (with the exception of far-related gene family, *twisted gastrulation*; *tsg* (Mason *et al.*, 1994)), which makes us impossible to apply this type of investigation. But the study of CCN expression pattern in vertebrate embryo would give some basic insight when integrated into the analysis of mutants of other genes. However, most of the studies of CCN gene family started in the context of cell proliferation or tumorigenesis and a few reports focused on its role in development or in embryogenesis. The Wnt-related subfamily of CCN such as Wise (Itasaki *et al.*, 2003) was well documented for their expression pattern in embryogenesis, and its role was analyzed as the determinant of dorsoventral polarity of axial structure. For the principal members of the CCN gene family, CCN1 and CCN2 were reported to express from early stage of development showing specific patterns (O'Brien and Lau, 1992) (Shimo *et al.*, 2002) and the loss of function experiments using the gene disruption technique or dominant negative constructs demonstrated that they play an important role in both skeletogenesis and vasuculogenesis (Mo *et al.*, 2002; Latinkic *et al.*, 2003; Abreu *et al.*, 2002; Lau and

Lam, 1999). Compared with them, CCN3 is not yet well studied, but several reports have indicated the gene expression pattern of *CCN3*, which will give some help to unveil its role not only in development but also in adult tissues.

CCN3 EXPRESSION DURING EARLY STAGE DEVELOPMENT

Several reports demonstrated the expression pattern of *CCN3* in embryonic stage, which indicates that the pattern of *CCN3* expression may differ between early and middle to late stage. In chicken embryo, the expression started in Hensen's node area from the beginning of gastrulation (Katsube *et al.*, 2001; Fig. 1). Histologically, principal expression at this area was in the forming tissue

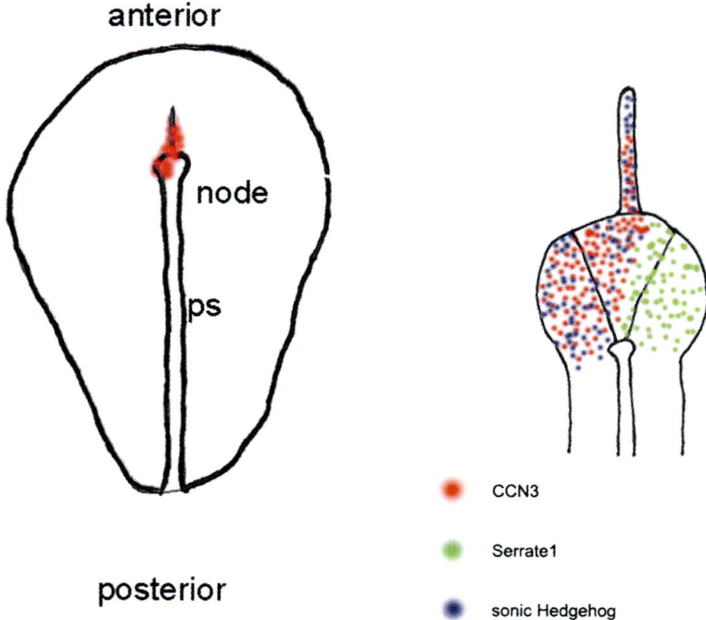

Fig. 1.   Schematical view of *CCN3* expression in early stage chicken embryo. Left: Overview of *CCN3* at the gastrulation stage. Right: Magnified view of Hensen's node area. Gene expression of *CCN3* was demonstrated with related genes.

   CCN3 expression starts from the beginning of gastrulation. Left-right asymmetry was observed at this stage. *CCN3* was expressed in the forming notochord and floor plate area in the left part. This is similar with other left deviated gene expression such as sonic Hedgehog, but was more limited. Serrate1, a ligand of Notch1 showed its expression in the right part of the node. ps: primitive streak, node: Hensen's node.

of the notochord (inter-discs of vertebral column in adult) and the floor plate (ventral medial part of the central nervous system). This expression pattern indicates that CCN3 may commit in the formation of axial structure affecting its ventral polarity. The expression also showed a typical pattern of left-right asymmetry in early stage of gastrulation. Several genes have been reported for their left-right asymmetrical expression such as secreting proteins, receptors, cytoskeletons, etc. (Meyers and Martin, 1999; Takeda *et al.*, 1999; Tsukui *et al.*, 1999; Monsoro–Burq and Le Douarin, 2001; Aruga, 2004) and the expression pattern *CCN3* is quite similar to that of sonic hedgehog and some other left side expressing genes. CCN3 association with Connexin43 was reported to regulate the function of gap junction and to suppress the cell proliferation (Gellhaus *et al.*, 2004). The gap junction is also important for the left-right asymmetry in embryogenesis (Levin and Mercola, 1999) and CCN3 may influence on it via Connexin43. Direct or indirect interaction of CCN3 with these genes would be expected.

CCN3 EXPRESSION IN MIDDLE TO LATE STAGE DEVELOPMENT

Until now, the expression of *CCN3* during middle to late stage was investigated by two groups. Mouse *CCN3* expression was detected principally in mesoderm-derived tissues and the central nervous system, in which it showed a diverged expression pattern (Fig. 2). In middle stage of mouse

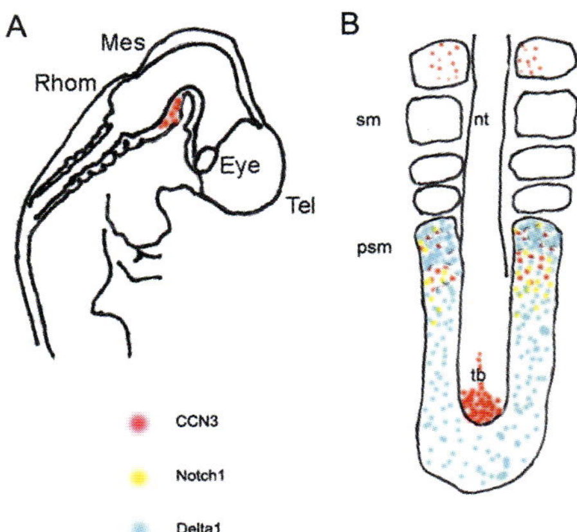

Fig. 2.   Schematical view of *CCN3* expression in middle stage. A: *CCN3* expression in the ventral side of Mesencephalon (midbrain). B: *CCN3* expression in the area of somitogenesis.

Rhom: Rhombencephalon, Mes: Mesencephalon, Tel: Telencephalon, Eye: eye ball, sm: somite, psm: presomitic mesoderm, tb: tail bud, nt neural tube.

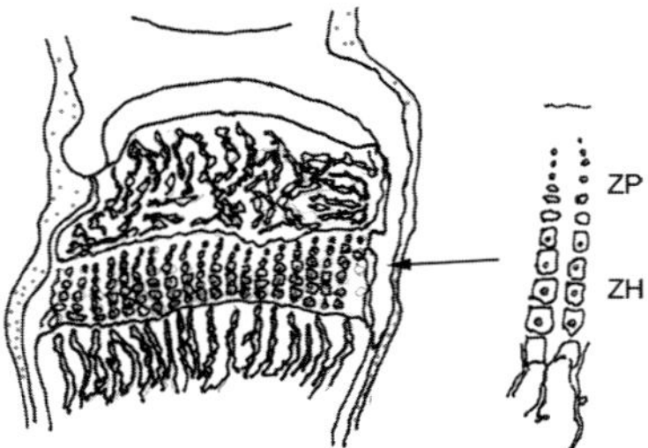

Fig. 3. CCN3 expression in developing bone and related tissues. It is observed principally in the periosteum and chondrocytes. Expression in chondrocytes is in the zone of hypertrophy (ZH) but not in the zone of proliferation (ZP). Immunoreactivity of CCN3 in chondrocytes is in the nucleus.

embryogenesis, hypaxial area of developing somite showed *CCN3* expression, which corresponded to the Myo-D and α-SMA positive region (Natarajan *et al.*, 2000). This region differentiates into the proximal muscle tissue (body wall muscle) and in part the costal bone tissue in late stage. CCN3 may have some role for myogenesis suppressing the migration of the progenitor cells. Its expression continued in more mature muscle tissues and even in adult muscle indicating the general importance of CCN3 in the regulation of muscle tissue. However, several muscle tissues such as tongue were reported not to express *CCN3* and it is necessary to verify temporal alternation of its expression. In other mesoderm-derived tissues, the expression in the developing great arteries and major blood vessels was remarkable. Bone tissue itself did not express *CCN3*, but some neighboring tissue such as tendon and periosteum were strongly positive for *CCN3* (Fig. 3). Later, the expression continued in various mesoderm-derived tissues including the glomerulus of kidney, the growth plate cartilage of bone and vascular smooth muscles, follicles of vibrissae. In general view, the expression of these areas continued in the postnatal stage indicating that they may be the prelude of adult expression.

In the central nervous system, the expression in the ventral horn of spinal cord, the motoneuronal area was reported in both mouse and human embryo (Su *et al.*, 1998; Kocialkowski *et al.*, 2001). This expression started in the mature stage motoneurons that initiated the formation of neuromuscular junctions in the axonal process. In chicken embryo, *CCN3* expression in the central nervous system was detected in the ventral region of midbrain, which will develop the oculomotor nucleus. This finding is interesting because the oculomotor nucleus develops faster than other motoneuronal regions in chicken embryo. The protein locality of CCN3 was not yet determined in these areas, but the commitment of CCN3 to the maturation or assembly of neuromuscular junction seems one of the possibilities. As mentioned previously, CCN3 commitment in the gap junction formation was reported, but there is no report for neuromuscular junction.

Cranial structures derived from the neural crest showed its expression in mouse (Natarajan *et al.*, 2000). Olfactory neuroepithelium and the sensory ganglia of cranial nerves had its expression during development. The expression in these areas was observed in mouse embryo, which was not verified in case of chicken embryo. So far, the meaning of *CCN3* expression in the neural crest derivatives is still not well understood.

Possible Role of CCN3 by its Analysis in Embryogenesis

By several different approaches, proteins that directly interact with CCN3 have been already reported and are mentioned in other chapters. Integrins were well investigated for their interaction with CCN1 and were also reported to interact with CCN3 in vasculogenesis (Lin *et al.*, 2003). CCN3 interacts with other extracellular proteins such as Fibulin (Perbal *et al.*, 1999). But commitment of these proteins during early stage is not clear since their expression pattern has little similarity. Some of Notch related genes are expressed in the same area with *CCN3* in a similar pattern at such stage. Notch and its ligands are classified as a member of Neurogenic genes in Drosophila that regulate the development of neural precursor cells (Wharton *et al.*, 1985). The mechanism of Notch signal in Drosophila has been explained by the model of "lateral inhibition". This model can tell how the neighboring cells mutually differentiate into different types of cells. "Inhibition" comes from the first findings in Notch hypomorphic mutant to suppress the neural development in the proneural clusters of epidermoblasts, but now it is convenient to call "lateral specification" because the Notch action induces any kind of repulsive differentiation between neighboring cells. Usually, Notch functions as a receptor of transmembrane type ligands such as Delta (Delta-like, Dll) or Serrate (Jagged, Jag) (Fehon *et al.*, 1990). We speculated an association of Notch/CCN3 because their expression pattern was quite similar in the caudal area. Defect of vascular remodeling in CCN1 and CCN2 knockout mouse in embryogenesis reminded us the similar pattern in Notch4 or Jagged 1 knockout mice, which prompted us to challenge this experiment (Gridley, 2001). Apparently, CCN3 could bind to the EGF motifs in the Notch extracellular domain and stimulated the Notch signal (Sakamoto *et al.*, 2002), which implies the importance of CCN3 for the regulation of stem cells. In *in vitro* experiments, myogenesis was suppressed by

CCN3 stimulating Notch signal. Recently, we investigated the expression pattern of Delta1 and Serrate1 in gastrulation and found that Serrate1 expression pattern also showed a left-right asymmetrical pattern, which expressed exclusively in right region. The commitment of CCN1 to gastrulation was confirmed in Xenopus (Latinkic *et al.*, 2003) and that similar effect of CCN3 is expected. Recently the Notch signal has effect on left-right asymmetry with association of calcium ion delivery (Raya *et al.*, 2003). Product of CCN gene family is a secreting molecule and its behavior may be different from the authentic ligands because they are membrane-anchored proteins. Notch signal regulation may be differently evolved in vertebrate, which needs to be verified by other aspect of CCN3. Calcium ion regulation by CCN3 may have some role in this function (Lombet *et al.*, 2003).

The expression of *CCN3* showed such a diverged pattern in embryogenesis, which makes us difficult to discover the proteins that interact with CCN3. Gene disruption experiment of *CCN3* will certainly give some key to understand, which is not yet published. Recently the commitment of CCN3 in some restitutive processes was reported (Ellis *et al.*, 2000). *CCN3* expression in smooth muscle tissue was enhanced in the recovery step after the injury of major arteries. Notch signal is important for vasculogenesis (Iso *et al.*, 2003) and this area was also reported for the augmented expression of Notch1 and its ligand, Jagged1 in injury (Lindner *et al.*, 2001). Notch family regulates stem cells in both vertebrate and invertebrate and its signal was stimulated by CCN3 association. This indicates the possible role of CCN3 in restitution. *CCN3* expression was extremely high in bone marrow-derived stem cells and downregulated with their differentiation (Kawashima *et al.*, 2004). In the restitutive process of major arteries, bone marrow derived stem cells contributed to the smooth muscle development (Sata *et al.*, 2002). Notch function in bone marrow derived stem cells has a strange aspect because its ligand's expression was very low or at a undetectable level. CCN3 may substitute for its authentic ligand function, which needs further investigation.

CCN is a family of secreting proteins, which has various counterparts other than Notch. Integrins are important to consider because their expression has similarity with CCN3. Some other factors may interact with CCN3 protein directly or indirectly. Systematic screening of candidate counterpart is still an important subject to clarify the role of CCN3 in development and in embryogenesis. Two-hybrid study of yeast is not easy to apply for the interaction of proteins that exist in the extracellular media and phenotypic analysis of embryo still has some importance to know its function using genetically mutated animals.

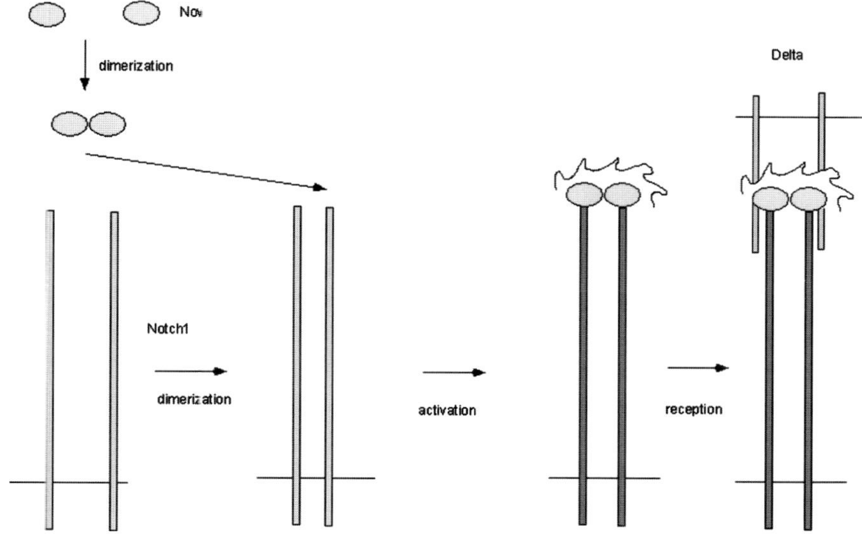

Fig. 4.   Hypothetical molecular interaction of CCN3 with Notch signal pathway.

REFERENCES

Abreu JG, Ketpura NI, Reversade B and De Robertis EM (2002) Connective-tissue growth factor (CTGF) modulates cell signalling by BMP and TGF-beta. *Nat Cell Biol* **4**:599–604.

Aruga J (2004) The role of Zic genes in neural development. *Mol Cell Neurosci* **26**: 205–221.

Ellis PD, Chen Q, Barker PJ, Metcalfe JC and Kemp PR (2000) Nov gene encodes adhesion factor for vascular smooth muscle cells and is dynamically regulated in response to vascular injury. *Arterioscler Thromb Vasc Biol* **20**:1912–1919.

Fehon RG, Kooh PJ, Rebay I, Regan CL, Xu T, Muskavitch MA and Artavanis–Tsakonas S (1990) Molecular interactions between the protein products of the neurogenic loci Notch and Delta, two EGF-homologous genes in Drosophila. *Cell* **61**:523–534.

Gellhaus A, Dong X, Propson S, Maass K, Klein–Hitpass L, Kibschull M, Traub O, Willecke K, Pe:bal B, Lye SJ and Winterhager E (2004) Connexin43 interacts with NOV: a possible mechanism for negative regulation of cell growth in choriocarcinoma cells. *J Biol Chem* **279**:36931–36942.

Gridley T (2001) Notch signaling during vascular development. *Proc Natl Acad Sci USA* **98**:5377–5378.

Iso T, Hamamori Y and Kedes L (2003) Notch signaling in vascular development. *Arterioscler Thromb Vasc Biol* **23**:543–553.

Itasaki N, Jones CM, Mercurio S, Rowe A, Domingos PM, Smith JC and Krumlauf R (2003) Wise, a context-dependent activator and inhibitor of Wnt signalling. *Development* **130**:4295–4305.

Katsube K, Chuai ML, Liu YC, Kabasawa Y, Takagi M, Perbal B and Sakamoto K (2001) The expression of chicken NOV, a member of the CCN gene family, in early stage development. *Gene Expr Patterns* **1**:61–65.

Kawashima N, Shindo K, Sakamoto K, Kondo H, Umezawa A, Kasugai S, Perbal B, Suda H, Takagi M and Katsube K (2004) Molecular and cell biological properties of mouse osteogenic mesenchymal progenitor cells, Kusa. *Bone Mineral Res* **23**(2): 123–133, 2005.

Kocialkowski S, Yeger H, Kingdom J, Perbal B and Schofield PN (2001) Expression of the human NOV gene in first trimester fetal tissues. *Anat Embryol (Berl)* **203**:417–427.

Latinkic BV, Mercurio S, Bennett B, Hirst EM, Xu Q, Lau LF, Mohun TJ and Smith JC (2003) Xenopus Cyr61 regulates gastrulation movements and modulates Wnt signalling. *Development* **130**:2429–2441.

Lau LF and Lam SC (1999) The CCN family of angiogenic regulators: the integrin connection. *Exp Cell Res* **248**:44–57.

Levin M and Mercola M (1999) Gap junction-mediated transfer of left-right patterning signals in the early chick blastoderm is upstream of Shh asymmetry in the node. *Development* **126**:4703–4714.

Lin CG, Leu SJ, Chen N, Tebeau CM, Lin SX, Yeung CY and Lau LF (2003) CCN3 (NOV) is a novel angiogenic regulator of the CCN protein family. *J Biol Chem* **278**:24200–24208.

Lindner V, Booth C, Prudovsky I, Small D, Maciag T and Liaw L (2001) Members of the Jagged/Notch gene families are expressed in injured arteries and regulate cell phenotype via alterations in cell matrix and cell-cell interaction. *Am J Pathol* **159**: 875–883.

Lombet A, Planque N, Bleau AM, Li CL and Perbal B (2003) CCN3 and calcium signaling. *Cell Commun Signal* **1**:1.

Mason ED, Konrad KD, Webb CD and Marsh JL (1994) Dorsal midline fate in Drosophila embryos requires twisted gastrulation, a gene encoding a secreted protein related to human connective tissue growth factor. *Genes Dev* **8**: 1489–1501.

Meyers EN and Martin GR (1999) Differences in left-right axis pathways in mouse and chick: functions of FGF8 and SHH. *Science* **285**:403–406.

Mo FE, Muntean AG, Chen CC, Stolz DB, Watkins SC and Lau LF (2002) CYR61 (CCN1) is essential for placental development and vascular integrity. *Mol Cell Biol* **22**:8709–8720.

Monsoro-Burq A and Le Douarin NM (2001) BMP4 plays a key role in left-right patterning in chick embryos by maintaining Sonic Hedgehog asymmetry. *Mol Cell* **7**:789–799.

Natarajan D, Andermarcher E, Schofield PN and Boulter CA (2000) Mouse Nov gene is expressed in hypaxial musculature and cranial structures derived from neural crest cells and placodes. *Dev Dyn* **219**:417–425.

Nusslein–Volhard C (1993) General Motors Cancer Research Prizewinner Laureates Lectures. Alfred P. Sloan, Jr. Prize. The formation of the embryonic axes in Drosophila. *Cancer* **71**:3189–3193.

O'Brien TP and Lau LF (1992) Expression of the growth factor-inducible immediate early gene cyr61 correlates with chondrogenesis during mouse embryonic development. *Cell Growth Differ* **3**:645–654.

Perbal B, Martinerie C, Sainson R, Werner M, He B and Roizman B (1999) The C-terminal domain of the regulatory protein NOVH is sufficient to promote interaction with fibulin 1C: a clue for a role of NOVH in cell-adhesion signaling. *Proc Natl Acad Sci USA* **96**:869–874.

Raya A, Kawakami Y, Rodriguez–Esteban C, Buscher D, Koth CM, Itoh T, Morita M, Raya RM, Dubova I, Bessa JG, de la Pompa JL and Belmonte JC (2003) Notch activity induces Nodal expression and mediates the establishment of left-right asymmetry in vertebrate embryos. *Genes Dev* **17**:1213–1218.

Sakamoto K, Yamaguchi S, Ando R, Miyawaki A, Kabasawa Y, Takagi M, Li CL, Perbal B and Katsube K (2002) The nephroblastoma overexpressed gene (NOV/ccn3) protein associates with Notch1 extracellular domain and inhibits myoblast differentiation via Notch signaling pathway. *J Biol Chem* **277**:29399–29405.

Sata M, Saiura A, Kunisato A, Tojo A, Okada S, Tokuhisa T, Hirai H, Makuuchi M, Hirata Y and Nagai R (2002) Hematopoietic stem cells differentiate into vascular cells that participate in the pathogenesis of atherosclerosis. *Nat Med* **8**:403–409.

Shimo T, Wu C, Billings PC, Piddington R, Rosenbloom J, Pacifici M and Koyama E (2002) Expression, gene regulation, and roles of Fisp12/CTGF in developing tooth germs. *Dev Dyn* **224**:267–278.

Su BY, Cai WQ, Zhang CG, Su HC and Perbal B (1998) A developmental study of novH gene expression in human central nervous system. *C R Acad Sci III* **321**:883–892.

Takeda S, Yonekawa Y, Tanaka Y, Okada Y, Nonaka S and Hirokawa N (1999) Left-right asymmetry and kinesin superfamily protein KIF3A: new insights in determination of laterality and mesoderm induction by kif3A−/− mice analysis. *J Cell Biol* **145**:825–836.

Tsukui T, Capdevila J, Tamura K, Ruiz–Lozano P, Rodriguez–Esteban C, Yonei–Tamura S, Magallon J, Chandraratna RA, Chien K, Blumberg B, Evans RM and Belmonte JC (1999) Multiple left-right asymmetry defects in Shh(−/−) mutant mice unveil a convergence of the shh and retinoic acid pathways in the control of Lefty-1. *Proc Natl Acad Sci USA* **96**:11376–11381.

Wharton KA, Johansen KM, Xu T and Artavanis–Tsakonas S (1985) Nucleotide sequence from the neurogenic locus notch implies a gene product that shares homology with proteins containing EGF-like repeats. *Cell* **43**:567–581.

CHAPTER 10

REGULATION OF CCN PROTEINS BY ALTERATIONS OF THE CYTOSKELETON

Brahim Chaqour and Margarete Goppelt-Struebe*

Department of Anatomy and Cell Biology
SUNY Downstate Medical Center
450 Clarkson Avenue Box 5, Brooklyn, NY 11203, USA
Brahim.Chaqour@downstate.edu

**Medical Clinic IV, University Erlangen-Nuremberg*
Loschgestrasse 8, 91054 Erlangen, Germany
Goppelt-Struebe@rzmail.uni-erlangen.de

CCN1/Cyr61 and CCN2/CTGF are highly expressed in several stress-related pathologies resulting from increased externally applied forces or from forces generated internally by the cellular components of tissue organs. In accordance with their *in vivo* distribution pattern, the CCN1/Cyr61 and CCN2/CTGF genes are strongly induced upon application of mechanical forces on cultured cells, or through manipulation of the cellular contractile apparatus. Mechanistically, alteration of the actin cytoskeleton tension and/or actin polymerization promoted by various components of the signaling machinery i.e. small Rho GTPases and MAP kinases, or by actin binding proteins seem to regulate the expression of the immediate early CCN genes. What could the significance be of enhanced CCN1/Cyr61 and CCN2/CTGF expression by mechanical stress? The role of these proteins in the cellular mechano-transduction process as response to environmental stress and their participation in long term modifications of the extracellular environment are issues of ongoing research.

1. INTERPLAY BETWEEN CELLULAR MECHANICS AND CCN IMMEDIATE EARLY GENE EXPRESSION

Cyr61/CCN1 and CTGF/CCN2 are co-induced upon exposure of the cells to various hormones, growth factors, inflammatory molecules and apoptotic

177

agents. In particular, co-induction of Cyr61/CCN1 and CTGF/CCN2 occurs upon stimulation of connective tissue type cells with TGF-beta1, VEGF, FGF, angiotensin II, prostaglandins, bioactive lipids, thrombin, factor IX, estrogens and apoptotic agents (Brigstock, 2002; Chaqour *et al.*, 2002; Kim *et al.*, 2003; Kireeva *et al.*, 1997; Liang *et al.*, 2003; Lin *et al.*, 2004; Pendurthi *et al.*, 2000; Sampath *et al.*, 2001). However, the observation that the immediate early CCN genes can also be strongly induced upon application of mechanical forces to the cells further rationalized, at least in our point of view, studies of their regulation and function in mechanically active environment (Kessler *et al.*, 2001; Tamura *et al.*, 2001). Indeed, under physiological and physio-pathological conditions, the primary signal sensed by many cells is mechanical. For instance, basal physiological processes ranging from blood circulation and the micturition reflex to the senses of hearing and touch are initiated by forces rather than molecules impinging on cell surface receptors and initiating a cascade of biochemical reactions. Experimentally, mechanical forces applied to and/or generated by the cells or the stiffness of substrata to which the cells are anchored affect normal cell phenotype and biological function and are suggested to largely contribute to pathological conditions *in vivo* (Gunst *et al.*, 2003; Janmey and Weitz, 2004; Knoll *et al.*, 2003; Silver *et al.*, 2003; Ingber, 2002). Under extremely or even moderately strained conditions, the cellular components of organ systems, particularly fibroblasts and smooth muscle cells, become subjected to mechanical inputs beyond a normally acceptable range and cells respond in an inappropriate fashion to altered types of mechanical signals. The transfer of such an excessive strain results in the production of various growth factors, cytokines, hormones ultimately leading to hypertrophic and fibrotic responses. Consistent with this is the observation that both Cyr61/CCN1 and CTGF/CCN2, while either not or minimally expressed in normal adult tissues, they are strongly up-regulated in several pathological conditions that result from increased mechanical strain including atherosclerosis, post-angioplastic restenosis, hypertension and bladder outlet obstruction (Chaqour *et al.*, 2002; Grzeszkiewicz *et al.*, 2002; Hilfiker *et al.*, 2002; Schober *et al.*, 2003; Wunderlich *et al.*, 2000). Cyr61/CCN1 was reported to be one of the earliest genes whose expression was turned on in smooth muscle-rich tissues (e.g. aorta and bladder) with the onset of hypertension or bladder outlet obstruction, and remained expressed at constant level thereafter (Chaqour *et al.*, 2002; Unoki *et al.*, 2003). The specialized cases of abnormal scarring, e.g. keloids, represent another example of situation in which the mechanical regulation of Cyr61/CCN1 and CTGF/CCN2 is of relevance since these apparently well-enriched lesions in both Cyr61/CCN1 and CTGF/CCN2, have long been known to develop in regions of the body that are subjected to relatively higher mechanical strain than others (Denton and Abraham, 2001; Igarashi *et al.*, 1996; Ito *et al.*, 1998). The scar that finally develops and subsists is itself a tissue under increased mechanical strain. Additionally, although not yet documented, over expression of these genes can also be predicted in the growing number of other pathologies now thought to be associated, to greater or lesser extents, with mechanical stress including pulmonary hypertension, reperfusion injury, bronchopulmonary dysplasia, excess pulmonary ventilation and acute respiratory distress syndrome. It is noteworthy, however, that in several cases, Cyr61/CCN1 and CTGF/CCN2 are not simultaneously induced suggesting that their genes may not systematically be coordinately regulated and that perhaps different etiological factors prompt their expression *in vivo*.

The current state of our knowledge purports Cyr61/CCN1 and CTGF/ CCN2 as precursors to profound remodeling within the tissues and early markers of pathologies initiated by stress conditions. We will first describe the interplay between cellular mechanics and Cyr61/CCN1 gene expression. Studies related to CTGF/CCN2 will be summarized further in detail below.

2. CYR61/CCN1 IS A MECHANO-SENSITIVE GENE

Muscle, cartilage and bone in particular, provide ideal tissues for the study of the mechanical regulation and function of Cyr61/CCN1 and CTGF/CCN2 because these tissues experience a wide range of strains during normal use due to both their own cytoskeletally generated tension and external loading. The types of stress defined as force per unit area, to which different tissues respond are legion and it is likely that the mechanical sensing mechanisms involved vary accordingly. However, despite the previously established role of Cyr61/CCN1 in chondrogenesis and its upregulation by factors critical for osteoblast function and differentiation such as 1α,25 dihydroxy vitamin D3, EGF, TNF-α and IL-1, there is still a paucity of information regarding the mechanical regulation and function of Cyr61/CCN1 in chondrocytes and osteoclasts (Schutze *et al.*, 1998). Similarly, the effects of other mechanical forces of compression and static stretch that either bone or cartilage tissues are more prone to experience have not been investigated in the context of mechanical regulation of Cyr61/CCN1. Studies by Hadjiargyrou *et al.* suggested that Cyr61/CCN1 plays a preponderant role in cartilage and bone formation and may serve as an important regulator of fracture healing (Hadjiargyrou *et al.*, 2002; Schild and Trueb, 2004). Whether Cyr61/CCN1 is involved in mechanical load-dependent growth and/or degeneration of bone is yet to be explored.

By using a model system of lung fibroblasts cultured in a three-dimensional collagen gel, Schild and Trueb found that while not being expressed when cells are in a relaxed state, Cyr61/CCN1 was strongly upregulated when the cells were in a stressed state suggesting an important role of intracellular force-generated tension in the modulation of Cyr61/CCN1 expression (Schild and Trueb, 2004). The work of Lau's group also demonstrated the involvement of Cyr61/CCN1 in wound repair prompted by skin fibroblasts and myofibroblasts which play a central role in closing the wound tissues through their capacity to produce strong contractile forces (Chen *et al.*, 2001b). In addition to responding to internally generated tension, the Cyr61/CCN1 gene is responsive to externally imposed forces particularly in smooth muscle and endothelial cells. Investigation of the biological effects of mechanical forces have been originally focused on these cell types as the layer of endothelial cells along the inside of blood vessels protects the smooth muscle from the direct shearing effects of the flowing blood whereas the pulsing blood clearly stretches the entire vascular wall. Studies using cultured smooth muscle cells from various tissue beds showed that the Cyr61/CCN1 gene is strongly but transiently up-regulated upon the application of a cyclic biaxial strain to cultured monolayer smooth muscle cells while that of CTGF/CCN2 was unaffected (Tamura *et al.*, 2001). Microarray screens for shear stress responsive genes in endothelial cells indicated that Cyr61/CCN1 was down-regulated after 6 hours of shearing deformation although an earlier burst of its gene expression cannot be ruled out given its immediate early gene expression type of pattern *in vitro* (McCormick *et al.*, 2001). Conversely, although skeletal muscle cells originate from tissues that continuously bear high tensile stress, they surprisingly express low amount of Cyr61/CCN1. Skeletal muscle are composed of terminally differentiated cells, unlike fibroblasts and smooth muscle cells, which suggests that perhaps terminal differentiation of the cells and production of Cyr61/CCN1 and/or CTGF/CCN2 are mutually exclusive.

3. MECHANO-TRANSDUCTION MECHANISMS INVOLVED IN CYR61/CCN1 GENE EXPRESSION

Research into mechano-transduction pathways that define the translation of mechanical stimulus into an intracellular biochemical process is necessary to effectively use mechanical therapies and/or pharmacological intervention to halt the deleterious effects of excessive strain. However, many of the mechano-sensing receptors, pathways and second messengers at work while the cells sense mechanical stimuli and/or their environment are yet to be defined. It has become clear, however, that the notion of separate and linear pathways linking mechanical stimuli to the expression of a mechano-sensitive gene is an oversimplification. Instead, complex and interdependent signaling networks are likely involved. The cellular structures that initially sense mechanical strain at the cell surface and convert it into Cyr61/CCN1 gene activation are yet to be identified although preliminary studies seem to indicate that integrins are, at least, partly involved (Chaqour B, unpublished data). Several intracellular signaling events such as increased intracellular Ca^{2+} concentration, activation of kinases such as PKC, p38 mitogen-activated kinases (MAPK) and phosphatidyl inositol 3-kinase and/or GTPases such as RhoA that are involved in stimulation of the Cyr61/CCN1 gene by chemical stimuli, seems to be implicated in the mechanical regulation of Cyr61/CCN1 as well suggesting that mechano-transduction mechanisms may be similar to those used by soluble factors (Han *et al.*, 2003; O'Brien *et al.*, 1990; Tamura *et al.*, 2001). However, strain-induced Cyr61/CCN1 expression is sensitive to mechanisms that sense actin dynamics i.e. increased actin polymerization and decreased globular actin levels and all signals require an intact actin cytoskeleton (Tamura *et al.*, 2001).

The role of the cytoskeleton is important in intracellular signaling as it serves as a major scaffolding element for signaling machinery components involved in intra-cellular communications and permits the compartmentalization of the cytoplasm and organization of specialized zones for sustained signaling between cell surface and nucleus. The importance of the cytoskeleton architecture in relaying PKC and PI 3-kinase signals for instance, has been reported and likely plays a role in mediating Cyr61/CCN1 gene activation. Studies have shown that direct activation of PKC isoforms by phorbol ester has dramatic effects on the assembly and disassembly of the filamentous actin (F-actin) network and there is abundant evidence showing an isoenzyme-specific translocation to the F-actin components of the cytoskeleton in intact cells (Kiley *et al.*, 1995; Slater *et al.*, 2003). Alternatively, either PKC or PI 3-kinase may affect reorganization of the actin cytoskeleton and gene expression through their interaction with Rho proteins although activation of PKC and/or PI 3-kinase can either be upstream or downstream of the Rho GTPases depending on the system (Ren and Schwartz, 1998; Wang and Bitar, 1998). The precise complex mechanism underlying cross talk between these signaling molecules and the actin cytoskeleton and its impact on stretch-induced Cyr61/CCN1 gene expression is yet to be unraveled. In contrast, the role of RhoA GTPase signaling on the actin cytoskeleton is well-established although the downstream elements of pathways through which

RhoA-mediated cytoskeletal organization regulates gene expression are poorly understood (Ridley, 2001b). Among RhoA targets, RhoA-associated kinase seemed to concomitantly alter actin stress fiber formation and Cyr61/CCN1 expression (Ridley, 2001b). Functionally, RhoA-associated kinase directly phosphorylates myosin light chains and negatively regulates myosin phosphatases and increases acto-myosin-based contractility (Chrzanowska–Wodnicka and Burridge, 1996; Kimura *et al.*, 1996; Yoshizaki *et al.*, 2004). The resulting contractile forces are thought to contribute to the formation of stress fibers and focal contacts. In addition, RhoA-kinase also activates LIM kinase which subsequently phosphorylates cofilin that inhibits actin-depolymerizing activity, thus contributing to actin fiber stabilization (Sotiropoulos *et al.*, 1999). However, whether these signaling pathways directly affect actin polymerization *per se* and F-actin rearrangement is unknown. Recent studies indicate that regulation of phosphatidyl inositol metabolism by RhoA GTPase is likely involved because the increase in phosphatidyl inositol turnover often correlates with the increase in F-actin levels within the cells (Hilpela *et al.*, 2004). The availability of adequate tools to evaluate not only total cellular phosphatidyl inositol but also local concentrations within the cells should enhance our knowledge in this field.

4. STRAIN OR "STRETCH" RESPONSIVE ELEMENTS IN THE CYR61/CCN1 GENE

Mechanical stimulus-specific gene transcription is based on the idea that mechanical strain causes the expression and/or activation of transcription factors which bind to mechano-responsive promoter elements suggesting that RhoA-actin signaling possibly affects the expression of Cyr61/CCN1 transcription in mechanically stimulated cells through such mechanisms. Studies have shown that activation of RhoA GTPase by G-protein coupled receptor agonists such as sphingosine 1-phosphate affected the expression of Cyr61/CCN1 gene by modulating the activity of the transcription factor AP-1 and CREB either directly or through the MAPK family member, p38 (Han *et al.*, 2003). Since the intracellular mechano-sensing mechanisms do not require that the intracellular and nuclear signals be distinct from those initiated by soluble factors, the AP-1 and CRE binding sites localized with the promoter region of the human Cyr61/CCN1 gene are potential mechanical stretch responsive elements. These elements seem to act as hypoxia-responsive elements as well (Kunz *et al.*, 2003). Additionally, the promoter region of both Cyr61/CCN1 and CTGF/CCN2 contains so-called shear stress-responsive elements (SSRE) representing the core sequence of NF-κB binding sites found previously in shear stress-responsive genes (Resnick *et al.*, 1993). However, specificity cannot be explained in terms of presence or absence of SSRE since as shown by microarray technology, numerous genes do not contain SSREs but were responsive to shearing deformation of endothelial cells and vice versa (Chen *et al.*, 2001a; McCormick *et al.*, 2001; Zhao *et al.*, 2002).

5. MODULATION OF CTGF/CCN2 EXPRESSION *IN VITRO* AND *IN VIVO*

5.1. CTGF/CCN2 Expression in Skin Disorders

TGF-beta mediates up-regulation of CTGF/CCN2 in wound healing and in various pathophysiological situations. In scleroderma, the initial transactivation of CTGF/CCN2 was mediated through the TGF-beta specific smad signaling pathway, whereas the maintenance of CTGF/CCN2 expression was independent of TGF-beta signaling (Leask *et al.*, 2003; Holmes *et al.*, 2003). CTGF/CCN2 expression was increased in patients with radiation enteritis with established fibrosis, without a concomitant up-regulation of TGF-beta (Vozenin–Brotons *et al.*, 2003). These examples indicate that even though TGF-beta is the major regulator of CTGF/CCN2, additional factors have to be considered to understand the physiological and pathophysiological relevance of this protein. In particular, mechanical alterations modulate the expression of CTGF/CCN2 as outlined in the introduction.

Activated fibroblasts, so called myofibroblasts are the cells associated with CTGF/CCN2 expression in fibrotic tissue or healing wounds. These cells generate the contractile forces required for wound closure and healing and play an essential role in the excessive matrix production during fibrosis (Lorena *et al.*, 2002; Badid *et al.*, 2000). Since *in vitro* studies provide a basis for the molecular understanding of stress-mediated regulation of CTGF/CCN2, three-dimensional collagen-1 matrices were used as model system to investigate the influence of mechanical stress on the cell biology of fibroblasts (Grinnell, 2003). Under stressed conditions, CTGF/CCN2 was observed to be up-regulated, whereas release of mechanical stress led to a rapid down-regulation of CTGF/CCN2 expression (Schild and Trueb, 2004; Schild and Trueb, 2002). Modulation of CTGF/CCN2 gene expression in response to mechanical stress suggests that CTGF is molecular regulator of stress adaptational states of the cells.

TGF-beta-mediated fibroblast differentiation is enhanced when mechanical tension is applied to cells (Arora *et al.*, 1999). Similarly, TGF-beta-mediated differentiation and subsequent matrix contraction are dependent on

CTGF/CCN2 expression, but they are not promoted by CTGF/CCN2 alone (Garrett *et al.*, 2004). Fibroblast differentiation may thus be an example of an effective cooperation of soluble mediators and environmental restraints.

5.2. CTGF/CCN2 Expression in Cardiovascular Diseases

Hemodynamic forces are involved in the initiation and localization of early atherogenetic lesions, which are located preferentially in specific regions of arterial wall, which experience non-uniform blood flow (Nerem, 1992). CTGF/CCN2 is strongly expressed in endothelial cells of atherosclerotic lesions although a causative role for CTGF/CCN2 in the initiation of atherosclerosis has not yet been established (Oemar *et al.*, 1997; Schober *et al.*, 2002). *In vitro* studies confirmed that CTGF/CCN2 belongs to the group of genes which are strongly up-regulated in endothelial cells exposed to non-uniform shear stress (Yoshisue *et al.*, 2002). Constant shear stress, in contrast, reduced CTGF/CCN2 expression in primary HUVEC (McCormick *et al.*, 2003). These data were in line with the hypothesis that physiological shear stress protects against fibrotic and atherosclerotic disease processes which are supported by turbulent flow. However, CTGF/CCN2 mRNA expression remained unaltered when laminar flow was applied for 24 hr in cultured HUVEC or bovine endothelial cells (Eskin *et al.*, 2004). In another study, CTGF/CCN2 was shown to be upregulated by turbulent as well as laminar flow (Garcia–Cardena *et al.*, 2001), which is in contrast to the *in vivo* situation, where CTGF/CCN2 is not expressed in normal vessels, exposed to constant laminar flow. Utilization of different types of cells and apparatus and various shear stress regimen and magnitudes may account for these apparently contradictory results. More data on CTGF/CCN2 protein and an improved molecular understanding of CTGF/CCN2 regulation will help to understand the role of CTGF/CCN2 in vessels exposed to changes in pressure and flow.

5.3. CTGF/CCN2 Expression in Kidney Disorders

Variations in hemodynamics have an impact on end organs such as the kidney. CTGF/CCN2 expression in renal diseases has been extensively studied (Goldschmeding *et al.*, 2000; Abdel and Mason, 2004; Riser and Cortes, 2001). Depending on the disease, high glucose and *TGF-beta* may be the major inducers of CTGF/CCN2. However, it is noteworthy that the synthesis of renal glomerular proteins is also modulated by mesangial cell stretch. Systemic arterial hypertension and conditions of impaired glomerular pressure autoregulation lead to excessive expansion and repetitive cycles of distension contraction of the elastic glomeruli (Cortes *et al.*, 1999). In animal models, an enhanced glomerular capillary pressure was associated with an increased synthesis of extracellular matrix proteins and inflammatory mediators (Ingram and Scholey, 2000). The up-regulation of CTGF/CCN2 in glomeruli of diabetic rats or patients, although being primarily related to increased glucose levels, may additionally be influenced by increases in capillary plasma flow rates (Wahab *et al.*, 2001; Zatz *et al.*, 1985).

Increased glomerular capillary pressure and wall tension are transmitted to resident glomerular cells. Exposure of mesangial cells to cyclic stress *in vitro*, transiently up-regulated CTGF/CCN2 (Riser and Cortes, 2001). In another study, hydrostatic pressure was related to sustained up-regulation of CTGF/CCN2 (Hishikawa *et al.*, 2001). Under these conditions, CTGF/CCN2 expression was associated with the apoptotic death of mesangial cells.

5.4. CTGF/CCN2 Expression in Bone and Cartilage

In the bone, CTGF/CCN2 is considered to be a hypertrophic chondrocyte-specific gene product, implicated in proliferation and differentiation of chondrocytes and in skeletal growth and modeling/remodeling (Takigawa *et al.*, 2003). Mechanical regulation is also implicated in cartilage biology. Cyclic tensile strains or shear promote cartilage growth and ossification (Wong and Carter, 2003). In an *in vitro* study, Wong *et al.* compared the effect of tensile strain and cyclic hydrostatic pressure on CTGF/CCN2 expression in primary chondrocytes (Wong *et al.*, 2003). Only application of tensile strain induced CTGF/CCN2, whereas hydrostatic pressure was without effect, in contrast to the report by Hishikawa *et al.* in mesangial cells (Hishikawa *et al.*, 2001). Continuous application of mechanical stimulation was also applied *in vivo* in experimental tooth movement, a model for mechanical-dependent bone growth (Yamashiro *et al.*, 2001). CTGF/CCN2 mRNA expression was increased in osteocytes at the compressed and the stretched side of the teeth, indicative of complex signaling pathways mediating the up-regulation.

Thus, both *in vitro* and *in vivo* data indicate that the CTGF/CCN2 gene is sensitive to mechanical stress in soft and hard tissues. The different *in vivo* and *in vitro* models, however, do not yet provide us with a consistent view of the essential requirements regarding type of applied mechanical force, strength or duration.

6. MOLECULAR MECHANISMS COUPLING MECHANICAL STRESS TO CTGF/CCN2 EXPRESSION

As outlined above, multiple signaling pathways are implicated in the transmission of mechanical stress to gene expression. Related to CTGF/CCN2, the experimental evidence presently available points to the small GTPase RhoA and alterations in the level of globular actin (G-actin) as major contributors to mechano-sensitive signal transmission. Therefore, we will focus on this aspect of CTGF/CCN2 regulation.

Mechanical stimuli such as tension or cyclic strain lead to rearrangement of the actin cytoskeleton, which is associated with increased contraction (Cooper, 1991). Pressure applied to cells *in vitro* increased the formation of actin stress fibers shifting the ratio of non polymerized G-actin to polymerized F-actin (Cipolla *et al.*, 2002). Changes in the ratio of G- and F-actin are also detectable *in vivo*. In diabetic glomeruli, which are exposed to increased mechanical strain, actin was found to be disorganized and the structure of the fibrillar F-actin was disrupted (Cortes *et al.*, 2000).

The most important sensors of mechanical stress are integrins linking extra cellular matrix proteins to intracellular signaling. Organization of integrins into focal complexes is dependent on the type of matrix molecule, but also modulated by the physical state of the matrix (Katz *et al.*, 2000). Via adaptor molecules such as integrin linked kinase, integrins are coupled to the actin cytoskeleton and to various regulatory proteins, among them kinases and small GTPases (Li and Xu, 2000; Attwell *et al.*, 2003). The small GTPases of the Rho family are central in mechano-transduction, mediating the formation of focal complexes (Balaban *et al.*, 2001) and also as transducers of signals leading to changes in gene expression and morphology (Ridley, 2001a). Activation of RhoA increases the formation of F-actin stress fibers via the down stream RhoA-associated kinase ROCK (Riento and Ridley, 2003).

Activation of the small GTPase RhoA significantly contributes not only to Cyr61/CCN1 expression as we previously mentioned, but also to the upregulation of CTGF/CCN2, linking the regulation of CTGF/CCN2 gene expression to changes of the actin cytoskeleton. Interference with RhoA signaling by toxin B or more specifically C3 exoenzyme prevented up-regulation of CTGF/CCN2 by lysophosphatidic acid, a known activator of RhoA (Hahn *et al.*, 2000; Heusinger–Ribeiro *et al.*, 2001). Similarly, disruption of microtubuli by colchicine, which activates RhoA in a receptor independent way, also activated CTGF/CCN2 in a toxinB-sensitive manner (Ott *et al.*, 2003). Involvement of RhoA in CTGF/CCN2 expression was confirmed by over-expression of constitutively active RhoA (RhoA V14). RhoA signaling pathways interact with other pathways involved in CTGF/CCN2 expression. Interference with RhoA activity reduced CTGF/CCN2 expression even if other pathways were stimulated. Inhibition of RhoA-associated kinase inhibited TGF-beta-mediated up-regulation of CTGF/CCN2, which was primarily mediated by the Smad 3/4 signaling pathway (Goppelt–Struebe *et al.*, 2001; Iwanciw *et al.*, 2003). Similarly, angiotensin-mediated induction of CTGF was due to stimulation of the MAP kinase pathway, but was still inhibited by interference with RhoA signaling (Iwanciw *et al.*, 2003). Interference with RhoA signaling thus allows inhibition of CTGF/CCN2 expression largely independent of the primary stimulus.

As a regulator of the cytoskeleton, the small GTPase RhoA is expressed in all types of cells. In the network of interacting signaling mediators it seems to play a key role in maintaining the basal turnover of CTGF/CCN2 mRNA and also in the stimulated expression of CTGF/CCN2. Furthermore, RhoA is a target for pharmacological interference with CTGF/CCN2 expression. By inhibition of the posttranslational modification of RhoA statins (HMG CoA reductase inhibitors) inhibit CTGF/CCN2 induction *in vitro* and *in vivo* (Eberlein *et al.*, 2001; Muehlich *et al.*, 2004; Watts and Spiteri, 2004; Song *et al.*, 2004, and unpublished results). Rho-kinase inhibitors, Y27632 or fasudil, which inhibit CTGF/CCN2 expression *in vitro*, may be another way to interfere with overexpression of CTGF/CCN2 *in vivo*.

As RhoA is a major regulator of the actin cytoskeleton it was obvious to investigate the effect of changes in actin organization on CTGF/CCN2 expression. Recruitment of G-actin into F-actin stress fibers by jasplakinolide increased CTGF/CCN2 expression, whereas disruption of F-actin by latrunculin B reduced CTGF/CCN2 expression (Ott *et al.*, 2003). Unexpectedly, cytochalasin D, which also rapidly disintegrated actin stress fibers, transiently increased CTGF/CCN2 (Goppelt–Struebe *et al.*, 2002). Both, cytochalasin D and latrunculin B enhance the cellular content of G-actin (Cipolla *et al.*, 2002), however, the availability of G-actin as modulator of gene expression seems to be different upon treatment with both agents: Cytochalasin D was shown to sequester and thus reduce the effective level of G-actin (Sotiropoulos *et al.*, 1999; Gineitis and Treisman, 2001). These data indicate that rather than being regulated by F-actin stress fibers, the expression of CTGF/CCN2 seems to be sensitive to changes in the level of G-actin. In line with this hypothesis, over-expression of mutant G-actin, which is no longer able to polymerize into F-actin (Posern *et al.*, 2002), significantly reduced the expression of CTGF/CCN2 in endothelial cells (unpublished result). The sensitivity of CTGF/CCN2 expression towards varying levels of G-actin may be the molecular mechanism by which exogenous pressure or changes in morphology are being translated into gene expression.

7. FUNCTIONAL SIGNIFICANCE OF CYR61/CCN1 AND CTGF/CCN2 EXPRESSION UNDER STRESS CONDITIONS

Mechanical stress experiments do help us understand how internally generated forces and/or externally imposed forces on the cells lead to changes in CCN gene expression. However, we need to be mindful that in these kinds of experiments, cells go from being static to being mechanically deformed. As reported for Cyr61/CCN1 and CTGF/CCN2, their expression declined rapidly and even disappeared after a short time period of mechanical deformation as the stretching environment becomes the cell's new normalcy. The rapid reestablishment of basal expression might be indicative of an adaptive mechanism in which compensatory signaling pathways are activated to allow gene transcription to return to normal levels in the stimulated cells. At these later time points, cells may more accurately represent those *in vivo* which exist normally in a mechanically active environment. However, such compensatory mechanisms do not seem to take place in pathological conditions since the up-regulation of Cyr61/CCN1 and CTGF/CCN2 appeared to be both rapid and long lasting in the affected tissues which theoretically would result in sustaining their effects (Chaqour *et al.*, 2002; Grzeszkiewicz *et al.*, 2002; Abdel and Mason, 2004). The *in vivo* environment and the strain magnitude within the tissue may account for the sustained expression. Nonetheless, the expression even transient of Cyr61/CCN1 or CTGF/CCN2 may have long term implications. Previous studies suggested that Cyr61/CCN1 can regulate the expression of genes involved in angiogenesis and matrix remodeling (Chen *et al.*, 2001b; Mo *et al.*, 2002). In agreement with this, interference with Cyr61/CCN1 expression in mechanically stimulated cells markedly reduced mechanical strain-induced VEGF, α_v integrin and smooth muscle alpha-actin gene expression but had no effect on type I collagen, fibronectin and myosin heavy chain isoform expression (Zhou *et al.*, 2005). An intact cytoskeleton is required for Cyr61/CCN1-dependent regulation of gene expression indicating that cytoskeleton integrity is required for both Cyr61/CCN1 expression and activity. Therefore, Cyr61/CCN1 may well be an integral part of the mechano-transduction process by promoting the expression of mechanosensors such as integrins and/or by propagating the mechanical signal to neighboring cells via the expression of autocrine and paracrine factors such as VEGF.

CTGF/CCN2 is a protein, which exerts its effects characteristically by interaction with other proteins in a synergistic or inhibitory manner (e.g. Abreu *et al.*, 2002; Inoki *et al.*, 2002). Regulation of CTGF/CCN2 by mechanical forces from the environment may thus add to the complexity and variability of the regulation of cellular communication. Whether CTGF/CCN2 assume these function in stress-related pathologies and what other functions CTGF/CCN2 may be associated with in various cell types remain to be investigated. It is clear, however, that both the expression and function of the immediate early CCN genes cannot be considered in isolation from the mechanically dynamic structures within and without the cells and tissues.

REFERENCES

Abdel WN and Mason RM (2004) Connective tissue growth factor and renal diseases: some answers, more questions. *Curr Opin Nephrol Hypertens* **13**:53–58.

Abreu JG, Ketpura NI, Reversade B and De Robertis EM (2002) Connective-tissue growth factor (CTGF) modulates cell signalling by BMP and TGF-beta. *Nat Cell Biol* **4**:599–604.

Arora PD, Narani N and McCulloch CA (1999) The compliance of collagen gels regulates transforming growth factor-beta induction of alpha-smooth muscle actin in fibroblasts. *Am J Pathol* **154**:871–882.

Attwell S, Mills J, Troussard A, Wu C and Dedhar S (2003) Integration of cell attachment, cytoskeletal localization, and signaling by integrin-linked kinase (ILK), CH-ILKBP, and the tumor suppressor PTEN. *Mol Biol Cell* **14**:4813–4825.

Badid C, Mounier N, Costa AM and Desmouliere A (2000) Role of myofibroblasts during normal tissue repair and excessive scarring: interest of their assessment in nephropathies. *Histol Histopathol* **15**:269–280.

Balaban NQ, Schwarz US, Riveline D, Goichberg P, Tzur G, Sabanay I, Mahalu D, Safran S, Bershadsky A, Addadi L and Geiger B (2001) Force and focal adhesion assembly: a close relationship studied using elastic micropatterned substrates. *Nat Cell Biol* **3**:466–472.

Brigstock DR (2002) Regulation of angiogenesis and endothelial cell function by connective tissue growth factor (CTGF) and cysteine-rich 61 (CYR61). *Angiogenesis* **5**:153–165.

Chaqour B, Whitbeck C, Han JS, Macarak E, Horan P, Chichester P and Levin R (2002) Cyr61 and CTGF are molecular markers of bladder wall remodeling after outlet obstruction. *Am J Physiol Endocrinol Metab* **283**:E765–E774.

Chen BP, Li YS, Zhao Y, Chen KD, Li S, Lao J, Yuan S, Shyy JY and Chien S (2001a) DNA microarray analysis of gene expression in endothelial cells in response to 24-h shear stress. *Physiol Genomics* **7**:55–63.

Chen CC, Mo FE and Lau LF (2001b) The angiogenic factor Cyr61 activates a genetic program for wound healing in human skin fibroblasts. *J Biol Chem* **276**:47329–47337.

Chrzanowska–Wodnicka M and Burridge K (1996) Rho-stimulated contractility drives the formation of stress fibers and focal adhesions. *J Cell Biol* **133**:1403–1415.

Cipolla MJ, Gokina NI and Osol G (2002) Pressure-induced actin polymerization in vascular smooth muscle as a mechanism underlying myogenic behavior. *FASEB J* **16**:72–76.

Cooper JA (1991) The role of actin polymerization in cell motility. *Annu Rev Physiol* **53**:585–605.

Cortes P, Mendez M, Riser BL, Guerin CJ, Rodriguez–Barbero A, Hassett C and Yee J (2000) F-actin fiber distribution in glomerular cells: structural and functional implications. *Kidney Int* **58**:2452–2461.

Cortes P, Riser BL, Yee J and Narins RG (1999) Mechanical strain of glomerular mesangial cells in the pathogenesis of glomerulosclerosis: clinical implications. *Nephrol Dial Transplant* **14**:1351–1354.

Denton CP and Abraham DJ (2001) Transforming growth factor-beta and connective tissue growth factor: key cytokines in scleroderma pathogenesis. *Curr Opin Rheumatol* **13**:505–511.

Eberlein M, Heusinger–Ribeiro J and Goppelt–Struebe M (2001) Rho-dependent inhibition of the induction of connective tissue growth factor (CTGF) by HMG CoA reductase inhibitors (statins). *Br J Pharmacol* **133**:1172–1180.

Eskin SG, Turner NA and McIntire LV (2004) Endothelial cell cytochrome P450 1A1 and 1B1: up-regulation by shear stress. *Endothelium* **11**:1–10.

Garcia–Cardena G, Comander J, Anderson KR, Blackman BR and Gimbrone MA, Jr (2001) Biomechanical activation of vascular endothelium as a determinant of its functional phenotype. *Proc Natl Acad Sci USA* **98**:4478–4485.

Garrett Q, Khaw PT, Blalock TD, Schultz GS, Grotendorst GR and Daniels JT (2004) Involvement of CTGF in TGF-beta1-stimulation of myofibroblast differentiation and collagen matrix contraction in the presence of mechanical stress. *Invest Ophthalmol Vis Sci* **45**:1109–1116.

Gineitis D and Treisman R (2001) Differential usage of signal transduction pathways defines two types of serum response factor target gene. *J Biol Chem* **276**:24531–24539.

Goldschmeding R, Aten J, Ito Y, Blom I, Rabelink T and Weening JJ (2000) Connective tissue growth factor: just another factor in renal fibrosis? *Nephrol Dial Transplant* **15**:296–299.

Goppelt–Struebe M, Hahn A, Iwanciw D, Rehm M and Banas B (2001) Regulation of connective tissue growth factor (ccn2; ctgf) gene expression in human mesangial cells: modulation by HMG CoA reductase inhibitors (statins). *Mol Pathol* **54**: 176–179.

Goppelt–Struebe M, Heusinger–Ribeiro J and Fischer B (2002) Dual role of cytochalasin D in the regulation of connective tissue growth factor. *Ann NY Acad Sci* **973**: 57–59.

Grinnell F (2003) Fibroblast biology in three-dimensional collagen matrices. *Trends Cell Biol* **13**:264–269.

Grzeszkiewicz TM, Lindner V, Chen N, Lam SC and Lau LF (2002) The angiogenic factor cysteine-rich 61 (CYR61, CCN1) supports vascular smooth muscle cell adhesion and stimulates chemotaxis through integrin alpha(6)beta(1) and cell surface heparan sulfate proteoglycans. *Endocrinology* **143**:1441–1450.

Gunst SJ, Tang DD and Opazo SA (2003) Cytoskeletal remodeling of the airway smooth muscle cell: a mechanism for adaptation to mechanical forces in the lung. *Respir Physiol Neurobiol* **137**:151–168.

Hadjiargyrou M, Lombardo F, Zhao S, Ahrens W, Joo J, Ahn H, Jurman M, White DW and Rubin CT (2002) Transcriptional profiling of bone regeneration. Insight into the molecular complexity of wound repair. *J Biol Chem* **277**:30177–30182.

Hahn A, Heusinger–Ribeiro J, Lanz T, Zenkel S and Goppelt–Struebe M (2000) Induction of connective tissue growth factor by activation of heptahelical receptors. Modulation by rho proteins and the actin cytoskeleton. *J Biol Chem* **275**:37429–37435.

Han JS, Macarak E, Rosenbloom J, Chung KC and Chaqour B (2003) Regulation of Cyr61/CCN1 gene expression through RhoA GTPase and p38MAPK signaling pathways. *Eur J Biochem* **270**:3408–3421.

Heusinger–Ribeiro J, Eberlein M, Wahab NA and Goppelt–Struebe M (2001) Expression of connective tissue growth factor in human renal fibroblasts: regulatory roles of RhoA and cAMP. *J Am Soc Nephrol* **12**:1853–1861.

Hilfiker A, Hilfiker–Kleiner D, Fuchs M, Kaminski K, Lichtenberg A, Rothkotter HJ, Schieffer B and Drexler H (2002) Expression of CYR61, an angiogenic immediate early gene, in arteriosclerosis and its regulation by angiotensin II. *Circulation* **106**:254–260.

Hilpela P, Vartiainen MK and Lappalainen P (2004) Regulation of the actin cytoskeleton by PI(4,5)P2 and PI(3,4,5)P3. *Curr Top Microbiol Immunol* **282**:117–163.

Hishikawa K, Oemar BS and Nakaki T (2001) Static pressure regulates connective tissue growth factor expression in human mesangial cells. *J Biol Chem* **276**:16797–16803.

Holmes A, Abraham DJ, Chen Y, Denton C, Shi-wen X, Black CM and Leask A (2003) Constitutive connective tissue growth factor expression in scleroderma fibroblasts is dependent on Sp1. *J Biol Chem* **278**:41728–41733.

Igarashi A, Nashiro K, Kikuchi K, Sato S, Ihn H, Fujimoto M, Grotendorst GR and Takehara K (1996) Connective tissue growth factor gene expression in tissue sections from localized scleroderma, keloid, and other fibrotic skin disorders. *J Invest Dermatol* **106**:729–733.

Ingber DE (2002) Mechanical signaling and the cellular response to extracellular matrix in angiogenesis and cardiovascular physiology. *Circ Res* **91**:877–887.

Ingram AJ and Scholey JW (2000) Stress-responsive signal transduction mechanisms in glomerular cells. *Curr Opin Nephrol Hypertens* **9**:49–55.

Inoki I, Shiomi T, Hashimoto G, Enomoto H, Nakamura H, Makino K, Ikeda E, Takata S, Kobayashi K and Okada Y (2002) Connective tissue growth factor binds vascular endothelial growth factor (VEGF) and inhibits VEGF-induced angiogenesis. *FASEB J* **16**:219–221.

Ito Y, Aten J, Bende RJ, Oemar BS, Rabelink TJ, Weening JJ and Goldschmeding R (1998) Expression of connective tissue growth factor in human renal fibrosis. *Kidney Int* **53**:853–861.

Iwanciw D, Rehm M, Porst M and Goppelt–Struebe M (2003) Induction of connective tissue growth factor by angiotensin II: integration of signaling pathways. *Arterioscler Thromb Vasc Biol* **23**:1782–1787.

Janmey PA and Weitz DA (2004) Dealing with mechanics: mechanisms of force transduction in cells. *Trends Biochem Sci* **29**:364–370.

Katz BZ, Zamir E, Bershadsky A, Kam Z, Yamada KM and Geiger B (2000) Physical state of the extracellular matrix regulates the structure and molecular composition of cell-matrix adhesions. *Mol Biol Cell* **11**:1047–1060.

Kessler D, Dethlefsen S, Haase I, Ploman M, Hirche F, Krieg T and Eckes B (2001) Fibroblasts in mechanically stressed collagen lattices assume a 'synthetic' phenotype. *J Biol Chem* **276**:36575–36585.

Kiley SC, Jaken S, Whelan R and Parker PJ (1995) Intracellular targeting of protein kinase C isoenzymes: functional implications. *Biochem Soc Trans* **23**:601–605.

Kim KH, Min YK, Baik JH, Lau LF, Chaqour B and Chung KC (2003) Expression of angiogenic factor Cyr61 during neuronal cell death via the activation of c-Jun N-terminal kinase and serum response factor. *J Biol Chem* **278**:13847–13854.

Kimura K, Itoh M, Amano M, Chihara K, Fukata Y, Nakafuku M, Yamamori B, Feng J, Nakano K, Nakano T, Okawa K, Iwamatsu A and Kaibuchi K (1996) Regulation of myosin phosphatase by Rho and Rho-associated kinase (Rho-kinase). *Science* **269**:221–223.

Kireeva ML, Latinkic BV, Kolesnikova TV, Chen CC, Yang GP, Abler AS and Lau LF (1997) Cyr61 and Fisp12 are both ECM-associated signaling molecules: activities, metabolism, and localization during development. *Exp Cell Res* **233**:63–77.

Knoll R, Hoshijima M and Chien K (2003) Cardiac mechanotransduction and implications for heart disease. *J Mol Med* **81**:750–756.

Kunz M, Moeller S, Koczan D, Lorenz P, Wenger RH, Glocker MO, Thiesen HJ, Gross G and Ibrahim SM (2003) Mechanisms of hypoxic gene regulation of angiogenesis factor Cyr61 in melanoma cells. *J Biol Chem* **278**:45651–45660.

Leask A, Holmes A, Black CM and Abraham DJ (2003) Connective tissue growth factor gene regulation. Requirements for its induction by transforming growth factor-beta 2 in fibroblasts. *J Biol Chem* **278**:13008–13015.

Li C and Xu Q (2000) Mechanical stress-initiated signal transductions in vascular smooth muscle cells. *Cell Signal* **12**:435–445.

Liang Y, Li C, Guzman VM, Evinger AJ, III, Protzman CE, Krauss AH and Woodward DF (2003) Comparison of prostaglandin F2alpha, bimatoprost (prostamide), and butaprost (EP2 agonist) on Cyr61 and connective tissue growth factor gene expression. *J Biol Chem* **278**:27267–27277.

Lin MT, Chang CC, Chen ST, Chang HL, Su JL, Chau YP and Kuo ML (2004) Cyr61 expression confers resistance to apoptosis in breast cancer MCF-7 cells by a mechanism of NF-kappaB-dependent XIAP up-regulation. *J Biol Chem* **279**: 24015–24023.

Lorena D, Uchio X, Costa AM and Desmouliere A (2002) Normal scarring: importance of myofibroblasts. *Wound Repair Regen* **10**:86–92.

McCormick SM, Eskin SG, McIntire LV, Teng CL, Lu CM, Russell CG and Chittur KK (2001) DNA microarray reveals changes in gene expression of shear stressed human umbilical vein endothelial cells. *Proc Natl Acad Sci USA* **98**:8955–8960.

McCormick SM, Frye SR, Eskin SG, Teng CL, Lu CM, Russell CG, Chittur KK and McIntire LV (2003) Microarray analysis of shear stressed endothelial cells. *Biorheology* **40**:5–11.

Mo FE, Muntean AG, Chen CC, Stolz DB, Watkins SC and Lau LF (2002) CYR61 (CCN1) is essential for placental development and vascular integrity. *Mol Cell Biol* **22**:8709–8720.

Muehlich S, Schneider N, Hinkmann F, Garlichs CD and Goppelt–Struebe M (2004) Induction of connective tissue growth factor (CTGF) in human endothelial cells by lysophosphatidic acid, sphingosine-1-phosphate, and platelets. *Atherosclerosis* **175**:261–268.

Nerem RM (1992) Vascular fluid mechanics, the arterial wall, and atherosclerosis. *J Biomech Eng* **114**:274–282.

O'Brien TP, Yang GP, Sanders L and Lau LF (1990) Expression of cyr61, a growth factor-inducible immediate-early gene. *Mol Cell Biol* **10**:3569–3577.

Oemar BS, Werner A, Garnier JM, Do DD, Godoy N, Nauck M, Marz W, Rupp J, Pech M and Luscher TF (1997) Human connective tissue growth factor is expressed in advanced atherosclerotic lesions. *Circulation* **95**:831–839.

Ott C, Iwanciw D, Graness A, Giehl K and Goppelt–Struebe M (2003) Modulation of the expression of connective tissue growth factor by alterations of the cytoskeleton. *J Biol Chem* **278**:44305–44311.

Pendurthi UR, Allen KE, Ezban M and Rao VM (2000) Factor VIIa and thrombin induce the expression Cyr61 and connective tissue growth factor, extracellular matrix signaling proteins that could act as possible downstream mediators in factor VIIa tissue factor-induced signal transduction. *J Biol Chem* **275**: 14632–14641.

Posern G, Sotiropoulos A and Treisman R (2002) Mutant actins demonstrate a role for unpolymerized actin in control of transcription by serum response factor. *Mol Biol Cell* **13**:4167–4178.

Ren XD and Schwartz MA (1998) Regulation of inositol lipid kinases by Rho and Rac. *Curr Opin Genet Dev* **8**:63–67.

Resnick N, Collins T, Atkinson W, Bonthron DT, Dewey CF, Jr and Gimbron MA, Jr (1993) Platelet-derived growth factor B chain promoter contains a cis-acting fluid shear-stress-responsive element. *Proc Natl Acad Sci USA* **90**:7908.

Ridley AJ (2001a) Rho family proteins: coordinating cell responses. *Trends Cell Biol* **11**:471–477.

Ridley AJ (2001b) Rho proteins: linking signaling with membrane trafficking. *Traffic* **2**:303–310.

Riento K and Ridley AJ (2003) Rocks: multifunctional kinases in cell behaviour. *Nat Rev Mol Cell Biol* **4**:446–456.

Riser BL and Cortes P (2001) Connective tissue growth factor and its regulation: a new element in diabetic glomerulosclerosis. *Ren Fail* **23**:459–470.

Sampath D, Winneker RC and Zhang Z (2001) Cyr61, a member of the CCN family, is required for MCF-7 cell proliferation: regulation by 17beta-estradiol and overexpression in human breast cancer. *Endocrinology* **142**:2540–2548.

Schild C and Trueb B (2002) Mechanical stress is required for high-level expression of connective tissue growth factor. *Exp Cell Res* **274**:83–91.

Schild C and Trueb B (2004) Three members of the connective tissue growth factor family CCN are differentially regulated by mechanical stress. *Biochim Biophys Acta* **1691**:33–40.

Schober JM, Chen N, Grzeszkiewicz TM, Jovanovic I, Emeson EE, Ugarova TP, Ye RD, Lau LF and Lam SC (2002) Identification of integrin alpha(M)beta(2) as an adhesion receptor on peripheral blood monocytes for Cyr61 (CCN1) and connective tissue growth factor (CCN2): immediate-early gene products expressed in atherosclerotic lesions. *J Lipid Mediat Cell Signal* **99**:4457–4465.

Schober JM, Lau LF, Ugarova TP and Lam SC (2003) Identification of a novel integrin alpha Mbeta 2 binding site in CCN1 (CYR61), a matricellular protein expressed in healing wounds and atherosclerotic lesions. *J Biol Chem* **278**:25808–25815.

Schutze N, Lechner A, Groll C, Siggelkow H, Hufner M, Kohrle J and Jakob F (1998) The human analog of murine cystein rich protein 61 [correction of 16] is a 1alpha,25-dihydroxyvitamin D3 responsive immediate early gene in human fetal osteoblasts: regulation by cytokines, growth factors, and serum. *Endocrinology* **139**: 1761–1770.

Silver FH, DeVore D and Siperko LM (2003) Invited review: role of mechanophysiology in aging of ECM: effects of changes in mechanochemical transduction. *J Appl Physiol* **95**:2134–2141.

Slater SJ, Cook AC, Seiz JL, Malinowski SA, Stagliano BA and Stubbs CD (2003) Effects of ethanol on protein kinase C alpha activity induced by association with Rho GTPases. *Biochemistry* **42**:12105–12114.

Song Y, Li C and Cai L (2004) Fluvastatin prevents nephropathy likely through suppression of connective tissue growth factor-mediated extracellular matrix accumulation. *Exp Mol Pathol* **76**:66–75.

Sotiropoulos A, Gineitis D, Copeland J and Treisman R (1999) Signal-regulated activation of serum response factor is mediated by changes in actin dynamics. *Cell* **98**:159–169.

Takigawa M, Nakanishi T, Kubota S and Nishida T (2003) Role of CTGF/HCS24/ ecogenin in skeletal growth control. *J Cell Physiol* **194**:256–266.

Tamura I, Rosenbloom J, Macarak E and Chaqour B (2001) Regulation of Cyr61 gene expression by mechanical stretch through multiple signaling pathways. *Am J Physiol Cell Physiol* **281**:C1524–C1532.

Unoki H, Furukawa K, Yonekura H, Ueda Y, Katsuda S, Mori M, Nakagawara K, Mabuchi H and Yamamoto H (2003) Up-regulation of cyr61 in vascular smooth muscle cells of spontaneously hypertensive rats. *Lab Invest* **83**:973–982.

Vozenin–Brotons MC, Milliat F, Sabourin JC, de Gouville AC, Francois A, Lasser P, Morice P, Haie–Meder C, Lusinchi A, Antoun S, Bourhis J, Mathe D, Girinsky T and Aigueperse J (2003) Fibrogenic signals in patients with radiation enteritis are associated with increased connective tissue growth factor expression. *Int J Radiat Oncol Biol Phys* **56**:561–572.

Wahab NA, Yevdokimova N, Weston BS, Roberts T, Li XJ, Brinkman H and Mason RM (2001) Role of connective tissue growth factor in the pathogenesis of diabetic nephropathy. *Biochem J* **359**:77–87.

Wang P and Bitar KN (1998) Rho A regulates sustained smooth muscle contraction through cytoskeletal reorganization of HSP27. *Am J Physiol* **275**:G1454–G1462.

Watts KL and Spiteri MA (2004) Connective Tissue Growth Factor expression and induction by Transforming Growth Factor β, is abrogated by Simvastatin via a Rho signalling mechanism. *Am J Physiol Lung Cell Mol Physiol* **287**:4323–4332.

Wong M and Carter DR (2003) Articular cartilage functional histomorphology and mechanobiology: a research perspective. *Bone* **33**:1–13.

Wong M, Siegrist M and Goodwin K (2003) Cyclic tensile strain and cyclic hydrostatic pressure differentially regulate expression of hypertrophic markers in primary chondrocytes. *Bone* **33**:685–693.

Wunderlich K, Pech M, Eberle AN, Mihatsch M, Flammer J and Meyer P (2000) Expression of connective tissue growth factor (CTGF) mRNA in plaques of human anterior subcapsular cataracts and membranes of posterior capsule opacification [In Process Citation]. *Curr Eye Res* **21**:627–636.

Yamashiro T, Fukunaga T, Kobashi N, Kamioka H, Nakanishi T, Takigawa M and Takano–Yamamoto T (2001) Mechanical stimulation induces CTGF expression in rat osteocytes. *J Dent Res* **80**:461–465.

Yoshisue H, Suzuki K, Kawabata A, Ohya T, Zhao H, Sakurada K, Taba Y, Sasaguri T, Sakai N, Yamashita S, Matsuzawa Y and Nojima H (2002) Large scale isolation of non-uniform shear stress-responsive genes from cultured human endothelial cells through the preparation of a subtracted cDNA library. *Atherosclerosis* **162**:323–334.

Yoshizaki H, Ohba Y, Parrini MC, Dulyaninova NG, Bresnick AR, Mochizuki N and Matsuda M (2004) Cell type-specific regulation of RhoA activity during cytokinesis. *J Biol Chem* **279**:44756–44762.

Zatz R, Meyer TW, Rennke HG and Brenner BM (1985) Predominance of hemody-
     namic rather than metabolic factors in the pathogenesis of diabetic glomerulopathy.
     Proc Natl Acad Sci USA **82**:5963–5967.
Zhao Y, Chen BP, Miao H, Yuan S, Li YS, Hu Y, Rocke DM and Chien S (2002)
     Improved significance test for DNA microarray data: temporal effects of shear
     stress on endothelial genes. *Physiol Genomics* **12**:1–11.
Zhou D, Herrick D, Rosenbloom J and Chaqour B (2005) Cyr61 mediates the expres-
     sion of VEGF, α_V-integrin, and α-actin genes through cytoskeletally-based mechan-
     otransduction mechanisms in bladder smooth muscle cells. *J Appl Physiol*, in press.

CHAPTER 11

PATHOGENESIS OF SYSTEMIC SCLEROSIS AND CCN2 (CONNECTIVE TISSUE GROWTH FACTOR)

Kazuhiko Takehara*

Professor
Department of Dermatology
Kanazawa University Graduate School of Medical Science
and School of Medicine

The pathogenesis of systemic sclerosis(SSc) is not clarified yet, although some cytokines seemed to be closely related to the fibrotic process of this disorder. Transforming growth factor β (TGF-β) has received many attentions as an essential factor in the pathogenesis of various fibrotic disorders. In addition, we have shown that connective tissue growth factor (CCN2) is closely related to the pathogenesis of SSc as follows: (i) CCN2 mRNA expression was observed in the fibrotic lesions but not in the early non-fibrotic lesions or atrophic lesions; (ii) Serum CCN2 protein concentrations were significantly elevated, and this was correlated with skin sclerosis and lung fibrosis; and (iii) In our animal model, TGF-β induced subcutaneous fibrosis and subsequent CCN2 application caused persistent fibrosis, which was correlated with upregulation of procollagene mRNA promotor activities. Based on these data, we hypothesize that a two-step process of fibrosis occurs in SSc, that is TGF-β induces fibrosis in the early stage, and afterwards CCN2 acts to maintain tissue fibrosis.

Keywords: Systemic sclerosis; fibrosis; transforming growth factor β; connective tissue growth factor.

*13-1 Takara-machi, Kanazawa, 920-8641, Japan, Phone: +81-76-265-2343, Fax: +81-76-234-4270, e-mail: takehara@med.kanazawa-u.ac.jp

1. PATHOGENESIS OF SYSTEMIC SCLEROSIS

Systemic sclerosis (SSc) is a multisystem disorder of connective tissue characterized by excessive fibrosis in the skin and various internal organs such as the lungs, kidneys, esophagus and heart etc. Although the pathogenesis of this disease remains unknown, three major abnormalities are emphasized and intricately connected: autoimmunity, abnormal connective tissue metabolism, and disturbed vascular systems.

Regarding abnormal connective tissue metabolism and subsequent fibrosis, many studies suggest that several growth factors and cytokines released from inflammatory cells, endothelial cells, fibroblasts, and other cells in the involved organs play important roles in the initiation and the maintenance of connective tissue fibrosis (LeRoy *et al.*, 1989; Varga *et al.*, 1987; Roberts and Sporn, 1993; Takehara *et al.*, 1987; Etoh *et al.*, 1990).

Among these factors, the importance of the roles of CCN2 is described in this chapter.

2. CONNECTIVE TISSUE GROWTH FACTOR AND FIBROTIC DISORDERS

Connective tissue growth factor (CCN2) has been suggested to play an important role generally in the development of various fibrosis. CCN2 is a cysteine-rich peptide originally identified from cultured human umbilical endothelial cell (HUVEC) supernatants that exhibitis PDGF-like chemotactic and mitogenic activities on mesenchymal cells, and appears to be antigenically related to PDGF A and B chain peptides. (Bradham *et al.*, 1991) Human foreskin fibroblasts produce high levels of CCN2 mRNA and protein after activation with TGF-β, but not other growth factors such as PDGF, EGF, or basic fibroblast growth factor (bFGF) (Igarashi *et al.*, 1993). Thus, CCN2 is a candidate autocrine stimulator released in response to TGF-β in skin fibroblasts as shown in Fig. 1, and also appears to participate in the pathologic process of fibrosis.

In fact CCN2 mRNA is overexpressed in a large number of fibrotic conditions, including SSc (Igarashi *et al.*, 1995), localized scleroderma (Igarashi *et al.*, 1996), keloid (Igarashi *et al.*, 1996), atherosclerosis (Oemar *et al.*, 1997), renal fibrosis (Ito *et al.*, 1998), inflammatory bowel disease (Dammeier *et al.*, 1998), chronic pancreatitis (di Mola *et al.*, 1999), lung fibrosis, (Jasky *et al.*, 1998) and liver fibrosis (Paradis *et al.*, 1999). CCN2 mRNA expression was also confirmed in other skin fibrotic disorders including cutaneous fibrohistocytic and vascular tumors (Igarashi *et al.*, 1998).

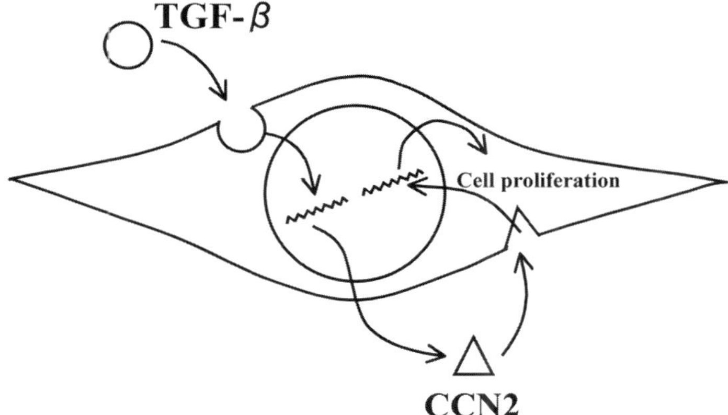

Fig. 1.   The relationship between TGF-β and CCN2. CCN2 acts as an autocrine growth factor when skin fibroblasts are stimulated with TGF-β.

Although it is likely that CCN2 expression is not specific for fibrosis in SSc, several lines of evidence indicated that CCN2 is closely involved in fibrosis in SSc.

3. CCN2 AND SSc

Our studies strongly suggest that CCN2 is involved in the pathogenesis of SSc as follows.

(1) CCN2 mRNA Expression in the Fibroblasts in SSc Sclerotic Lesions

When tissues from patients with SSc were examined by *in situ* hybridization with the antisense CCN2 probe, dermal fibroblasts were all positive in all 12 cases that showed histologic sclerosis (Igarashi *et al.*, 1993). Positive signals were more abundant in the sclerotic stage than in the inflammatory stage. Moreover, no CCN2 mRNA expression was observed in tissue from the atrophic stage of SSc or from the presclerotic stage. Our finding of elevated CCN2 mRNA expression in SSc lesions was confirmed using cultured fibroblasts derived from SSc lesions (Holemes *et al.*, 2001; Shi-wen *et al.*, 2000).

(2) Serum Concentrations of CCN2 in SSc

By ELISA, we examined serum samples from patients with SSc (Sato *et al.*, 2000), and serum concentrations of CCN2 were found to be elevated in these

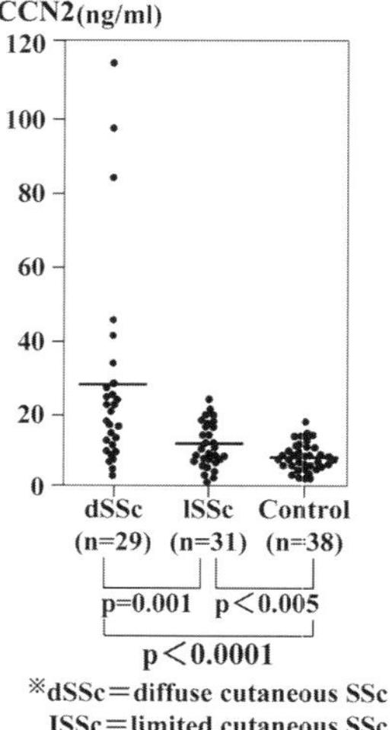

Fig. 2.   Concentrations of CCN2 in sera from patients with diffuse cutaneous SSc(dSSc), limited cutaneous SSc(lSSc), and healthy controls. Serum concentration of CCN2 was determined by ELISA. Horizontal lines show median values.

patients (Fig. 2). Since the elevated CCN2 levels were not observed in patients with systemic lupus erythematosus and polymyositis/dermatomyositis, it is likely that the elevation of CCN2 levels is specific for SSc among various autoimmune connective tissue diseases. Furthermore, the findings that CCN2 levels correlated with the extent of skin sclerosis and the severity of pulmonary fibrosis suggest that CCN2 plays a critical role in the development of fibrosis in SSc (Sato *et al.*, 2000).

(3) An Animal Model of Skin Fibrosis by Exogenous Injection of TGF-β and CCN2

In 1986, Roberts *et al.* reported that TGF-β injection into newborn mice caused granulation tissue formation and skin fibrosis (Roberts *et al.*, 1986). We initially tried to establish an animal model of skin fibrosis by TGF-β injection using a similar method. We injected 800 ng of TGF-β 1, 2 or 3 into the subcutaneous tissue of newborn mice for 7 days. All types of TGF-β caused strong granulation formation and fibrotic changes after 3 consecutive injections; however, a single injection of TGF-β alone did not cause persistent fibrosis, and fibrosis disappeared after 7 days (Shinozaki *et al.*, 1986). Therefore, we tried to establish persistent fibrosis by combination of TGF-β and other growth factors, including CCN2, as described (Shinozaki *et al.*, 1997; Mori *et al.*, 1999).

Table 1. Histological responses to single, simultaneous, and serial injections of different growth factors into newborn mice

	Day 4	Day 8	Day 14
TGF-β alone	++	−	−
CCN2 alone	±	−	−
bFGF alone	+	−	−
TGF-β + **CCN2**	+++	+++*	+++*
TGF-β + hFGF	+++	+++*	+++*
TGF-β → **CCN2**	++	++*	++
TGF-β → bFGF	++	++*	++
CCN2 → TGF-β	±	+	+
b-FGF → TGF-β	+	+	−

−: no change; ±: slight edema and some cell infiltration; +: edematous granulation tissue; ++: granulation tissue consisting of lymphocytes, histiocytes, and fibroblasts; + + +: fibrotic tissue consisting of fibroblast aggregation and extracellular matrix deposition, and asterisks denotes marked fibrosis. →: In the first 3 days and next 4 days, different growth factors were injected.

The results of these experiments are summarized in Table 1. CCN2 injection caused slight edema and some cell infiltration. Similarly, b-FGF injection caused slight edematous granulated tissue formation. Simultaneous injection of TGF-β plus CCN2 or TGF-β plus b-FGF resulted in fibrotic tissue formation, consisting of fibroblast aggregation and ECM deposition, persisted for up to 14 days, even though the injections were discontinued on Day 7. To examine the tissue response further, two different growth factors were injected serially; TGF-β on Days 1–3 followed by CCN2 on Days 4–7. Serial injections of CCN2 or b-FGF after TGF-β caused fibrotic tissue formation (Fig. 3). Injection of CCN2 or b-FGF before TGF-β did not cause any significant change compared with TGF-β alone (Table 1).

There results clearly demonstrate that a single application of any growth factor is not sufficient to induce persistent fibrosis, despite continuous injections. Instead, interaction of multiple growth factors seems to be necessary for the induction of persistent fibrosis in this animal model. Our findings on serial application of different growth factors suggest that TGF-β plays an important role in inducing granulation and fibrotic tissue formation (Mori *et al.*, 1999).

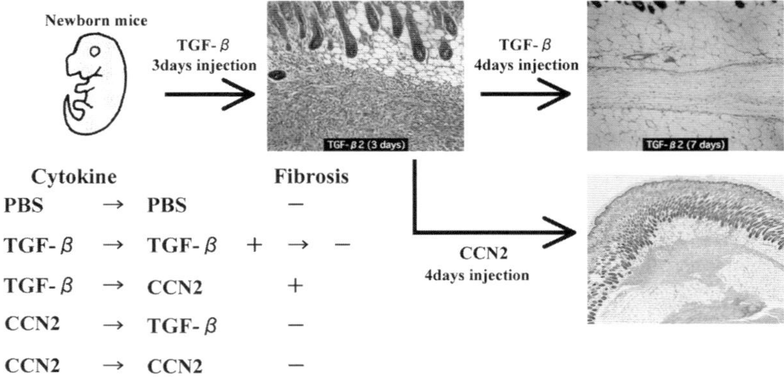

Fig. 3.   TGF-β caused strong granulation formation and fibrotic changes after 3 consecutive injection;however, this fibrotic changes disappeared ever after 7 consecutive injections. In contrast, CCN2 injection between 4–7 days caused persistent fibrotic changes.

4. MECHANISM OF PERSISTENT FIBROTIC CHANGES IN THE ANIMAL MODEL

To further define the mechanisms of skin fibrosis induced by TGF-β and CCN2 *in vivo*, in this study, we investigated the effects of growth factors on the promoter activity of pro $\alpha2$ (I) collagen (COL1A2) gene in skin fibrosis. For this purpose, we utilized transgenic reporter mice harboring the -17 kb promoter sequence of the mouse COL1A2 linked to either a firefly luciferase gene or a bacterial β-galactosidase gene. Serial injections of CCN2 after TGF-β resulted in a sustained elevation of COL1A2 mRNA expression and promoter activity compared with consecutive injection of TGF-β alone on Day 8. We also demonstrated that the number of fibroblasts with activated COL1A2 transcription was increased by serial injections of CCN2 after TGF-β in comparison with injection of TGF-β alone. Furthermore, the serial injections recruited mast cells and macrophages. The number of mast cells reached the maximum on Day 4 and maintained relatively high levels up till Day 8. In contrast to the kinetics of mast cells, the number of macrophages was increased on Day 4 and continued to rise during the following consecutive CCN2 injections until Day 8. These results suggested that CCN2 maintains TGF-β induced skin fibrosis by sustaining COL1A2 promoter activation and increasing the number of activated fibroblasts. The infiltrated mast cells and macrophages may also contribute to the maintenance of fibrosis (Chujo *et al.*; Fig. 4).

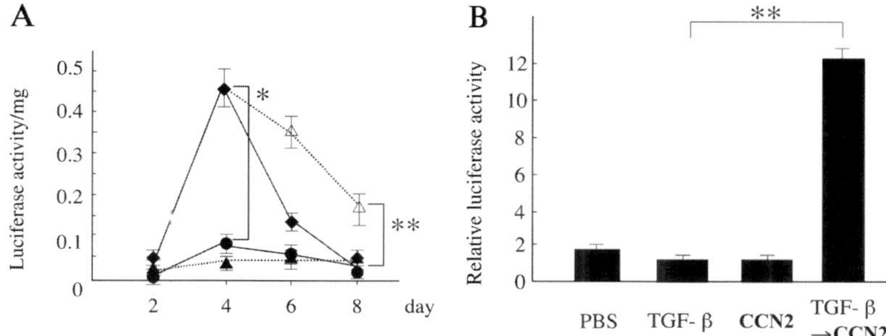

Fig. 4. CCN2 maintains the TGF-β-induced activation of the COL1A2 promoter *in vivo*. A: Newborn transgenic mice were injected with TGF-β3, CCN2 or PBS as a control into the neck subcutaneous tissue once a day for 3 or 7 consecutive days. For some of them, we injected CCN2 for 4 days after injecting TGF-β for the first 3 days. Luciferase activity, which represents the activity of pro α2(I) collagen promoter in granulation tissue, was determined on Day 2, 4, 6 and 8 using a luminometer. Closed circles, squares and triangles express control, TGF-β and CCN2, respectively. Open triangles represent serial injections of CCN2 after TGF-β. Luciferase activity of the TGF-β injected mouse reached the maximum levels on Day 4 and fell down to the control level on Day 8. In contrast high levels of COL1A2 promoter activity was maintained after serial injections of TGF-β and CCN2. B: The *bar graph* showed the levels of COL1A2 promoter activity of consecutive or serial injections on Day 8. Serial injections of CCN2 after TGF-β showed a 12-fold increase in the luciferase activity on Day 8 compared with consecutive injections of PBS or TGF-β. Values are the mean of 5 mice in each group. *p < 0.01, **p < 0.0001.

5. THE TWO-STEP FIBROSIS HYPOTHESIS IN SSc

Based on the results with TGF-β and CCN2 described above, we hypothesized that a 2-step process of fibrosis occurs in SSc. We think that TGF-β induces fibrosis in the early stage of SSc, and then CCN2 acts to maintain tissue fibrosis. TGF-β induces CCN2 mRNA, but some additional factor is required for continuous CCN2 mRNA expression, because CCN2 induced in an autocrine manner disappeared after 3 days injection of TGF-β. The mechanism of continuous CCN2 expression is a key to this disorder, and efforts to reveal it are further required (Takehara, 2003).

REFERENCES

Bradham DM, Igarashi A, Potter RL and Grotendorst GR (1991) Connective tissue growth factor: a cysteine-rich mitogen secreted by human vascular endothelial cells is related to the SRC induced immediate early gene product CEF-10. *J Cell Biol* **114**:1285–1294.

Chujo S, Shirasaki F, Kawara S, Inagaki Y, Kinbara T, Inaoki M, Takigawa M and Takehara K. Connective tissue growth factor causes persistent Pro α(1) collagen gene expression induced by transforming growth factor-β in a mouse fibrosis model. *J Cell Physiol*, in press.

Dammeier J, Brauchle M, Falk W, Grotendorst GR and Werner S (1998) Connective tissue growth factor; A novel regulator of mucosal repair and fibrosis in inflammatory bowel disease? *Int J Biochem Cell Biol* **30**:909–922.

di Mola FF, Friess H, Martignoni ME, *et al.* (1999) Connective tissue growth factors is a regulator for fibrosis in human chronic pancreatitis. *Ann Surg* **230**:63–71.

Etoh T, Igarashi A, Ishibashi Y and Takehara K (1990) The effects of scleroderma sera on endothelial cell survival *in vitro*. *Arch Dermatol Res* **282**:516–519.

Holemes A, Abraham DJ, Sa S, Shiwen X, Black CM and Leask A (2001) CTGF and SMADs, maintenance of scleroderma phenotype is independent of SMAD signaling. *J Biol Chem* **276**:10594–10601.

Igarashi A, Okachi H, Bradham DM and Grotendorst GR (1993) Regulation of connective tissue growth factor gene expression in human skin fibroblasts and during wound repair. *Mol Cell Biol* **4**:637–645.

Igarashi A, Nashiro K, Kikuchi K, *et al.* (1995) Significant correlation between connective tissue growth factor gene expression and skin sclerosis in tissue sections from patients with systemic sclerosis. *J Invest Dermatol* **105**:280–284.

Igarashi A, Nashiro K, Kikuchi K, *et al.* (1996) Connective tissue growth factor gene expression in tissue sections from localized scleroderma, keloid and other fibrotic skin disorders. *J Invest Dermatol* **106**:729–733.

Igarashi A, Hayashi N, Nashiro K and Takehara K (1998) Differenential expression of connective tissue growth factor gene in cutaneous fibrohistiocytic and vascular tumors. *J Cutan Pathol* **25**:729–733.

Ito Y, Aten J, Bende RJ, *et al.* (1998) Expression of connective tissue growth factor in human renal fibrosis. *Kidney Int* **53**:853–861.

Jasky JA, Ortiz LA, Tonthat B, *et al.* (1998) Connective tissue mRNA expression is upregulated in bleomycin-induced lung fibrosis. *Am J Physiol* **275**:L365–L371.

LeRoy EC, Edwin A, Smith M, Kahaieh B, Trojanowska M and Silver RM (1989) A strategy for determining the pathogenesis of systemic sclerosis: is transforming growth factor β the answer? *Arthritis Rheum* **32**:817–825.

Mori T, Kawara S, Shinozaki M, *et al.* (1999) Role and interaction of connective tissue growth factor with transforming growth factor-13 in persistent fibrosis; a mouse fibrosis model. *J Cell Physiol* **181**:153–159.

Oemar BS, Werner A, Garnier J-M, *et al.* (1997) Human connective tissue growth factor is expressed in advanced atherosclerotic lesions. *Circulation* **251**:748–752.

Paradis V, Dargere D, Vidaud M, *et al.* (1999) Expression of connective tissue growth factor in experimental rat and human liver fibrosis. *Hepatology* **30**:968–976.

Roberts AB and Sporn MB (1993) Physiological actions and clinical applications of transforming growth factor-β. *Growth Factors* **8**:1–9.

Roberts AB, Sporn MB, Assoian JK, *et al.* (1986) Transforming growth factor type β; rapid induction of fibrosis and angiogenesis *in vivo* and stimulation of collagen formation *in vitro*. *Proc Natl Acad Sci USA* **83**:4167–4171.

Sato S, Nagaoka T, Hasegawa M, *et al.* (2000) Serum levels of connective tissue growth factor are elevated in patients with systemic sclerosis: association with extent of skin sclerosis and severity of pulmonary fibrosis. *J Rheumatol* **27**:149–154.

Shinozaki M, Kawara S, Hayashi N, Kakinuma T, Igarashi A and Takehara K (1997) Induction of subcutaneous tissue fibrosis in newborn mice by transforming growth factor-beta-simultaneous application with basic fibroblast growth factor causes persistent fibrosis. *Biochem Biophys Res Commun* **237**:292–297.

Shi-wen X, Pennington D, Holmes A, *et al.* (2000) Autocrine over expression of CTGF maintains fibrosis; RDA analysis of fibrosis genes in systemic sclerosis. *Exp Cell Res* **259**:213–224.

Takehara K, LeRoy EC and Grotendorst GR (1987) TGF-β inhibition of endothelial cell proliferation: alteration of EGF binding and EGF-induced growth regulatory (competence) gene expression. *Cell* **49**:415–422.

Takehara K (2003) Hypothesis: pathogenesis of systemic sclerosis. *J Rheumatol* **30**:755–759.

Varga J, Rosenbloom J and Jimenez SA (1987) Transforming growth factor β causes a persistent increase in steady-state amounts of type I and type III collagen and fibronectin mRNAs in normal human dermal fibroblasts. *Biochemistry* **297**: 597–604.

CHAPTER 12

FUNCTION AND REGULATION OF CCN5

Mark R. Gray and John J. Castellot Jr.

Department of Anatomy and Cellular Biology
Tufts University School of Medicine
136 Harrison Avenue
Boston, Massachusetts 02459 USA

I. INTRODUCTION

The CCN family of proteins consists of six highly conserved, cysteine-rich cell surface and extracellular matrix (ECM)-associated proteins (Brigstock, 2003). CCN proteins help regulate cell proliferation and motility, as well as numerous other biological properties; they have been implicated in numerous pathophysiologic models. With one exception — CCN5 — all CCN proteins have a distinctive sequence consisting of four polypeptide modules: Insulin-Like Growth Factor-Binding Protein (IGFBP), von Willebrands Factor-C (VWC), thrombospondin-1 (TSP1), and a carboxy-terminal (CT) domain. The IGFBP domain includes a consensus sequence found in all known IGF-binding proteins. The von Willebrand factor type C repeat of the VWC domain is found in some collagens and mucins, and the domain might have a role in the formation of protein complexes and protein oligomerization. The thrombospondin type I repeat (TSP1) domain has been implicated in promotion of cell attachment and binding to matrix proteins and sulfated glycoconjugates. The carboxy-terminal (CT) domain is thought to have a role in oligomerization and/or receptor binding and has been shown to be mitogenic in some cell types.

The CCN5 gene, known previously as Cop-1 (Delmolino *et al.*, 1997; 2001), rCop-1 (Zhang *et al.*, 1998), Wisp-2 (Pennica *et al.*, 1998), and CTGF-L (Kumar *et al.*, 1999), encodes the only CCN protein that does not include the CT domain. Because of this major structural difference, it is expected that many functions of CCN5 should be significantly different from those mediated by the four-domain CCN proteins, reviewed in the other chapters of this volume.

In some model systems, CCN5 seems to have opposite or complementary functions to those of the other CCN proteins, leading to the notion that CCN5 is the antagonist (either pharmacologically or biologically) of the four-domain CCN proteins.

The purpose of this chapter is to review all biochemical data on functions of CCN5 published to date. It is interesting that several investigators independently found the CCN5 gene in screens for genes that were overexpressed significantly in well-defined cell-based assays in a wide variety of research areas. The findings of each of these groups are reviewed in the following paragraphs, and their results are then tabulated together and compared. Finally, we provide our thoughts on the future directions for CCN5 research in the context of the CCN family as a whole.

II. SUMMARY OF DISCOVERIES AND INVESTIGATIONS OF CCN5

A. Studies of Vascular Smooth Muscle Cells

In 1997, Delmolino and Castellot reported the discovery of a new gene in the CCN family that had features of a growth arrest gene (Delmolino *et al.*, 1997). This gene, first named *hicp*, and then later re-named *Cop-1*, was found using the subtractive hybridization method to compare arrays of mRNA molecules from heparin-treated rat aortic vascular smooth muscle cells (VSMC) to that of untreated cells (Delmolino *et al.*, 2001). Previous work had shown that VSMC proliferation is down-regulated by heparin (Castellot *et al.*, 1981). The new experiments were designed to identify heparin-regulated genes that directly caused VSMC growth arrest. CCN5 was identified as a heparin-induced gene that was expressed at high levels in quiescent (serum-starved; G_o) VSMC and at very low levels in proliferating (serum stimulated) VSMC. The CCN5 pattern of expression is characteristic of growth arrest-specific genes. Experiments with cultured VSMC also revealed that addition of serum to growth-arrested cells causes a rapid fall in CCN5 mRNA expression within two hours, and a five-fold decrease within 6 hours. Further studies demonstrated that platelet-derived growth factor (PDGF), but not epidermal growth factor (EGF), also reduced CCN5 expression five-fold, similar to serum. TGF-β and INF-β, known inhibitors of VSMC proliferation, did not induce CCN5 expression. Conditioned medium from cells that overexpressed CCN5 inhibited proliferation of VSMC, demonstrating that CCN5 is a secreted protein (Delmolino *et al.*, 2001). Induction of CCN5 was shown to be specific to VSMC and not vascular endothelial cells, or VSMC lines that were resistant to the antiproliferative effect of heparin. However, it should be noted that while heparin did not induce CCN5 expression in vascular endothelial cells, this cell type produces significant amounts of the protein in the absence of heparin.

Because VSMC hyperplasia has a key role in chronic and acute vascular pathological processes, and high levels of CCN5 were found in the normal rat aorta VSMC, further CCN5 experiments were based on the rat vascular injury model (Lake *et al.*, 2003). Following arterial injury, VSMC actively migrate through the basement membrane of the intimal endothelial cell layer, proliferate, and form a myointimal lesion. Immunohistochemical analysis of transverse arterial sections using an antibody to CCN5 showed that abundant CCN5 protein staining in the medial quiescent VSMC layer of uninjured rat carotid arteries disappeared within two days after balloon injury and returned in both the medial and myointimal VSMC layers 14 days after injury. To produce this injury repair phenotype, VSMC must acquire both increased rates of growth and motility. Tests for motility, including scratch wound, Boyden chamber, and Matrigel invasion assays, demonstrated that VSMC had increased motility and invasiveness during the proliferative state, correlating in time with reduced CCN5 expression. Further experiments were done using VSMC forced to express higher levels of CCN5. VSMC were transfected with a recombinant adenovirus that carried a CCN5 gene driven by a cytomegalovirus (CMV) promoter (AdCCN5). These studies revealed dose-dependent decreases in VSMC growth rates, motility, and invasiveness (Lake *et al.*, 2003). Similar testing of VSMC transfected with AdCCN5 showed that adhesiveness and apoptosis were not affected by increased levels of CCN5. Transfection with small inhibitory RNA (siRNA) molecules specific for CCN5 added further evidence that CCN5 regulates the heparin effect on VSMC proliferation and motility (Lake and Castellot, 2003). Reduction of CCN5 protein levels using RNA interference (RNAi) abrogated the heparin-mediated decrease in VSMC proliferation, and increased VSMC proliferation rates and motility. Knockdown of CCN5 using RNAi caused an increased production of matrix metalloproteinase-2 (MMP-2), an enzyme used by actively motile VSMC (and other cells) to digest extracellular matrix. MMP-2 activity is reduced in VSMC transfected with AdCCN5, providing more evidence for CCN5 regulation of MMP-2 expression. Reduction of CCN5 protein levels by RNAi also resulted in the 50% reduction of expression of smooth muscle cell α-actin and caused an altered cytoskeletal morphology of VSMC. The results of caspase-3 assays suggested that knockdown of CCN5 by RNAi does not cause apoptosis in VSMC.

B. Uterine Smooth Muscle Cells

After initial results from the studies of CCN5 in VSMC were obtained, the role of CCN5 was investigated in another smooth muscle cell model involving aberrant proliferation. Human leiomyomas are benign neoplasias of uterine smooth muscle cells that are clinically significant in approximately 20% of women of reproductive age. It was previously demonstrated that heparin can inhibit both mitogenesis and motility of human myometrial and leiomyoma SMC (Mason *et al.*, 2003). Matched pairs of uterine SMC samples taken from normal uterine smooth muscle (myometrium) and adjacent leiomyoma smooth muscle from the same individuals were analyzed for CCN5 expression using Western blot analysis, real-time PCR, and immunohistochemistry (Mason *et al.*, 2004a; Mason *et al.*, 2004b). Although expression of CCN5 mRNA was easily detected in both myometrial and leiomyoma growth-arrested SMC it was not maintained by heparin after serum stimulation, in sharp contrast to VSMC (Lake *et al.*, 2003; Mason *et al.*, 2004a). In a group of ten autologous pairs of myometrial and leiomyoma SMC, CCN5 expression was lower in all ten leiomyoma SMC samples in comparison to their normal myometrial counterparts. Overexpression of CCN5 by transfection of the AdCCN5 retroviral expression vector used in the VSMC experiments into the uterine SMC (Lake *et al.*, 2003) resulted in significant reductions of both proliferation and motility in both myometrial and leiomyoma SMC. This data suggests that CCN5 might be necessary to maintain the normal uterine SMC phenotype, and that loss of CCN5 might result in formation of leiomyomas. Furthermore, it demonstrates that leiomyoma SMC, despite losing the ability to express CCN5, are still responsive to it.

Further study of human myometrial SMC showed that levels of CCN5 mRNA and protein expression are normally 5-fold higher during the proestrous (high estrogen) phase of the menstrual cycle than in the metestrous (low estrogen) phase, suggesting CCN5 regulation by estrogen. This hypothesis was verified in ovariectomized rats treated with exogenous estrogen ($\sim$90 pg/mL; Mason *et al.*, 2004b). Estrogen-treated rats had 4-8-fold higher levels of CCN5 mRNA and protein in uterine tissue than untreated animals. Immunohistochemical analysis showed that during the proestrous phase of the cycle, CCN5 localizes throughout the uterus, including the endometrium, endometrial glands, myometrium, and serosa. In support of the hypothesis that estrogen regulates CCN5 levels, Mason *et al.* (2004a) observed much higher levels of CCN5 mRNA and protein in uterine tissue in women during the proliferative (high estrogen) phase of their cycle compared to the menstrual (low estrogen) phase.

C. Studies of Rodent Embryonic Fibroblasts

The CCN5 gene independently emerged from experiments designed to identify downstream target genes that are activated during cell transformation caused by activated H-ras in the presence of an inactive P53 gene (Zhang *et al.*, 1998). The differential display PCR method was used to compare mRNA from normal rat embryonic fibroblasts (REF) to that of H-ras-transformed REF. Expression of CCN5 mRNA, originally named rCop-1, was absent after cell transformation. Further experiments testing mouse embryonic fibroblasts (MEF) transformed by a variety of methods, including the alkylating agents (BPA31 or DA31), Kirsten and SV40 viruses, and spontaneous mutant backgrounds, showed that MEF transformed by any method undergo the loss of CCN5 mRNA expression. Expression of CCN5 mRNA was absent in murine fibroblasts during growth arrest and for 12 hours after release from growth arrest by serum supplementation, and peaked at 20 hours during S phase. Forced expression of CCN5 in transformed REF by transfection with retroviral CCN5 expression vectors led to a 10-fold reduction in cell number, caused in part by cell death, and reduced tumorigenicity.

D. Studies of Transformed Mammary Epithelial Cells and Human Colon Tumors

CCN5, named Wnt-1 induced signal protein-2 (Wisp-2) in these experiments, emerged from a study to identify downstream target genes after Wnt-1 transformation of mouse mammary epithelial cells (C57MG) (Pennica *et al.*, 1998). Using the suppression subtraction hybridization (SSH) method, CCN5 mRNA was found to be up-regulated 5-fold at two or three days after transformation induced by Wnt-1 expression. C57MG cells that over-express Wnt-1 demonstrate induction of CCN5 mRNA. In the breast tumors of Wnt-1 transgenic mice, CCN5 expression was observed in fibrovascular tumor stroma cells. As part of the same study, CCN5 mRNA expression was examined in a series of primary human colon tumors. Although the chromosomal region that includes the CCN5 gene was amplified in some of the tumors, CCN5 mRNA expression was lower in tumor cells compared to normal mucosal cells by 2–30-fold in 79% cases tested. The data suggests strongly that CCN5 is a downstream target of Wnt-1 signaling.

E. Studies of Human Bone Cells

Analysis of a human osteoblast cDNA library using expressed sequence tag (EST) analysis revealed a highly expressed CCN5 gene (Kumar *et al.*, 1999).

It was first named CTGF-L because of its ~60% sequence identity with the CTGF gene (CCN2). CCN5 mRNA expression was detected in primary human osteoblasts and synovial fibroblasts, and in numerous other cell types found in developing adult and fetal bone, but not in osteosarcoma cells. CCN5 expression in primary osteoblast cells is not affected by osteotropic agents such as parathyroid hormone, TGF-β, vitamin D, and estrogen. A recombinant human CCN5 was made and tested for direct biochemical assays to characterize CCN5 physical interactions. These experiments demonstrated that CCN5 binds with high specificity to both IGF-I and IGF-II, blocks binding of fibrinogen to integrin receptors, and promotes adhesion of osteoblast cells to surfaces coated with CCN5. Recombinant human CCN5 also inhibited the production of osteocalcin by rat osteoblast-like cells.

F. Studies of Human Breast Cancer Cells

CCN5 was identified in human breast cancer cells (MCF7) in experiments designed to find new estrogen-responsive genes (Inadera *et al.*, 2000). Serial analysis of gene expression (SAGE) was used to show that CCN5 was one of four genes with a markedly increased level (16-fold) of mRNA in MCF7 cells treated with estrogen (10 nM). The three other genes found using SAGE, cathepsin, pS2, and high mobility group 1 protein, have been previously characterized as estrogen-inducible. Further experiments showed that the increase in CCN5 mRNA was completely prevented by co-incubation of estrogen-treated cells with the pure anti-estrogen agonist ICI182,780, strongly suggesting that CCN5 is directly regulated by estrogen signaling via the estrogen receptor. CCN5 mRNA induction by estrogen was reduced by 75% in the presence of actinomycin D and not affected by cycloheximide, suggesting that the estrogen signaling pathway for CCN5 induction depends primarily on new transcription. Western blot analysis revealed that CCN5 protein induced by estrogen treatment was also present in conditioned medium from estrogen-treated MCF7 cells (Inadera *et al.*, 2002). After addition of 10 nM estradiol to MCF7 cells, CCN mRNA began to increase at 4 hrs, peaked at 24 hrs, and remained at the peak level (5-fold higher than unstimulated cells) for up to 3 days (Inadera *et al.*, 2000; Banerjee *et al.*, 2003). Levels of CCN5 mRNA were estradiol-dose-dependent, with peak levels at 1 nM, and EC_{50} between 10–100 pM. Induction of CCN5 mRNA in MCF7 cells was partially dissected using phosphokinase (PK) intracellular signaling pathway activators (Inadera, 2003). Treatment with the protein kinase A (PKA) activator cholera toxin plus 3-isobutyl-1-methylxanthine (CT/IBMX) induced a five-fold increase in CCN5 mRNA expression in MCF7 cells without estrogen stimulation, with or without the estrogen receptor. In contrast, treatment with 12-O-tetradecanoylphorbol-13-acetate (TPA), a stimulator of protein kinase C (PKC) activity, completely blocked CCN5 mRNA expression in MCF7 cells stimulated by estrogen. The blocking by TPA was independent from cell growth.

Other investigators have shown that CCN5 mRNA is minimally detectable in normal human mammary epithelial cells and over-expressed in serum-stimulated (including estrogen) MCF7 cells (Zoubine *et al.*, 2001). In another series of experiments, human mammary epithelial cells up-regulated CCN5 mRNA expression after transfection with an estrogen receptor α gene construct, providing additional evidence for CCN5 regulation via estrogen signaling (Banerjee *et al.*, 2003).

III. CCN5 GENE, mRNA, AND PROTEIN STRUCTURE

A. Tools Developed for Studies of CCN5

1. *Antibodies to CCN5*

Most investigators of CCN5 function have generated polyclonal antibodies in rabbits against portions of the rat and human CCN5 protein. All but one of them are directed against short synthetic polypeptide fragments derived from the CCN5 sequence (Table 1). Many of the antibodies are specific for certain modular domains of CCN5. Kumar *et al.* (1999) reported that

Table 1.   Rabbit polyclonal antibodies made against CCN5.

Species	Affinity purified	Amino acid residues	Added epitopes	CCN module*	Reference
rat	no	24–250	His6	includes all three modules	Zhang *et al.*, 1998
rat	yes	103–117	none	VWC	Castellot (unpublished)
human		57–69	none	IGFBP	Inadera *et al.*, 2002
human		88–108	none	IGFBP-VWC	Zoubine *et al.*, 2001
human	yes	103–117	none	VWC	Lake *et al.*, 2003
human		135–151	none	VWC	Kumar *et al.*, 1999
human		234–248	none	5 amino acid residues of TSP	Kumar *et al.*, 1999
human		237–250	none	2 amino acid residues of TSP	Inadera *et al.*, 2002
human		239–250	none	none; carboxy-terminal peptide	Zoubine *et al.*, 2001

*CCN5 domain polypeptide sequence coordinates: IGFBP, 24–93; VWC, 98–164; variable domain, 165–193; TSP, 194–238 (Brigstock, 2003).

antibodies against VWC domain peptides and the C terminus of CCN5 were non-neutralizing and therefore not useful for blocking experiments. However, Lake *et al.* (2003) reported that the antibody against the human CCN5 amino acid residues 103–117 (within the VWC module) was capable of being completely neutralized by the immunizing peptide. In many of the immunohistochemistry and Western blot experiments, polyclonal antibody preparations were first affinity-purified using columns made with the immunizing peptide. None of these antibodies have been reported to cross-react with other members of the CCN family of proteins, or any other proteins. One of the antibodies was raised against the full-length CCN5 protein (not including the signal peptide) with an added His6 epitope tag (Zhang *et al.*, 1998).

2. *Viral and plasmid CCN5 expression constructs*

Several rat CCN5 expression constructs have been developed to enable overexpression of CCN5 in cultured cells during experimental or non-physiologic situations (Table 2). All of them have been shown to produce CCN5-mediated biological activity. Although two of them have CCN5 protein-coding regions with attached epitopes, there has been no evidence for either gene that the epitope tag had any biologic affect in the assays performed. None of these expression constructs have been used in animal models.

3. *Inhibitory RNA molecules for CCN5*

RNA interference (RNAi) is a simple and inexpensive method to reduce the expression of any nucleus-encoded gene, such as CCN5. A 21 nucleotide siRNA was made *in vitro* for a sequence 306 bp downstream from the CCN5 translation start codon (Lake and Castellot, 2003). This CCN5 siRNA was transfected into VSMC and caused an 80% reduction in endogenous CCN5 protein levels. The transfection efficiency was monitored by observing the number of cells

Table 2.   Viral and plasmid CCN5 expression constructs.

Construct name	Vector type	CCN5 insert	Added markers	Reference
pMFG-S-rCop-1	retrovirus	complete rat cDNA	none	Zhang *et al.*, 1998
pBabe-Puro-rCop-1	retrovirus	complete rat cDNA	none	Zhang *et al.*, 1998
pA3M/COP-1	mammalian expression	complete rat cDNA	C-terminal myc	Delmolino *et al.*, 2001
AdCCN5	adenovirus	complete rat cDNA	C-terminal HA; separate GFP gene	Lake *et al.*, 2003

that displayed fluorescence after transfection of fluorescein labeled siRNA; the fraction transfected was routinely >90%.

4. *Recombinant CCN5*

Pure recombinant CCN5 would enable a wide variety of new experiments to dissect its structure and function. Unfortunately, the low solubility and high "stickiness" of CCN5, probably caused at least in part by its high (11%) cysteine content and the presence of several protein-binding sequence motifs, has made large-scale preparations of recombinant CCN5 very difficult. Only two examples of recombinant CCN5 have been reported.

To obtain enough rat CCN5 to prepare a polyclonal antibody against the full-length protein, a 741 bp cDNA fragment that contained the entire CCN5 coding region without the N-terminal 23 amino acid signal peptide was made using PCR and inserted into the His6-tag expression vector pQE32 (Zhang *et al.*, 1998). The recombinant pQE32 CCN5 was transformed into bacteria to produce the His6-tagged CCN5 fusion protein. This protein was insoluble when produced in bacteria, and it was necessary to purify it on a nickel-Sepharose column under denaturing conditions to obtain enough protein to immunize a rabbit. No other experimental uses have been described for this recombinant rat CCN5 protein.

A recombinant human CCN5 protein was made using a cDNA clone with the complete coding sequence (Kumar *et al.*, 1999). Appropriate 5′- and 3′-untranslated regions were attached to the cDNA fragment and the gene construct was then inserted into the CDN expression vector. In this vector, CCN5 expression was driven by a CMV promoter. In addition, two epitope tags were added, including an HIV polypeptide and His6 tag after the signal sequence with an enterokinase cleavage site between the epitope tags and CCN5. This CCN5 expression plasmid was transfected into CHO cells by electroporation and cells expressing CCN5 were bulk-selected in nucleoside-free medium. Conditioned medium from a large-scale culture of CHO-rhCCN5 cells was used to purify rhCCN5 on a TALON metal affinity column. As in the previous example of recombinant CCN5, poor expression of soluble secreted CCN5 was reported. However, the amount of rhCCN5 obtained was enough to perform several biochemical experiments as summarized above.

B. CCN5 Gene Location, mRNA Structure, and Promoter

The human CCN5 gene was mapped to human chromosome 20 using the radiation hybrid mapping technique (Pennica *et al.*, 1998). The CCN5 gene is in the middle of the long arm, at cytogenetic position 20q12-q13.1, approximately 63 kb distal from the adenosine deaminase (ADA) gene, the gene affected in severe combined immune deficiency (SCID). The other five known CCN genes are on different chromosomes (chromosomes 1, 6, and 8). The recently concluded human genome sequence project has produced the complete sequence of chromosome 20, as well as the sequence of the entire CCN5 gene. CCN5 has four exons of 96, 216, 254, and 842 bp separated by three introns of 4543, 4225, and 2095 bp. Several different CCN5 cDNA clones have been sequenced and taken together strongly suggest that the processed mRNA sequence includes 1708 nucleotides, including a 261 nucleotide 5' untranslated region, the 750 nucleotide open reading frame encoding the 250 amino acid residues, a 730 nucleotide 3' untranslated region beginning with a TAA stop codon at nucleotide 1011, and an mRNA poly-A addition consensus site at nucleotide 1710 (Zhang *et al.*, 1998; Delmolino *et al.*, 2001). The rat and human CCN5 amino acid sequences are ~74% identical (Kumar *et al.*, 1999). As discussed above, the CCN5 polypeptide sequence includes the IGFBP, VWC, and TSP1 modules that are homologous to the same modules in the 4-domain CCN family proteins.

Approximately three kb of DNA sequence upstream from the CCN5 translation start site has been analyzed (John Castellot, unpublished observations). This region includes three TATA box motifs, a CCAAT box, AP-1, AP-2, and AP-4 recognition sites; motifs for MyoD, motifs for heat shock factors, and stress response elements (recognized by p53). The presence of stress response elements supports the designation of CCN5 as a growth arrest gene that is expressed as part of the rescue response of cells subjected to potentially lethal environmental influences. An expression plasmid construct was generated that included the CCN5 3 kb upstream promoter region ligated to the luciferase reporter gene (John Castellot, unpublished observations). This was transfected into rat VSMC and COS-7 fibroblasts. Both cell types demonstrated increased luciferase activity, and the VSMC had 3-5 fold higher luciferase activity levels than the COS-7 cells. Functional analysis of the CCN5 promoter has been done using deletion derivatives of the CCN5 upstream region to drive the pGL3 basic control luciferase gene promoter. Rat VSMC were transfected with these constructs and grown in medium containing 10% serum with or without 300 ug/mL heparin or chondroitin sulfate (glycosoaminoglycan control). The background level of luciferase activity was 12–18 units, and the full length CCN5 upstream fragment resulted in 360 units of activity. All deletion constructs, except for one that included a heparin response element, produced low levels of luciferase when the transfected cells were grown with heparin stimulation. The heparin response element in the CCN5 promoter caused a 20-fold increased stimulation of the luciferase reporter gene. To our knowledge, this is the first demonstration of a glycosaminoglycan-response element in a promoter.

C. Analysis of the Molecular Size of the CCN5 Protein

The cDNA sequence predicts a molecular weight of CCN5 of 27.5 kDa. Western blot analysis has been used to monitor the molecular weight and abundance of the CCN5 protein in many of the studies described above. In some studies, the CCN5 protein under study has been modified by the addition of short epitope polypeptide markers. The molecular weights of CCN5 reported vary from 28 kDa (in growth arrested VSMC conditioned medium and cell lysates made from AdCCN5-transfected VSMC; Lake *et al.*, 2003), 26 kDa (in conditioned medium from osteoblasts; Kumar *et al.*, 1999), 28 kDa (in 10 nM estrogen-induced MCF7 cell lysates, conditioned medium, and ECM; Inadera *et al.*, 2002), 31 kDa (in MCF7 cell lysates; Zoubine *et al.*, 2001), 28 kDa (HA-tagged CCN5 from AdCCN5 in myometrial uterine SMC using anti-HA; Mason *et al.*, 2004a), 31 kDa (conditioned medium from VSMC transfected with an expression vector encoded CCN5 with a myc tag; Delmolino *et al.*, 2001), and 31 kDa (conditioned medium from Rat-1 embryonic fibroblast transfected with a CCN5 gene that had no leader peptide and a His6 tag; Zhang *et al.*, 1998). Most of the examples listed here are analyses of CCN5 secreted into the medium, expression vector encoded CCN5, or CCN5 produced by tumor cells; none are from lysates made from unperturbed animal cells.

Recent Western blot analyses of lysates from unperturbed cells and tissues reveal a more complex array of CCN5 size variants (Mark Gray and John Castellot, unpublished results). Organs and tissues (aorta, uterus, heart, liver, pancreas, spleen, stomach, skeletal muscle, ovary, and kidney) were carefully dissected from adult male and female mice and rats and separated from surrounding tissues immediately post-mortem and frozen. Tissue lysates examined on Western blots after electrophoresis in 15% polyacrylamide-SDS gels revealed a prominent band at 47 kDa in all samples, and a more modest 28 kDa band in some. Additional bands of approximately 45 kDa, 58 kDa, and several other larger species were also found among some of the tissues analyzed. CCN5 bands of lower molecular mass than 28 kDa were not detected. Comparison of lysates made from the same tissues in both rat and mouse demonstrated that patterns of tissue-specific variants were the same in both animals. The sum of the band intensities of all CCN5 variants in each tissue lysate was proportional to the intensity of CCN5 staining in immunohistochemical analyses of adult tissues (discussed below; Mark Gray and John Castellot, unpublished results). These data suggest that CCN5 is modified *in vivo* by tissue-specific post-translational modification(s) in many, if not all, organs and tissues.

IV. MODELS USED FOR THE STUDY OF CCN5 EXPRESSION

A. Tissue and Animal Models

CCN5 has been studied in living animals and in animal tissues, as briefly described earlier (Table 3). Animal models include observational studies of normal cycling female rats (Mason *et al.*, 2004a), and rats subjected to experimental treatments such as ovariectomy followed by exogenous estrogen administration (Mason *et al.*, 2004b) and balloon injury to the carotid artery (Lake *et al.*, 2003). Only a small number of normal human tissues have been examined for CCN5 expression. No general conclusions about CCN5 tissue-specific expression can be made with certainty from this extremely short list of studies of normal tissues and tumors, except that CCN5 expression tends to be decreased or absent in neoplastic cells. Because of the cost and complexity of analyzing CCN5 expression in animals and tissues, most studies to date have been directed toward cultured cells.

B. Cultured Cell Models

Most of the information known about the CCN5 protein was obtained through the study of cultured mammalian cells. A variety of normal and neoplastic cell types have been examined and experimentally challenged to provide clues about CCN5 expression and function (Table 4). Normal cell types analyzed include SMC from rat aorta, SMC from rat and human uterus, mouse skeletal muscle, endothelial cells from rat epididymis, mammary epithelial cells from the mouse and humans, rat embryonic fibroblasts, and several cell types from human bone. In some experiments, normal cells were manipulated by addition of regulators such as serum, heparin, and estrogen; these results are summarized below. Neoplastic cells analyzed for CCN5 expression include mouse and human breast tumor cells, human ovarian carcinoma cells, human osteosarcoma cells, and rat embryonic fibroblasts transformed using *H-ras* and *P53* gene mutations. Normal and transformed cells have been induced to over- and under-express CCN5 using either biochemical inducers, transfected CCN5

Table 3.   Analysis of CCN5 expression in animal models.

Organ or tissue	Experimental conditions	Methods of analysis	CCN5 expression	Reference
rat uterus	normal growth	Q-PCR; Western; IHC	mRNA and protein increased 5-fold during pro-estrus	Mason *et al.*, 2004b
ovariectomized rat uterus	exogenous estrogen	Q-PCR; IHC	mRNA increased 5.5-fold; protein increased 7.7-fold	Mason *et al.*, 2004b
injured rat common carotid artery	balloon injury	IHC	protein present in media pre- and post-injury; in intima post-injury	Lake *et al.*, 2003
human colon mucosa	normal growth	RT-PCR	mRNA present	Pennica *et al.*, 1998
human colon adenocarcinoma	normal growth	RT-PCR	mRNA under-expressed 2–30-fold in 79% of tumors	Pennica *et al.*, 1998
stromal fibroblasts in murine breast carcinoma	Wnt-1 over-expression	IHC	mRNA present	Pennica *et al.*, 1998
human osteosarcoma	normal growth	Northern	mRNA absent	Kumar *et al.*, 1999
human osteoclastoma	normal growth	Northern	mRNA absent	Kumar *et al.*, 1999
human fetal bone	normal growth	IHC; Northern	mRNA and protein present	Kumar *et al.*, 1999

IHC: immunohistochemistry; Q-PCR: quantitative (real-time) PCR; RT-PCR: reverse transcriptase PCR.

Table 4.   Analysis of CCN5 expression in cultured cell models.

Cell type	Experimental conditions	Methods of analysis	CCN5 expression found	Location	Biological effects of CCN5 expression	Reference
rat vascular smooth muscle cells	heparin treatment	Differential display; Northern; Western; IHC	mRNA and protein high in quiescence; low during proliferation	cell surface; perinuclear (Golgi); vesicular	decreased cell proliferation and motility	Delmolino *et al.*, 2001 Lake *et al.*, 2003
rat vascular smooth muscle cells with exogenous CCN5 gene	transfected with AdCCN5	IHC; Western	7-fold increase in CCN5	cell surface; perinuclear (Golgi); vesicular	reduces MMP-2 expression	Lake *et al.*, 2003 Lake and Castellot, 2003
rat vascular smooth muscle cells with reduced CCN5 expression	transfected with RNAi-CCN5	Q-PCR; Western; IHC; zymography	mRNA and protein decreased by ~80%		MMP-2 increased ~2-fold; reduced actin filaments	Lake and Castellot, 2003
rat epididymal endothelial cells	growth arrested	Northern	mRNA absent			Delmolino *et al.*, 2001
murine fibroblasts (3T3 and COS-7)	growth arrested	Northern	mRNA generally absent; tiny signal for 3T3			Delmolino *et al.*, 2001
COS-7 murine fibroblasts that over-express CCN5	transfection with pA3M-CCN5	IP from conditioned medium	protein in conditioned medium	secreted	secreted CCN5 inhibited proliferation of VSMCs	Delmolino *et al.*, 2001
human myometrial uterine smooth muscle cells	growth arrested	IHC; Western; Q-PCR	mRNA and protein expressed in proestrus			Mason *et al.*, 2004a
human leiomyoma uterine smooth muscle cells	growth arrested	IHC; Western; Q-PCR	high in G_0; mRNA and protein down-regulated throughout cell cycle		maintains quiescence	Mason *et al.*, 2004a
human uterine smooth muscle cells over-expressing CCN5	transfected with AdCCN5	Western; Q-PCR	protein and mRNA up-regulated		reduced cell proliferation and motility	Mason *et al.*, 2004a
human leiomyoma smooth muscle cells over-expressing CCN5	transfected with AdCCN5	Western; Q-PCR	protein and mRNA up-regulated		reduced cell proliferation and motility	Mason *et al.*, 2004a
murine skeletal muscle (C2C12)	growth arrested	Northern	mRNA absent			Delmolino *et al.*, 2001

Table 4.   (*Continued*)

Cell type	Experimental conditions	Methods of analysis	CCN5 expression found	Location	Biological effects of CCN5 expression	Reference
mouse fibroblasts (3T3-A31)	growth arrest and serum stimulated	Northern	mRNA throughout cell cycle (peaks in S-phase); no mRNA in quiescence			Zhang *et al.*, 1998
rat embryonic fibroblasts	serum stimulated	differential display Northern	mRNA in continuously growing cells			Zhang *et al.*, 1998
rat embryo fibroblasts that over-express CCN5	transfected with retroviral CCN5	Northern; Western; FACS	mRNA and protein produced		no dead cells	Zhang *et al.*, 1998
primary human osteoblasts from trabecular bone	serum stimulated	EST analysis; IHC; Northern; Western; ISH;	mRNA, protein	secreted		Kumar *et al.*, 1999
primary human synovial fibroblasts	serum stimulated	Northern; IHC; ISH; Western	mRNA, protein	secreted		Kumar *et al.*, 1999
mouse mammary epithelial (C57MG)	none	SSH; Northern	low mRNA levels			Pennica *et al.*, 1998
human mesangial cells	serum stimulated	Northern	no mRNA			Kumar *et al.*, 1999
human stromal cells (TF274)	serum stimulated	Northern	no mRNA			Kumar *et al.*, 1999
human mammary epithelial cells (HUMEC)	serum stimulated	RT-PCR Western	little mRNA; no protein			Zoubine *et al.*, 2001
CHO-recombinant that expresses rhCCN5	cells transfected with rhCCN5	Western; N-terminal protein sequencing	increased protein	secreted		Kumar *et al.*, 1999

(*Continued*)

Table 4.   (*Continued*)

Cell type	Experimental conditions	Methods of analysis	CCN5 expression found	Location	Biological effects of CCN5 expression	Reference
H-ras-transformed mouse embryonic fibroblasts (MEFs)	transformed by mutations (activated ras and loss of p53), chemicals, and viruses	Northern	no mRNA			Zhang *et al.*, 1998
transformed rat embryonic fibroblasts	serum stimulated	differential display Northern	no mRNA			Zhang *et al.*, 1998
transformed rat embryonic fibroblasts that overexpress CCN5	transfected with retroviral CCN5-His6 tag	Northern; Western; IHC; FACS; tumorigenicity tests	mRNA and protein produced	cell surface; perinuclear (Golgi) and vesicular	10-fold reduced cell number at 3 days; many dead cells; reduced tumorigenicity	Zhang *et al.*, 1998
mouse mammary epithelial cells (C57MG) overexpressing Wnt-1	transfected with Wnt-1 retrovirus	SSH; Northern	mRNA induced 5-fold			Pennica *et al.*, 1998
human breast tumor cells (MCF7)	serum stimulated	RT-PCR; Northern; Western	mRNA 5.3-fold higher than growth arrested cells; protein present			Zoubine *et al.*, 2001
human breast tumor cells (MCF7)	estrogen stimulated	SAGE; Northern blot; Western blot	16-fold increased mRNA with 10 nM estrogen	protein in conditioned medium and in ECM		Inadera *et al.*, 2000 Inadera *et al.*, 2002
human osteosarcoma cells	serum stimulated	Northern; ISH	no mRNA			Kumar *et al.*, 1999
human ovarian carcinoma cells (HeLa)	serum stimulated	Northern	no mRNA			Kumar *et al.*, 1999

ECM: extracellular matrix; IHC: immunohistochemistry; IP: immunoprecipitation; ISH: *in situ* hybridization; Q-PCR: quantitative (real-time) PCR; RT-PCR: reverse-transcriptase PCR; SAGE: serial analysis of gene expression; SSH: sequential subtractive hybridization; FACS: fluorescence-activated cell sinting.

expression vectors, or CCN5 small inhibitory RNAs. Cells were examined for mRNA or protein expression using a wide variety of standard molecular techniques. Most of the data from these experiments consists of variations in the level of CCN5 expression, with a few examples of subcellular localization data. Biological effects of CCN5 expression in these models are discussed further below.

C. Other Cells and Tissues Examined for CCN5 Expression

The experiments summarized in Tables 3 and 4 suggest that CCN5 is expressed in a wide variety of mammalian cell types. In some of the studies, other tissues were surveyed for the presence (or absence) of CCN5 mRNA expression (Table 5). All available data is grouped in Table 5 by organ/tissue tested. No experimental manipulations to alter CCN5 expression were deliberately imposed on the tissues and organs included in these surveys. Almost all of them are tests for the presence of CCN5 mRNA in mRNA preparations made from whole organs and displayed on Northern blots. Three investigators used commercially-prepared Northern blots that included 2 µg of poly(A)+ RNA from various organs/tissues loaded in each lane (all from Clontech, Inc.). All of the data reported in one of the studies is negative, i.e., no CCN5 mRNA was found in any rat and mouse tissue tested by Northern blot (Zhang *et al.*, 1998). For some organs (leukocytes, spleen, pancreas, liver, stomach, kidney, and placenta), no CCN5 mRNA was detected in any of the studies. For other organs/tissues (such as heart, lung, intestine, testis, ovary, skeletal muscle, and bone), one or more studies reported the presence of CCN5 mRNA.

There is much variability in the Northern blot data, even when the same commercial blots were tested by different investigators (Pennica *et al.*, 1998; Kumar *et al.*, 1999). In some cases, results for the same tissue (lung, intestine, skeletal muscle) ranged from negative (−) to strong (+++) expression levels. Analysis of organ/tissue mRNA localization by Northern blots might reveal where high levels of CCN5 transcription take place, but cannot reveal where the secreted protein is normally localized. The Northern blot experiments are limited by mRNA abundance and the balance of cell/tissue types represented in the organ mRNA preparations and thus may not have enough sensitivity to detect CCN5 reliably in some instances.

The sensitivity of Northern blot analysis for detection of CCN5 mRNA is called into question further when compared with CCN5 protein localization data. CCN5 distribution in both adult and embryonic animals has not been mapped either spatially or temporally in a systematic manner. In recent

Table 5.   Survey of tissues and organs for expression of CCN5 mRNA.

Organ	Expression[*]	Method	Reference
Brain			
mouse	−	ISH; Northern	Zhang *et al.*, 1998
rat	−	ISH; Northern	Zhang *et al.*, 1998
rat	+	Northern[a]	Delmolino *et al.*, 2001
human	−	Northern[a]	Kumar *et al.*, 1999
human	−	Northern[a]	Pennica *et al.*, 1998
human fetal	−	Northern[a]	Pennica *et al.*, 1998
Heart			
mouse	−	ISH; Northern	Zhang *et al.*, 1998
rat	−	ISH; Northern	Zhang *et al.*, 1998
rat	++	Northern[a]	Delmolino *et al.*, 2001
human	+	Northern[a]	Kumar *et al.*, 1999
human	−	Northern[a]	Pennica *et al.*, 1998
human fetal	−	Northern[a]	Pennica *et al.*, 1998
Lung			
mouse	−	ISH; Northern	Zhang *et al.*, 1998
rat	−	ISH; Northern	Zhang *et al.*, 1998
rat	+++	Northern[a]	Delmolino *et al.*, 2001
human	++	Northern[a]	Kumar *et al.*, 1999
human	−	Northern[a]	Pennica *et al.*, 1998
human fetal	++	Northern[a]	Pennica *et al.*, 1998
Blood vessels			
rat aorta	++	Northern	Delmolino *et al.*, 2001
human giant cell bone tumor	−	ISH	Kumar *et al.*, 1999
Blood			
human leukocytes	−	Northern[a]	Kumar *et al.*, 1999
human leukocytes	−	Northern[a]	Pennica *et al.*, 1998
Thymus			
human	−	Northern[a]	Kumar *et al.*, 1999
human	−	Northern[a]	Pennica *et al.*, 1998
human fetal	−	Northern[a]	Pennica *et al.*, 1998
Spleen			
mouse	−	ISH; Northern	Zhang *et al.*, 1998
rat	−	ISH; Northern	Zhang *et al.*, 1998
rat	−	Northern[a]	Delmolino *et al.*, 2001
human	−	Northern[a]	Kumar *et al.*, 1999
human	−	Northern[a]	Pennica *et al.*, 1998
human fetal	−	Northern[a]	Pennica *et al.*, 1998

Table 5. (*Continued*)

Organ	Expression*	Method	Reference
Pancreas			
mouse	−	ISH; Northern	Zhang *et al.*, 1998
rat	−	ISH; Northern	Zhang *et al.*, 1998
human	−	Northern[a]	Kumar *et al.*, 1999
human	−	Northern[a]	Pennica *et al.*, 1998
Liver			
rat	−	Northern[a]	Delmolino *et al.*, 2001
human	−	Northern[a]	Kumar *et al.*, 1999
human	−	Northern[a]	Pennica *et al.*, 1998
human fetal	−	Northern[a]	Pennica *et al.*, 1998
Stomach			
mouse	−	ISH; Northern	Zhang *et al.*, 1998
rat	−	ISH; Northern	Zhang *et al.*, 1998
Intestine			
mouse	−	ISH; Northern	Zhang *et al.*, 1998
rat	−	ISH; Northern	Zhang *et al.*, 1998
human small	−	Northern[a]	Kumar *et al.*, 1999
human small	−	Northern[a]	Pennica *et al.*, 1998
human colon	+	Northern[a]	Kumar *et al.*, 1999
human colon	+++	Northern[a]	Pennica *et al.*, 1998
Kidney			
mouse	−	ISH; Northern	Zhang *et al.*, 1998
rat	−	ISH; Northern	Zhang *et al.*, 1998
rat	−	Northern[a]	Delmolino *et al.*, 2001
human	−	Northern[a]	Kumar *et al.*, 1999
human	−	Northern[a]	Pennica *et al.*, 1998
human fetal	−	Northern[a]	Pennica *et al.*, 1998
Prostate gland			
human	+	Northern[a]	Kumar *et al.*, 1999
human	−	Northern[a]	Pennica *et al.*, 1998
Testis			
rat	+	Northern[a]	Delmolino *et al.*, 2001
human	++	Northern[a]	Kumar *et al.*, 1999
human	−	Northern[a]	Pennica *et al.*, 1998
Ovary			
human	+++	Northern[a]	Kumar *et al.*, 1999
human	+++	Northern[a]	Pennica *et al.*, 1998
Uterus			
rat	+++	Q-PCR	Mason *et al.*, 2004b
human	+++	Q-PCR	Mason *et al.*, 2004a

Table 5.   (*Continued*)

Organ	Expression*	Method	Reference
Skeletal muscle			
mouse	−	ISH; Northern	Zhang *et al.*, 1998
rat	−	ISH; Northern	Zhang *et al.*, 1998
rat	+	Northern[a]	Delmolino *et al.*, 2001
human	+	Northern[a]	Kumar *et al.*, 1999
human	+++	Northern[a]	Pennica *et al.*, 1998
human fetal	−	Northern[a]	Pennica *et al.*, 1998
Placenta			
human	−	Northern[a]	Pennica *et al.*, 1998
human	−	Northern[a]	Kumar *et al.*, 1999
Embryo			
whole mouse	−	ISH; Northern	Zhang *et al.*, 1998
Bone			
human osteoblasts-1° spongiosa	+++	ISH	Kumar *et al.*, 1999
human fetal osteoblasts-1° spongiosa	++	ISH	Kumar *et al.*, 1999
human osteoblasts-2° spongiosa	+	ISH	Kumar *et al.*, 1999
human fetal osteoblasts-2° spongiosa	+	ISH	Kumar *et al.*, 1999
human synovial macrophages	++	ISH	Kumar *et al.*, 1999
human bone marrow cells	+ to +++	ISH	Kumar *et al.*, 1999
human osteoclasts	+/−	ISH	Kumar *et al.*, 1999
human fetal bone chondrocytes	++	ISH	Kumar *et al.*, 1999
human fetal bone myocytes	+	ISH	Kumar *et al.*, 1999
human tumor stromal cells	+	ISH	Kumar *et al.*, 1999
human tumor osteoblasts	+	ISH	Kumar *et al.*, 1999
human tumor osteocytes	+	ISH	Kumar *et al.*, 1999
human tumor osteoclasts	−	ISH	Kumar *et al.*, 1999
human tumor bone marrow cells	+/−	ISH	Kumar *et al.*, 1999
human tumor macrophages	−	ISH	Kumar *et al.*, 1999
human osteoclastoma tissue	−	Northern	Kumar *et al.*, 1999

*Levels of expression: (−) negative; (+) weak; (++) moderate; (+++) strong;
[a]Clontech Northern blot tissue array; ISH: *in situ* hybridization; Q-PCR: quantitative (real-time) PCR.

immunohistochemistry experiments (Mark Gray *et al.*, unpublished results), adult organ and tissue samples from male and female mice and rats were isolated and frozen sections were analyzed by immunohistochemistry using the affinity purified, anti-CCN5 (amino acid residues 103–117 in the VWC domain; Table 1). Intense CCN5 expression was detected in the heart, aorta, uterus, bronchi, and digestive tract organs, as predicted by the Northern blot data (Table 5). However, CCN5 protein was also detected in many other adult tissues analyzed, including the pancreas, spleen, skeletal muscle, ovary, testis, thymus, brain, and kidney. Although CCN5 expression was often found in the smooth muscle-like tissues within an organ (for example, in the mesangial cells of the kidney glomerulus), strong expression of CCN5 protein in epithelial and endothelial tissues was also observed. Expression in the epithelial lining of ducts in the kidney and uterine endometrium was particularly striking. It is possible that CCN5 is expressed in all tissues and cell types at some time during development and adult life. In other experiments, the distribution of CCN5 protein in embryos has been analyzed by immunohistochemistry (Jones *et al.*, unpublished results). High levels of CCN5 protein expression were observed in all organs and tissues of mouse embryos (8–14 dpc). Ubiquitous embryonic distribution of CCN5 protein might be caused by exposure to high levels ($\sim$2 ng/mL or 1×10^{-8} M) of estrogen *in utero*. The wide distribution of CCN5 expression in embryos and adult organs and tissues suggests that CCN5 might have many additional biological functions beyond those previously identified in the highly-focused studies reviewed here.

V. BIOLOGICAL EFFECTS OF ALTERED CCN5 EXPRESSION

A. Arrest of Cell Proliferation

Changes in rates of cell proliferation after alterations of CCN5 expression have been reported (Table 6). In rat VSMC, CCN5 expression causes dose-dependent abrogation of cell proliferation (Delmolino *et al.*, 2001; Lake *et al.*, 2003). Increased expression of CCN5 in rat embryonic fibroblasts transformed by *H-ras* in the presence of a *P53* gene loss-of-function mutation results in a sudden arrest of cell proliferation and an induction of "cytotoxic" killing of cells (Zhang *et al.*, 1998). Although the magnitude of CCN5 induction was the same in both cell types, the mechanisms of growth arrest may be very different. It is also possible that the induction of apoptosis in the transformed rat embryonic fibroblasts is caused by the inability of neoplastic cells to enter the quiescent (G_o) state. Since the cells could undergo apoptosis if they are unable to enter quiescence, the apparent induction of apoptosis by CCN5 may be indirect and separate from its induction of growth arrest.

B. Promotion of Cell Proliferation

In contrast to CCN5-mediated growth arrest, serum induction of CCN5 in MCF7 breast tumor cells caused a three-fold increased number of cells (Zoubine *et al.*, 1998). Neoplastic breast epithelial cells have many differences

Table 6.   Biological effects following altered CCN5 expression.

Biological effect	Cell or tissue	Method of modifying CCN5 expression	CCN5 level	Modifications to CCN5	Magnitude of biological effect	Reference
Proliferation	rat vascular smooth muscle cells	serum stimulated + exogenous CCN5	5-fold increase	none	50% reduced number of cells	Delmolino *et al.*, 1998
Proliferation	rat vascular smooth muscle cells	transfection with AdCCN5	7-fold increase	C-terminal HA tag	50% reduced number of cells (300 moi AdCCN5)	Lake *et al.*, 2003
Proliferation	human uterine myometrial smooth muscle cells	transfection with AdCCN5	7-fold increase	C-terminal HA tag	65% reduced number of cells (200 moi AdCCN5)	Mason *et al.*, 2004a
Proliferation	human uterine leiomyoma smooth muscle cells	transfection with AdCCN5	7-fold increase	C-terminal HA tag	65% reduced number of cells (200 moi AdCCN5)	Mason *et al.*, 2004a
Proliferation	transformed rat embryonic fibroblasts	transfection with pBabe-CCN5	increased; levels not reported	His6 tag	90% fewer cells	Zhang *et al.*, 1998
Proliferation	human breast carcinoma cells (MCF7)	serum stimulation	5.3-fold increase	none	~3-fold increased cells 1d after serum induction	Zoubine *et al.*, 2001
Motility	rat vascular smooth muscle cells	transfection with AdCCN5	7-fold increase	C-terminal HA tag	3-fold reduced motility (300 moi AdCCN5)	Lake *et al.*, 2003
Motility	human uterine myometrial smooth muscle cells	transfection with AdCCN5	7-fold increase	C-terminal HA tag	49% reduced motility	Mason *et al.*, 2004a
Motility	human uterine leiomyoma smooth muscle cells	transfection with AdCCN5	7-fold increase	C-terminal HA tag	71% reduced motility	Mason *et al.*, 2004a
Motility	rat vascular smooth muscle cells	transfection with CCN5 siRNA	5-fold reduction	none	~2-fold increased motility	Lake and Castellot, 2003
Invasiveness	rat vascular smooth muscle cells	transfection with AdCCN5	7-fold increase	C-terminal HA tag	5-fold reduced invasiveness; 4-fold reduced MMP-2 protein level	Lake *et al.*, 2003

Table 6. (*Continued*)

Biological effect	Cell or tissue	Method of modifying CCN5 expression	CCN5 level	Modifications to CCN5	Magnitude of biological effect	Reference
Adhesion	rat vascular smooth muscle cells	transfection with AdCCN5	7-fold increase	C-terminal HA tag	no significant effect	Lake *et al.*, 2003
Adhesion	human osteoblasts, osteosarcoma cells, and osteoblast-like cells	cell binding to rhCCN5-coated surface	3–300 ng/mL rhCCN5 bound to well surface	none	3–20-fold increased adhesion	Kumar *et al.*, 1999
Apoptosis	transformed rat embryonic fibroblasts	transfection with pBabe-CCN5	increased; levels not reported	His6 tag	elevated FACS sub-G1 peak (more DNA fragmentation)	Zhang *et al.*, 1998
Apoptosis	rat vascular smooth muscle cells	transfection with AdCCN5	6–7-fold increase	C-terminal HA tag	no significant effect	Lake *et al.*, 2003
Apoptosis	rat vascular smooth muscle cells	transfection with CCN5 siRNA	5-fold reduction	none	no effect	Lake and Castellot, 2003
Morphology	rat vascular smooth muscle cells	transfection with CCN5 siRNA	5-fold reduction	none	fewer stress fibers; 50% reduced α-actin mRNA	Lake and Castellot, 2003
Tumorigenicity	transformed rat embryonic fibroblasts	transfection with retroviral vector expressing CCN5	increased; levels not reported	his	markedly reduced tumor size	Zhang, 1998
Osteocalcin secretion	rat Ros 17/2.8 osteoblast like cells	incubation in 1% FCS + rhCCN5 + vitamin D	1–3000 ng/mL rhCCN5 protein	none	~50% reduction of osteocalcin production	Kumar, 1999

from other cell types in the activity of intracellular signaling pathways; these differences might cause CCN5 to produce apparently opposite effects to those observed in an unperturbed cell types such as VSMC.

C. Inhibition of Cell Motility

Rat VSMC transfected with AdCCN5 have a 7-fold increased level of CCN5 mRNA and a 70–80% decrease in motility (Table 6; Lake *et al.*, 2003). In concordance with this observation, human uterine SMC — both normal and fibroid — also display significantly decreased motility when CCN5 is over-expressed (Mason *et al.*, 2004a). Conversely, VSMC transfected with CCN5 siRNA have a 2-fold increase in motility (Lake and Castellot, 2003). Motility regulation by CCN5 expression has not been studied in any other cell type.

D. Cellular Invasiveness

In the only test of the ability of CCN5 to promote invasiveness, rat VSMC transfected with AdCCN5 were compared to untreated VSMC in their ability to invade Matrigel-coated chambers (a surrogate model for basement membranes; Lake *et al.*, 2003). CCN5 over-expression caused an 80% reduction in VSMC invasiveness and a 75% reduction in matrix metalloproteinase-2 (MMP-2) protein expression (Lake and Castellot, 2003). CCN5 regulation of cellular invasiveness has not been reported in any other cell type.

E. Adhesion of Cells to Substrates

CCN5 mediation of the ability of cells to adhere to substrates has been tested in two model systems (Table 6). In experiments that utilize recombinant human CCN5, several types of bone cells demonstrate 3–20-fold increased adherence to plastic wells coated with CCN5 compared to uncoated plastic wells (Kumar *et al.*, 1999). In contrast, rat VSMC that over-express CCN5 after transfection of AdCCN5 do not adhere to plastic or collagen I-coated plastic any differently from untreated VSMC (Lake *et al.*, 2003). The apparent difference in substrate adhesion between these two cell types might be caused by experimental differences since the methods for increasing CCN5 levels and the assays used were very different.

F. Induction of Apoptosis

Apoptosis has been tested in rat VSMC by measuring levels of caspase-3 activity after induction or blocking of CCN5 mRNA (Table 6). Approximately

5–6-fold increased or decreased CCN5 mRNA does not affect levels of caspase-3 activity in rat VSMC (Lake *et al.*, 2003; Lake and Castellot, 2003). In contrast to rat VSMC, over-expression of CCN5 by transfecting a CCN5 expression vector into H-ras-transformed REF cells results in a 10-fold loss of cells 3 days after transfection and the appearance of many dead cells (Zhang *et al.*, 1999). Fluorescence-activated cell sorting (FACS) revealed a significant sub-G_1 DNA content peak from fragmented cells in transformed REF cells that over-express CCN5 compared to untreated REF. Although apoptosis has no role in CCN5 expression in VSMC, transformed cells may have defective cell cycle pathways for exiting the cycle and entering G_0, and thus may have no choice but to undergo apoptosis when they are stimulated to enter quiescence by CCN5.

G. Morphology

The only example of cytoskeletal changes caused by alterations of CCN expression is in rat VSMC that experience an 80% reduction in CCN5 expression after transfection with a CCN5 siRNA (Lake and Castellot, 2003). These cells do not form the typical hill-and-valley confluent growth-arrested phenotype of untreated VSMC. The reduced-CCN5 cells have a more cuboidal or epitheliod shape and have fewer prominent actin stress fibers compared to untreated cells. In support of the morphologic observation is a 50% reduction in α-actin mRNA as demonstrated by real-time PCR. No other examples of CCN5-mediated morphological changes have been reported.

H. Tumorigenicity

H-ras-transformed rat embryonic fibroblasts (REF) that over-express CCN5 were compared to untreated transformed REF in their ability to cause large tumors in athymic nu/nu mice (Zhang *et al.*, 1999). Tumors induced by untreated cells were much larger that those caused by transformed REF that over-express CCN5, suggesting that CCN5 suppresses tumor growth. No other *in vivo* tumorigenicity experiments have been reported using cells with modified CCN5 expression.

I. Osteocalcin Production

Rat Ros 17/2.8 osteoblast-like cells were incubated with recombinant human CCN5 and tested for the secretion of osteocalcin, a marker of osteoblast function (Kumar *et al.*, 1999). Compared with cells not exposed to recombinant

CCN5, the Ros 17/2.8 cells produced up to 50% less osteocalcin in a CCN5-dose-dependent manner. It is likely that many other genes down-regulated by CCN5 can be identified using similar simple assays with recombinant CCN5.

VI. REGULATORY MOLECULES FOR CCN5 EXPRESSION (TABLE 7)

A. Heparin

As discussed above, heparin can induce CCN5 mRNA in rat VSMC, but not in rat vascular endothelial cells, human uterine SMCs, or MCF7 human breast carcinoma cells (Delmolino *et al.*, 2001; Mason *et al.*, 2004a; Inadera *et al.*, 2002). CCN5 expression after heparin treatment of other cell types has not been reported. The ability of heparin to induce CCN5 expression in VSMC appears to be specific for heparin, since other well-characterized SMC proliferation inhibitors like INF-γ and TGF-β did not induce CCN5 in VSMC.

B. Estrogen

Estrogen up-regulates CCN5 mRNA 4–8-fold in human uterine SMC, and five-fold in MCF7 human breast carcinoma cells (Mason *et al.*, 2004a; Inadera *et al.*, 2000). In recent experiments (John Castellot, unpublished results), the effect of estrogen on CCN5 expression levels in rat VSMC has been tested using physiological doses of 17-β-estradiol (10^{-10} to 10^{-8} M) over the course of four days. Western blot analysis demonstrated an estrogen-dose-dependent induction of CCN5 protein. Cell proliferation tests revealed an estrogen-dose-dependent inhibition of VSMC cell number. Tests of other cell types for estrogen regulation of CCN5 have not been reported.

C. Serum

Fetal calf serum is a complex mixture of many growth factors and hormones. In rat VSMC, addition of serum to growth-arrested cells resulted in an 80% decrease in CCN5 mRNA levels (Delmolino *et al.*, 2001). The addition of serum could be mimicked by addition of PDGF, bFGF, or, to a lesser extent, EGF. In contrast, serum-stimulation of MCF7 human breast carcinoma cells caused a 5.3-fold increased level of CCN5 mRNA (Zoubine *et al.*, 2001).

D. Other CCN5 Regulatory Molecules

Several growth factors and PK activators have been tested for their effect on CCN5 mRNA expression in rat VSMC, MCF7 breast carcinoma cells,

Table 7. Tests of regulator molecules for CCN5 expression.

Regulator	Cell type tested	Effect on CCN5 expression	Reference
heparin	rat vascular smooth muscle cells	mRNA increased ∼5-fold	Delmolino *et al.*, 2001
heparin	rat aortic endothelial cells	no effect on CCN5 mRNA	Delmolino *et al.*, 2001
estrogen	rat uterine smooth muscle cells	mRNA increased 4–8-fold	Mason *et al.*, 2004
estrogen	human breast carcinoma cells (MCF7)	mRNA increased ∼5-fold	Inadera *et al.*, 2000
serum	rat vascular smooth muscle cells	5-fold decrease after G_0	Delmolino *et al.*, 2001
serum	human breast carcinoma cells (MCF7)	mRNA increased 5.3-fold	Zoubine *et al.*, 2001
Wnt-1	mouse mammary epithelial cells (C57MG)	up	Pennica *et al.*, 1998
activated H-ras, p53	rat embryonic fibroblasts	complete loss of mRNA	Zhang *et al.*, 1998
IGF-I and IGF-II	binds with rhCCN5 *in vitro*	not tested *in vivo*	Kumar *et al.*, 1999
IGF-I	human breast carcinoma cells (MCF7) stimulated with estrogen and blocked with estrogen receptor agonist	no effect on CCN5 expression	Inadera, 2003
PDGF	rat vascular smooth muscle cells	strongly suppresses CCN5 expression	Delmolino *et al.*, 2001
EGF	rat vascular smooth muscle cells	no effect on CCN5 expression	Delmolino *et al.*, 2001
TGF-β1	rat vascular smooth muscle cells	suppressed CCN5 expression completely	Delmolino *et al.*, 2001
IL-lα	human breast carcinoma cells (MCF7) stimulated with estrogen and blocked with estrogen receptor agonist	no effect on CCN5 expression	Inadera, 2003
INF-β	rat vascular smooth muscle cells	suppressed CCN5 expression completely	Delmolino *et al.*, 2001
CT/IBMX (PKA activator)	human breast carcinoma cells (MCF7) stimulated with serum	mRNA increased 6-fold; not dependent on estrogen receptor	Inadera, 2003
TPA (PKC activator)	human breast carcinoma cells (MCF7) stimulated with estrogen	mRNA blocked; dependent on new protein synthesis; not dependent on estrogen receptor	Inadera, 2003

and transformed rat embryonic fibroblasts (Table 7). Although this data is far from comprehensive, it strongly suggests that CCN5 signaling employs a wide variety of intracellular signaling pathways in a cell-type specific manner.

VII. SUMMARY AND FUTURE DIRECTIONS FOR CCN5 RESEARCH

A. Do General Pathophysiologic Roles Exist for CCN5?

Initially, we and others believed that the functions of CCN5 could be delineated from studies of single cell types, for example, VSMC and uterine SMC. However, the notion that CCN5 might have selectivity for SMC quickly breaks down when the data from normal and neoplastic cell types are considered. So far, only the surface has been scratched in describing CCN5 function. Many more normal cell types must be studied to better identify general functions for CCN5. Standard cell function tests need to be performed for many more cell and tissue types. Detailed studies examining the temporal and spatial expression patterns are needed. In addition, elucidation of the receptors and intracellular signaling pathways used by CCN5 lags behind that of other CCN family members, notably CCN1 and CCN2. Cell- and tissue-specific CCN5-mediated properties and functions will probably be demonstrated. For example, stimulation of cell proliferation by CCN5 in one cell type and induction of growth arrest by CCN5 in another might be the outcomes of signaling processes that share 90% of intermediary steps but differ at the final 10% of steps that lead to a binary selection decision (e.g. growth or arrest).

B. Does the Data Support the Hypothesis that CCN5 Is the Anti-4-Domain CCN?

This question cannot be answered at this time because of the issues outlined above. There are insufficient data for CCN5 expression and function in normal cells. For some models discussed above, there are examples in which CCN5 confers phenotypes that are opposite to phenotypes identified for other CCN proteins in the same cell type. There are also examples of CCN5-mediated outcomes that resemble those mediated by 4-domain CCN proteins. Experiments are needed that will test CCN5 and the 4-domain CCNs under the same conditions and in the same cell types.

C. The Need for Genetically Tractable Animal Models for CCN5 Expression

CCN5 gene knockout mice and transgenic CCN5 over-expressor mice could provide an opportunity to study CCN5 function in every cell type. However, if CCN5 is a protein needed for survival of every cell at some time during development, then it might not be possible to obtain a CCN5 knockout animal. Similarly, if CCN5 over-expression is not tolerated by a crucial cell type or organ system at any time during development, it might not be possible to obtain an over-expressing transgenic animal. It might become necessary to develop conditional CCN5 knockout and CCN5 over-expressing animals. Conditions that can be imposed include developmental timing and tissue/organ specificity. In any of these reconstructed animals, it will be necessary to demonstrate that the modified CCN5 gene is expressed in the cell normally, i.e. it is translated and the protein is processed (post-translationally modified) correctly for the cell type under investigation.

D. CCN5 — A Versatile Protein Whose Function and Regulation Varies by Cell Type?

The data to date suggest that CCN5 is multifunctional, mediating many diverse biological functions that vary according to cell type and tissue environment. So far, the data are limited to the functions chosen for study by each investigating group. Other functions outside the research scope of each laboratory are often ignored, resulting in very large information gaps. Most of the investigations began the study of CCN5 with a very specific potential clinical application in mind, or the hope for a simple molecular solution for a long-standing and difficult clinical disease challenge, such as vascular disease and cancer. Now that the complexity of CCN5 function and regulation is apparent, it is now necessary to broaden the research focus away from focused clinical applications to basic and systematic biochemical studies. Once these studies are completed, it will be much easier to ascertain the role of CCN5 in the pathogenesis and treatment of particular diseases, and to select potential therapeutic modalities for CCN5.

REFERENCES

Banerjee S, Saxena N, Sengupta K, Tawfik O, Mayo MS and Banerjee SK (2003) WISP-2 gene in human breast cancer: estrogen and progesterone inducible expression and regulation of tumor cell proliferation. *Neoplasia* **5**:63–73.
Brigstock DR (2003) The CCN family: A new stimulus package. *J Endocr* **178**:169–175.

Castellot JJ Jr, Addonizio ML, Rosenberg R and Karnovsky MJ (1981) Cultured endothelial cells produce a heparin-like inhibitor of smooth muscle cell growth. *J Cell Biol* **90**:372–379.

Delmolino LM, Stearns NA and Castellot JJ Jr (1997) Heparin induces a member of the CCN family which has characteristics of a growth arrest gene. *Mol Biol Cell* **8**:287a.

Delmolino LM, Stearns NA and Castellot JJ Jr (2001) COP-1, a member of the CCN family, is a heparin-induced growth arrest specific gene in vascular smooth muscle cells. *J Cell Physiol* **188**:45–55.

Inadera H, Hashimoto S, Dong HY, Suzuki T, Nagai S, Yamashita T, Toyoda N and Matsushima K (2000) WISP-2 as a novel estrogen responsive gene in human breast cancer cells. *Biochem Biophys Res Commun* **275**:108–114.

Inadera H, Dong H-Y and Matsushima K (2002) WISP-2 is a secreted protein and can be a marker of estrogen exposure in MCF-7 cells. *Biochem Biophys Res Commun* **294**:602–608.

Inadera H (2003) Estrogen-induced genes, WISP-2 and pS2, respond divergently to protein kinase pathway. *Biochem Biophys Res Commun* **309**:272–278.

Kumar S, Hand AT, Connor JR, Dodds RA, Ryan PJ, Trill JJ, Fisher SM, Nuttall ME, Lipshutz DB, Zou C, Hwang SM, Votta BJ, James IE, Rieman DJ, Gowen M and Lee JC (1999) Identification and cloning of a connective tissue growth factor-like cDNA from human osteoblasts encoding a novel regulator of osteoblast functions. *J Biol Chem* **274**:17123–17131.

Lake AC, Bialik A, Walsh K and Castellot JJ Jr (2003) CCN5 is a growth arrest-specific gene that regulates smooth muscle cell proliferation and motility. *Am J Pathol* **162**:219–231.

Lake AC and Castellot JJ Jr (2003) CCN5 modulates the antiproliferative effect of heparin and regulates cell motility in vascular smooth muscle cells.*Cell Commun Signal* **1**:5–17.

Mason HR, Nowak RA, Morton CC and Castellot JJ Jr (2003) Heparin inhibits the motility and proliferation of human myometrial and leiomyoma smooth muscle cells. *Am J Pathol* **162**:1895–1904.

Mason HR, Lake AC, Wubben JE, Nowak RA and Castellot JJ Jr (2004a) The growth arrest-specific gene CCN5 is deficient in human leiomyomas and inhibits the proliferation and motility of cultured human uterine smooth muscle cells. *Mol Hum Reprod* **10**:181–187.

Mason HR, Grove–Strawser D, Rubin BS, Nowak RA and Castellot JJ Jr (2004b) Estrogen induces CCN5 expression in the rat uterus *in vivo*. *Endocrinology*, published online Nov. 6, 2003.

Pennica D, Swanson TA, Welsh JW, Roy MA, Lawrence DA, Lee J, Brush J, Taneyhill LA, Deuel B, Lew M, Watanabe C, Cohen RL, Melhem MF, Finley GG, Quirke P, Goddard AD, Hillan KJ, Gurney AL, Botstein D and Levine AJ (1998) WISP genes are members of the connective tissue growth factor family that are up-regulated in wnt-1-transformed cells and aberrantly expressed in human colon tumors. *Proc Natl Acad Sci USA* **95**:14717–14722.

Zhang R, Averboukh L, Zhu W, Zhang H, Jo H, Dempsey PJ, Coffey RJ, Pardee AB and Liang P (1998) Identification of rCop-1, a new member of the CCN protein family, as a negative regulator for cell transformation. *Mol Cell Biol* **18**:6131–6141.

Zoubine MN, Banerjee S, Saxena NK, Campbell DR and Banerjee SK (2001) WISP-2: a serum-inducible gene differentially expressed in human normal breast epithelial cells and in MCF-7 breast tumor cells. *Biochem Biophys Res Commun* **282**:421–425.

CHAPTER 13

CCN3: A MULTIFUNCTIONAL SIGNALING REGULATOR

Nathalie Planque, Anne-Marie Bleau and Bernard Perbal*

Laboratoire d'Oncologie Virale et Moléculaire
Tour 54, Case 7048, Université Paris 7-D.Diderot,
2 Place Jussieu, 75005 Paris
perbal@ccr.jussieu.fr

The CCN3 protein is one of the three founding members of the CCN family (see Chapter 1). As a prototypic CCN protein, it is constituted by the assembly of the IGFBP, VWC, TSP1, and CT modules. The N-terminal signal peptide which is present in the neosynthesized protein allows efficient secretion of CCN3. Studies aimed at identifying the biological roles of CCN3 have been hampered by the low levels of CCN3 expressed in normal conditions, and by the lack of a simple biological system in which to study the properties of the normal CCN3 protein. Despite these limitations, the bulk of information that has been obtained over the past decade has permitted the identification of CCN3 as a key regulator of cell growth and differentiation. Recent results have indicated that CCN3 is a genuine signaling factor and that alterations in CCN3 expression constitute a marker for tumor typing. Furthermore, measurement of CCN3 amounts and inhibition or forced expression of CCN3 constitute potential therapeutic tools.

1. INTRODUCTION

The ccn3 gene (previously designated nov) was discovered as an integration site of the Myeloblastosis Associated Virus (MAV) in tumor DNA isolated from one MAV-induced chicken nephroblastoma (Joliot *et al.*, 1992). MAV is a tumorigenic retrovirus competent for replication, that does not contain oncogenic sequences from cellular origin (Perbal, 1995). MAV-induced avian nephroblastomas represent a unique model of the Wilms' tumor (Perbal, 1994), a pediatric tumor of the kidney affecting approximatively one in 10,000 children

(Beckwith, 1990). The integration of MAV within the ccn3 gene resulted in the elevated expression of a chimeric mRNA encoding an amino-truncated CCN3 protein. This gene was originally designated "nov" (Nephroblastoma OVerexpressed) because it was highly expressed in several other MAV-induced nephroblastomas, whereas, in normal post-natal kidney, it was expressed at low levels (Joliot *et al.*, 1992). Since no viral sequences were detected nearby ccn3 in these tumors, it was proposed that high expression was driven by the enhancer sequences contained in MAV Long Terminal Repeats that would be at a longer distance from ccn3. Fluorescent *in situ* hybridization (FISH) and use of bacterial artificial chromosomes (BACs) permitted us to establish that the CCN3 gene is not a preferential integration site of MAV in tumor DNA (Coullin *et al.*, 2002; Li *et al.*, submitted for publication). Furthermore, the expression of the human version of ccn3 was not increased in Wilms' tumors. *In situ* hybridization and immunocytochemistry studies established that the target cells for MAV are blastemal cells committed to epithelial differentiation, and that these cells express high levels of CCN3 (Cherel *et al.*, unpublished). Increased expression of ccn3 in nephroblastoma might therefore result from the clonal expansion of these target cells. Since proviral genomes are known to integrate in transcriptionally active genes, integration of MAV in ccn3 might have occurred in target cells in which it is expressed actively.

These observations suggested for the first time that the expression of ccn3 was high in cells undergoing differentiation and paved the road for the study of ccn3 expression during development.

2. CCN3 IN NORMAL CONDITIONS

2.1. Expression Patterns of CCN3 During Development

Early studies performed in chicken established that the expression of ccn3 was subjected to a very tight spatio-temporal regulatory control both at the embryonic and the adult stages (Joliot *et al.*, 1992). Very few studies have investigated the regulation of ccn3 expression in normal cells. The expression of ccn3 was reported to be downregulated indirectly by the Wilms' tumor protein WT1 (Martinerie *et al.*, 1996) and sequencing of the human and chicken promoters identified potential binding sites for SP1, AP1 and AP2 transcription factors (Perbal, 2001). Recently, the transcriptional activity of the ccn3 promoter was reported to be downregulated by parathyroid hormone related peptide (PTHrP), which is known to inhibit chondrocyte differentiation (see below).

Cloning of the human and mouse orthologues of ccn3 and preparation of specific antibodies raised against ccn3 proteins from three different species helped to establish that CCN3 was conserved throughout evolution and is expressed at different levels in tissues originating from the three germ layers. Major sites of expression include the nervous system, the musculo-skeletal system, the adrenal, the cartilage and the urogenital system. Gut mucosa, bronchial epithelium and pancreatic ducts are also positive for CCN3 expression but at a lower level (Perbal, 2001).

In general, a good match is observed between sites of RNA and protein detection. However, there are exceptions that suggest remote action of CCN3 at a distance from its site of biosynthesis (Su *et al.*, 2001; Perbal, 2001). In most cases, expression of CCN3 correlated with cellular differentiation and tissue remodeling.

2.1.1. *Nervous system*

In the chicken embryo, the neuroepithelium (early stage of development: embryonic Day 3) and the neural tube (latter stages of development: E3 to E7) show a strong staining for CCN3 (Katsube *et al.*, 2001; Ayer–Lelièvre *et al.*, unpublished). Likewise, the central nervous system (CNS) is a major site of CCN3 expression in the first trimester human fetus (Kocialkowski *et al.*, 2001) and at later developmental stages (Kocialkowski *et al.*, 2001; Su *et al.*, 2001). Both spinal cord and brain strongly express CCN3. The floorplate and the ventral horns of the spinal cord, the spinal nerves, and the dorsal root ganglia (DRG) exhibit an abundant expression of CCN3. In the ventral horn of the spinal cord, anatomical features suggest that CCN3 might be expressed in somatic motoneurons (Su *et al.*, 2001). The expression pattern of ccn3 in the nervous system strongly suggested that the CCN3 protein was playing a role in terminal differentiation of the nervous system and might be involved in the acquisition of cognitive functions (Su *et al.*, 2001; unpublished observations).

2.1.2. *Muscle*

Expression of CCN3 is detected both in embryonic and adult skeletal muscles. In the chicken embryo, from E4 onwards, developing myotome and skeletal muscle stain positive for CCN3. In first trimester human embryo, fusing myoblasts and myotubes of skeletal muscle are major sites of expression. CCN3 is also detected in cardiomyocytes and smooth muscle (Perbal, 2001; Kocialkowski *et al.*, 2001). Smooth muscle also expresses CCN3, but less than striated muscle (Kocialkowski *et al.*, 2001). In adult rat, the smooth muscle cells of aorta, and to a lesser extent, lung cells also stain positive for CCN3 (Ellis *et al.*, 2000) Northern blot analysis revealed that the amount of CCN3 mRNA is about 10 fold higher in aortas of adult rats than in aortas of the 3- and 14- day-old neonatal rats. Futhermore, CCN3 expression correlates with the expression of vascular smooth muscle cell (VSMC) differentiation markers, such as smooth muscle myosin heavy chain and smooth muscle actin (SMa22), during postnatal development. Altogether, these results suggested that CCN3 contributes to both skeletal and smooth muscle differentiation.

2.1.3. *Kidney*

Several observations suggested that CCN3 might be involved in kidney development. The chicken embryonic kidney expresses high levels of CCN3, whereas low amounts are detected at post-hatching stages (Perbal, 1994). At E3 and 4, CCN3 expression is detected in the mesonephric mesenchyme, epithelial vesicles and glomeruli. At later stages (E6 to E14), CCN3 expression is detected in differentiated tubules, metanephric epithelial vesicles, glomeruli and ureteric buds. The metanephric mesenchyme shows an increased staining for CCN3 as differentiation proceeds in the S-shaped bodies and a dramatic decrease is observed once differentiation has been achieved (Ayer–Lelièvre *et al.*, unpublished). In 8 to 10 week-old human embryos, the CCN3 protein is highly expressed in both metanephric and mesonephric glomeruli (Chevalier *et al.*, 1998; Kocialkowski *et al.*, 2001). Again, these observations suggested that CCN3 expression correlated with tissue differentiation.

2.1.4. *Cartilage*

Involvement of CCN3 in chondrocytic differentiation was also established both in chicken and human. In the developing chicken limb, at E6, CCN3 is detected in proliferating chondrocytes and in the perichondrium, whereas at E11/12, it is essentially found in the hypertrophic chondrocytes of the central diaphyseal region. In the developing wing at E14, the staining of CCN3 is greatly reduced and restricted to few hypertrophic chondrocytes. Northern blot performed on chicken limb buds established that CCN3 expression increases between E8 and E10, and then progressively decreases until E14, to reach a basal level at E15 (http://www.ccr.jussieu.fr/lovm/developpement.html; Perbal, 2001). The time course of limb development is completed after 10 days of incubation under normal conditions. The involvement of CCN3 in cartilage formation was also studied in chicken micromass cell cultures. In this system, mesenchymal cells of E3.5-chicken embryo seeded at high density undergo chondrocytic differentiation and cartilaginous nodules appear within 3 days of culture. Under these conditions, ccn3 levels increase progressively for two days, and decrease at Day 3, whereas the expression of collagen II — a marker of differentiation of progenitors that undergo chondroblastic differentiation — increases from Day 1 onwards (Perbal, 2001). Mesenchymal cells do not express ccn3 at the time they are seeded and nodules do not form when the cells are seeded at low density. Interestingly, CCN3 expression was not detected in the absence of cartilaginous nodule formation. Altogether, these observations and the results of immunohistochemical studies suggested that CCN3 was required over an extended period of time during limb development and is needed at late stages of chondrogenesis. Indeed, our unpublished observations established that CCN3 is required after CCN1 and CCN2 during the chondrocytic differentiation process (Ayer–Lelièvre *et al.*, unpublished; Perbal, 2001). In the early-stage human embryo, chondrocytes express low levels of CCN3. At later stages, CCN3 protein is detected in the perichondrium and in proliferating chondrocytes of the skull, ribs and lower limbs (Kocialkowski *et al.*, 2001; Perbal, 2001).

Immunohistochemistry and semi-quantitative RT-PCR analysis performed on distal femoral growth plates from E16.5 mice established that CCN3 was expressed in the pre-hypertrophic and early hypertrophic chondrocytes in the growth plate (Yu *et al.*, 2003). Hypertrophic, non-proliferating chondrocytes that have undergone differentiation were negative for ccn3 expression. In explant cultures, ccn3 expression was significantly reduced following treatment with PTHrP, which inhibits chondrocyte terminal differentiation, as shown by decreased type X collagen expression. Use of this system also helped to establish that the transcriptional activity of the ccn3 promoter was reduced by 30% in the presence of PTHrP. Since ccn3 positive pre-hypertrophic growth plate chondrocytes and early hypertrophic chondrocytes undergo the transition from proliferating to terminally differentiated cells, the restricted profile of ccn3 expression is in agreement with a role for CCN3 in chondrocyte differentiation in the growth plate.

2.2. Subcellular Localization of CCN3

The presence of a signal peptide at the N-terminus of CCN3, and other members of the CCN family, was indicative of export and secretion outside of the cells that produce these proteins. Indeed, CCN3 can be detected in conditioned cell culture medium. Use of specific antibodies permitted the identification of two CCN3-related proteins released into the medium. The apparent molecular weights of these proteins were found to slightly vary with the source of the producing cells, suggesting that specific post-translational modifications might depend upon the cell type. A large species that migrates as a 54–48 kDa protein, corresponds to the full length secreted CCN3 protein, whereas a shorter one whose apparent molecular weight of 32 kDa corresponds to an amino truncated CCN3 protein lacking the IGFBP and VWC modules. The existence of amino truncated isoforms is not unique to CCN3 (Perbal, 2004). A truncated CCN2 protein with an N-terminus identical to that of the 32 kDa CCN3 form, was also described in biological fluids and cell culture supernatants. A growing body of evidence suggests that these truncated forms are generated by proteolysis of the full length CCN proteins and are playing important biological functions that can synergize or antagonize those of the prototypic proteins.

In addition to the secreted forms that are released in the medium, various amounts of CCN3 proteins are also detected in the extracellular matrix, in the cytoplasm and in the nucleus of cells in culture (Perbal, 1999; Kyurkchiev *et al.*, 2004).

2.3. CCN3 Partners

The variety of biological processes in which CCN3 was implicated, and the detection of CCN3 proteins in different cellular compartments suggested that this protein might be involved in several regulatory pathways. The search for potential targets of CCN3 was first conducted by means of the two-hybrid system (Perbal, 1999; Perbal *et al.*, 1999).

The great number and variety of proteins that were found to interact with CCN3 in the two hybrid system first raised suspicion, as false positive are often observed with this powerful technique. However, the use of GST pull-down assays, co-immunoprecipitations and confocal microscopy soon confirmed that CCN3 was physically interacting with many different types of proteins. The localization and nature of these proteins was in agreement with the existence of various subcellular isoforms of CCN3 and suggested an implication of CCN3 in several key signaling pathways, and adhesion regulation (Fig. 1). The different partners of CCN3 included ECM proteins such as fibulin 1C and integrins, the membrane receptor Notch1, intracellular proteins such as S100A4 and the intracellular tail of connexin 43, and the nuclear rpb7-subunit of RNA polymerase II. Our results also established that truncated isoforms of CCN3 could bind specific targets and pointed to the CT domain of CCN3 as a critical determinant for protein interaction.

2.4. Functions of CCN3

2.4.1. *Control of cell adhesion*

The first clue for CCN3 being involved in the regulation of cell adhesion was drawn from the physical interaction of CCN3 with the matricellular protein fibulin 1C (Perbal *et al.*, 1999).

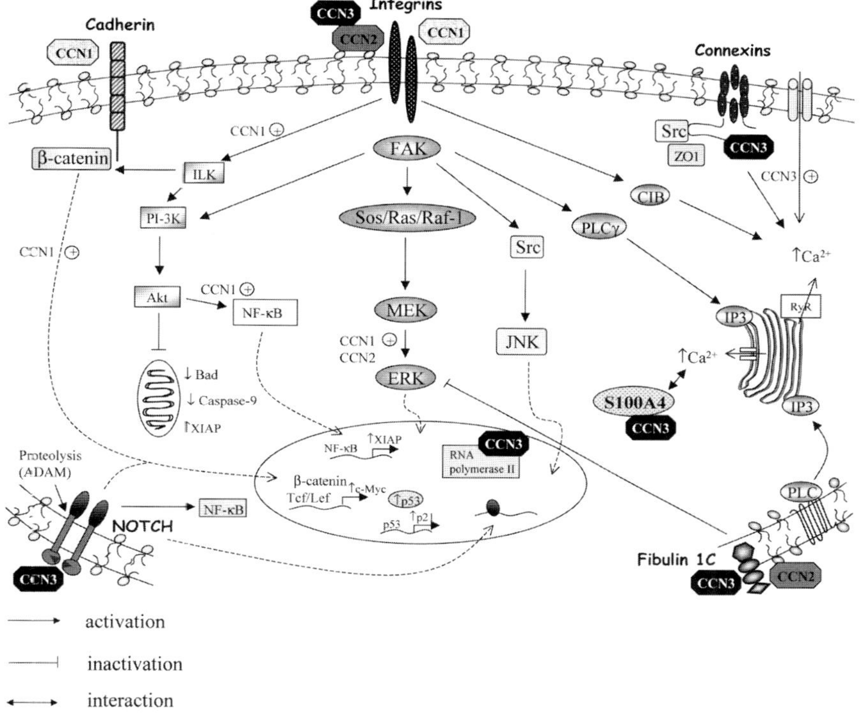

Fig. 1.   CCN3 and cell signaling.

The figure depicts the various signaling pathways in which CCN3 is thought to be involved and tentative relationships with CCN1 and CCN2 signaling pathways.

CCN3 is known to physically interact with proteins localized in the extracellular matrix ECM (integrins and fibulin 1C), at the plasma membrane (Notch receptor and connexins), in the cytoplasm (S100A4), and in the nucleus (rpb7 of RNA polymerase II). The regulation of intracellular levels of calcium ion is the first biological function attributed to CCN3.

The combined actions of CCN1, 2 and 3 (and probably of other CCN proteins) appear to be critical in the control of proliferation and apoptosis through their interaction with integrins and cadherins. The control of intracellular calcium ion concentration by CCN3 might reflect another key aspect of their biological properties.

Interaction with integrins constitutes a major feature in the biological action of the full length CCN3 and other CCN proteins on cell biology. Binding of CCN proteins to integrins was reported to result in increased angiogenesis, cell adhesion and migration. The interaction of CCN proteins with cadherin, integrins and Notch might permit to modulate and coordinate signaling pathways involved in cell survival and proliferation. Interaction of CCN1 with integrins stimulates cell survival through Akt-mediated inhibition of the Bad apoptotic factor and Caspase-9. In parrallel, CCN1 can induce, either through its binding to cadherin or through the integrins receptor, the translocation of β-catenin to the nucleus and subsequent accumulation of p21$^{\text{WAF1}}$. The activation of NK-κB, which is mediated by integrins, also leads to an increased synthesis of the

(*Continued*)

Fibulin 1C is a calcium binding ECM glycoprotein expressed in a great variety of cells. Interestingly, fibulin 1C expression pattern shows similarities with that of CCN3 during cartilage and bone development. Fibulin 1C is expressed in precartilage condensations of the phalanges in the developing mouse limb. In human embryos of gestational week (gw) 4, fibulin 1C staining was observed in the early mesenchymal bone anlagen. In embryos of gw 6.5 and 8, all perichondrial structures were positive for fibulin 1C expression but the chondrocytes themselves were negative. In embryos of gw 10, fibulin 1C expression is prominent in the matrix that surrounds the hypertrophic chondrocytes.

Fibulin 1C interacts with nidogen, fibronectin, laminin, aggrecan, versican, tropoelastin, and amyloid precursor protein. It associates with ECM structures such as elastic fibres, microfibrils, and basement membranes and it is thought to play a role in the regulation of cell migration as well as in hemostasis and thrombosis. It has been shown to promote attachment of platelets via a bridge of fibrinogen, and to inhibit the fibronectin-dependent haptotactic motility of breast carcinoma (MDA MB231) cells, epidermal carcinoma (A431), melanoma (A375 SM), rat pulmonary aortic smooth muscle cells

---

Fig. 1.   (*Continued*)

anti-apoptotic factor XIAP. In this context, it is worth noting that the NOTCH receptor, which is a partner of CCN3, was shown to activate NK-κB, increase p21$^{\text{WAF1}}$, and inhibit the β-catenin pathway. Therefore, it will be of interest to determine whether the interaction of Notch with CCN3 modulates its ability to induce NK-κB.

The activation of the FAK (focal adhesion kinase) pathway by most of the integrins induces modifications of the cell cytoskeleton and activation of the mitogen-activated protein kinase cascade (MAPK) and Ras extracellular signal-regulated kinase (ERK). A functional interaction between the CCN proteins along this signaling pathway is suggested by the observations that CCN1 and CCN2 can activate ERK and CCN3 can increase FAK phosphorylation. In the same line, Fibulin 1C, which is known to physically interact with CCN3, can negatively regulate ERK.

Activation of the MAPK c-Jun NH2-terminal kinase (JNK) pathways by integrins also appears to be critical for CCN proteins action in the control cell cycle progression.

Evidence also accumulates for a direct involvement of CCN3 isoforms in the regulation of transcription. CCN3 might interact with the rpb7 subunit of RNA polymerase II and proteins involved in chromatin remodeling, or bind to specific promoter sequences.

Considering the essential role of calcium ion in various levels of cell signaling, the modulation of the intracellular calcium concentration by CCN2 and CCN3 proteins might be of prime importance. The interaction of CCN3 with the cytoplasmic C-terminal tail of connexin Cx43, might account for the effects of CCN3 on intracellular calcium. It is worth noting that integrins binding can also induce an increase of the intracellular calcium concentration through activation of phospholipase C (PLC) and calcium and integrin binding protein (CIB). Whether other CCN proteins are also potent regulators of intracellular calcium levels remains to be established. The dual interaction of CCN1 with integrins and cadherins, and of CCN3 with connexins and integrins is supporting our proposal that the CCN proteins act as scaffolding factors which permit the coordination of various signaling pathways.

(PAC1) and Chinese hamster ovary (CHO) cells. The interaction of CCN3 and fibulin 1C was shown to involve the C-terminal module of CCN3. Whether this interaction modulates the regulation of cell adhesion by fibulin 1C is under current investigation.

Control of cell cell-adhesion and migration by CCN3 might also occur through a direct binding to various integrin receptors and to heparan sulfate proteoglycans (HSPG). In human umbilical vein cells (HUVECs), adhesion promoted by CCN3 involves integrins $\alpha_V\beta_3$, $\alpha_5\beta_1$, $\alpha_6\beta_1$ and HSPG. CCN3 stimulates migration through integrins $\alpha_V\beta_3$ and $\alpha_5\beta_1$ pathways, but not through $\alpha_6\beta_1$ pathway. CCN3 physically interacts with integrin receptors *in vitro*, suggesting that CCN3 functions in cell-adhesion and migration might also involve direct binding of the secreted CCN3 protein to integrin receptors. Recently, CCN3 was reported to show pro-angiogenic activity in a rat corneal assay (Lin *et al.*, 2003). The involvement of integrins in angiogenesis is well established. The relationships existing between CCN proteins, integrins and control of cell adhesion and migration are extensively reviewed in Chapter 3.

2.4.2. *Modulation of developmental pathways*

The interaction of CCN3 with the extracellular domain of the transmembrane Notch1 protein provides another level of signaling control, since CCN3 activates downstream effectors of the Notch signaling pathway (Sakamoto *et al.*, 2002). The role of Notch1 during embryonic development is well documented. In chicken the expression of CCN3- and Notch1 overlap in the pre-somitic mesoderm during early development (Katsube *et al.*, 2001) and recent data suggest that the binding of CCN3 to Notch1 may be of particular importance during normal muscle differentiation. The activation of Notch signaling pathway by CCN3 may provide new clues for understanding the pro- and anti-tumorigenic effects of Notch. These different aspects are discussed in Chapter 9.

The repair processes that are initiated after injury involve similar events than those happening during the embryonic development. In this context, the expression of CCN3 is dynamically regulated in response to blood vessel lesions (Ellis *et al.*, 2000). In a rat carotid injury model (balloon catheter injury), *in situ* hybridization analysis performed 7 days and 14 days after injury showed that CCN3 mRNA is first down-regulated in the media of the injured artery. Expression was barely detectable in the developing intima 7 days after injury, with the exception of the luminal surface which exhibited a strong expression.

At Day 14 after injury, CCN3 expression was increased throughout the intima. CCN3 was shown to promote VSMC adhesion, but did not seem to act on their proliferation and their differentiation.

2.4.3. *Calcium signaling*

Inasmuch as fibulin 1C, integrins, and Notch functions were highly dependent upon calcium ion distribution, the interaction of CCN3 with connexin43 and S100A4 shed a new light on the possible interconnections of these different signaling pathways and suggested that CCN3 might participate directly in calcium-related signaling.

Connexins are components of the gap junctions which constitute channels that allow the diffusion of small molecules such as ions, carbohydrates, amino acids, nucleotides, etc. The interaction of CCN3 with connexin 43 provided a biological support for the association of connexins with cancer development (see Chapter 14). This interaction also provided another relationship with calcium signaling pathways. Since the shape and functioning of the connexin channels have been reported to be dependent upon calcium ion concentrations, it is tempting to propose that calcium is a common theme in CCN3-mediated interactions. In support of this model, CCN3 was shown to physically interact with S100A4 and have the capacity to directly modulate the intracellular calcium concentration of responsive cells.

The S100A4 protein is thought to modulate the propagation of calcium signals. CCN3 and S100A4 expression pattern show a partial overlap. The human S100A4 gene is often rearranged in cancers, either by deletions, translocations or amplifications. High amounts of S100A4 are detected in metastasis, but S100A4 itself is unable to induce tumorigenesis. To date, S100A4 is known to be involved in cell motility, cell adhesion, cell cycle progression, and angiogenesis. It is tempting to postulate that the biological properties of S100A4 and CCN3 are closely related and are dependent upon the interaction of these two proteins.

Clues for a direct interaction of CCN3 with calcium channels were first obtained in studies performed with neuronal cells (SK-N-SH) that can undergo differentiation in culture (Li *et al.*, 2003). In these cells, CCN3 was shown to transiently increase the intracellular calcium concentration through a voltage-independent type of channel and mobilization of internal stores. The use of two tumor cells lines (G59 glioma cells and adrenocortical NCI295R cells) helped to establish that G59 responsive cells did not express this channel, and that NCI295 cells did not respond to exogenous CCN3. The identification of the mechanisms underlying these phenomena is underway (see review by Lombet *et al.*, 2003). At any rate, the effects of CCN3 on intracellular concentration levels permitted to attribute, for the first time, a biological function to this protein. They also open several avenues for exploring the relationships that might exist between calcium signaling and the subcellular localization of CCN3 in the context of normal and tumor cells in which calcium signaling is known to play a critical role.

2.4.4. *CCN3: a scaffolding regulator*

With the accumulation of evidence indicating that CCN3 and other members of the CCN family can interact with many different bioregulators involved in various signaling pathways, we have proposed that CCN3 acts as a scaffolding protein to permit an efficient coordination of regulatory circuits governing cell life and death.

It is well known that multifunctional complexes, assembled via interactions of several proteins, govern critical fundamental biological activities in the cells. The nature and number of partners constituting these complexes dictates the resulting activities and their regulation. Combinatorial events thus appear to be essential for the adaptation of the cells to changing environmental signals. Along this line, CCN3 is proposed to permit a functional cross talk between regulators that belong to different, complementary or antagonist, pathways through its multiple interaction capacities. The bioavailability of each component participating in the assembly of the functional complexes, including CCN3, would modulate the resulting signaling and cellular responses.

3. CCN3 IN PATHOLOGICAL CONDITIONS

For historical reasons, most of the studies performed to establish the potential involvement of CCN3 in pathology were conducted in tumor samples and cell lines. Although evidence suggesting that CCN3 alterations are also associated with non proliferative pathologies, we will focus in this chapter on the role and functions of CCN3 in cancer development.

Over the past decade, the expression levels of CCN3 have been shown to be impaired in a panel of human tumors representing the most common types of abnormal cellular proliferation. The current view is that increased CCN3 expression can be associated with either good or poor prognostic tumor features, reflecting the ability of this protein to potentially interact with several different partners.

3.1. Association of CCN3 Expression with Poor Prognosis

Increased expression of CCN3 was detected both at the RNA and protein levels in tumor samples and cell lines derived from osteosarcomas, renal cells carcinomas, and prostate carcinomas (Perbal, 2001; Planque and Perbal, 2003). In the case of osteosarcomas, an inverse relationship has been drawn between the level of CCN3 expression and alkaline phosphatase, an early marker of osteoblastic differentiation (Manara *et al.*, 2002). Inasmuch as alkaline phosphatase expression is correlated to the loss of aggressiveness of osteosarcoma cells, the expression of CCN3 in these tumors is considered as a marker of poor prognosis. Along the same line, the highest levels of CCN3 were secreted by renal cell carcinoma cells isolated from fast growing tumors. An inverse relationship could be drawn between the amount of CCN3 secreted by these cells and the capacity of these cells to establish and develop tumors following injection in nude mice (Glukhova *et al.*, 2001).

CCN3 was also highly expressed in cell lines derived from prostate metastases to bone (PC3), brain (DU145) and lymph node (LNCap) (Maillard *et al.*, 2001).

A study of Ewing's sarcomas provided the first clue for the use of CCN3 as a long term prognostic factor. Indeed, the expression of CCN3 in the primary tumors was found to correlate with a higher risk to develop metastasis (Manara *et al.*, 2002). Based on this observation, a thorough investigation of CCN3 expression in these tumors has been initiated.

3.2. Association of CCN3 Expression with Good Prognosis

In contrast with the above examples, elevated expression of CCN3 was associated with a favourable outcome in several other types of tumors.

The first evidence establishing a correlation between CCN3 expression and good prognosis came from the analysis of brain tumors (Li *et al.*, 1996). In this study, it was found that glioma and astrocytoma cells freshly explanted from a panel of tumors representing different grades expressed quite different levels of CCN3. From the results obtained, an inverse relationship was drawn between tumorigenicity and CCN3 expression. Neuroblastomas with good prognosis also express high levels of CCN3, whereas in tumors with bad prognosis, CCN3 staining was low to moderate (Perbal, 2001). Along the same line the highest levels of CCN3 expression were detected in enchondromas and low-grade chondrosarcomas (Yu *et al.*, 2003). In the case of Wilms' tumors and rhabdomyosarsomas, an elevated expression of CCN3 was associated with striated muscular differentiation (Chevalier *et al.*, 1998; Manara *et al.*, 2002), a situation reminiscent to the role of CCN3 in normal tissue differentiation.

CCN3 expression was also down-regulated upon induction of the BCR-ABL protein in a murine model for chronic myeloid leukemia (CML) and in blood cells from patients at different stages of their illness. Interestingly, levels returned to normal in patients undergoing remission (Gilmour *et al.*, submitted).

3.3. Antiproliferative Activity of CCN3: A Common Trait

To better understand the basis for association of CCN3 expression with both poor and favourable prognosis, the potential functions of CCN3 in the regulation of cell proliferation were investigated.

For this purpose, a series of plasmids which allowed constitutive and inducible expression of the human CCN3 protein were used to transfect cell lines derived from the two groups of tumors: G59 glioblastoma cells, and TC71 Ewing's tumor cells. In both cases, the expression of CCN3 resulted in a marked decrease in cell proliferation, which was correlated with the level of CCN3 protein expressed by the various transfectants that were used (Gupta *et al.*, 2001; Benini *et al.*, 2005). The isolation of TC71 transfectants in which the expression of CCN3 was driven by an inducible promoter helped to establish that the reduction of cell proliferation was indeed provoked by the production of secreted CCN3 protein.

In both cases, the expression of CCN3 was also found to reduce the tumorigenicity of the tumor cells. Interestingly, the secretion of CCN3 did not inhibit the implantation of the tumor cells, but reduced the expansion of the tumor. In the case of glioma cells, the tumors did not show a high level of vascularization, as compared to those induced by parental G59 cells. This observation, which is in contrast with the reported pro-angiogenic activity of CCN3 (see above), suggests that the CCN3 protein is interfering with tumor maintenance and expansion, rather than tumor establishment.

Recent results obtained with glioblastoma and choriocarcinoma cells (see Chapter 14) point to the importance of the interaction of CCN3 with connexin as a critical factor for inhibition of cell growth. Briefly, forced expression of CCN3 in JEG3 choriocarcinoma cells resulted in a reduction of cell proliferation and tumorigenicity (Gelhaus *et al.*, 2004), as previously established with Ewing's and glioma cells. New data established a link between the inhibition of cell proliferation and the interaction of CCN3 with connexin 43, both in

JEG3 cells and rat C6 glioma cells (Fu *et al.*, 2004), where CCN3 appears to be a key element in the connexin-mediated growth suppression.

Altogether, the bulk of data obtained thus far establishes CCN3 as a genuine antiproliferative protein, and reinforces the idea that the content of CCN3 in biological samples and the addressing of recombinant CCN3 to tumor cells might constitute new tools for tumor typing and therapy. These two studies also confirmed the existence of nuclear CCN3 isoforms in cancer cells, as previously established in the case of HeLa and 143 osteosarcoma cells (Perbal, 1999).

3.4. CCN3 Isoforms and Cancer

The first evidence suggesting that the production of truncated CCN3 isoforms might be associated with cancer development was obtained with the MAV-induced nephroblastomas (see above).

In one tumor, the integration of the MAV proviral genome into the second intron of CCN3 resulted in the elevated synthesis of a protein deprived of signal peptide. The truncated form (deleted of the first 63 aa) of the CCN3 protein was shown to morphologically transform primary chicken embryonic fibroblasts (CEFs) in culture (Joliot *et al.*, 1992), whereas the full length secreted protein showed an antiproliferative activity in a variety of chicken and human cell lines (Joliot *et al.*, 1992; Gupta *et al.*, 2001; Gellhaus *et al.*, 2004; Benini *et al.*, 2005; our unpublished observations). Since the expression of the truncated protein was expected to be altered, we hypothesized that modifications in CCN3 subcellular expression could uncover potential oncogenic activities (Perbal, 2004).

In support of this hypothesis, an aminotruncated form of CCN3 has been detected in the nucleus of HeLa and osteosarcoma cells (Perbal, 1999). The JEG3 human choriocarcinoma cells and the C6 rat glioma cell also show a nuclear staining for CCN3. Furthermore, stable expression of Cx43 in the JEG3 cells resulted in a modification of CCN3 labelling which was then detected in the cytoplasm and at the plasma membrane where it co-localized with Cx43 (Gellhaus *et al.*, 2004; Fu *et al.*, 2004). These observations reinforced the idea that an aberrant subcellular localization of CCN3 isoforms may be involved in the tumorigenic process.

Interestingly, an interaction between CCN3 and the rpb7 subunit of the RNA polymerase II was observed with the yeast two-hybrid system and the nuclear CCN3 labelling was found to co-localize with the transcriptional machinery but not with the replication machinery (Perbal, 2001). The C-terminal domain of CCN3 has also been reported to bind specific sequences in the promoter of the plasminogen activator inhibitor type-2 (PAI-2) (Mahony *et al.*, 1999). Taken together, the results suggested that nuclear forms of CCN3 may directly regulate gene transcription. Ongoing studies in our laboratory have indeed established that CCN3 truncated forms can transactivate transcription (Planque *et al.*, submitted).

The bulk of these results therefore established that CCN3 may exert different effects on cell growth, depending on its subcellular localization and modular composition; the full length secreted CCN3 showing antiproliferative activity, and the nuclear CCN3 isoforms acting as cell growth stimulators.

4. CONCLUSIONS

The variety of biological functions in which CCN3 is known to participate has long been a puzzle until the multiplicity of regulatory molecules that physically interact with CCN3 was uncovered. In agreement with the high levels of CCN3 expression detected in cells undergoing differentiation, the full length CCN3 protein shows an antiproliferative activity in normal fibroblastic cells and in a variety of tumor cells. On the other hand, recent evidence also established that an elevated expression of CCN3 is associated with the acquisition of more advanced tumor phenotype and metastatic potential. Although these two sets of observations may appear contradictory at first glance, they likely reflect the different levels of activity in which CCN3 is involved through its multiple interactions with regulators acting at various stages of cell growth control.

Even though expression of CCN3 in Ewing cells triggers a reduction of cell adhesion and increased motility, there are no examples, as yet, of CCN3 increasing the proliferation rate of cells. These observations therefore indicate a disconnection between proliferative activity and metastatic potential. Furthermore, these results highlight the importance of cellular context, and reinforce the idea that the ultimate functions of multimolecular complexes in which CCN3 participates may be highly dependent upon the nature and number of interacting partners. The physical coupling of multiple enzyme activities with their regulatory subunits is well documented in the case of metabolic pathways in eukaryotes. Recent evidence also suggests that tri-molecular complexes involving CCN3 may be required for osteogenic differentiation (unpublished observations).

One could argue that the effects on cell growth that are induced by ectopic expression of recombinant CCN3 in glioma cells, that do not express this protein, result in the activation of "abnormal" or "irrelevant" processes. However, this is unlikely because similar effects are observed in choriocarcinoma and Ewing cells in which forced expression of CCN3 results in an increase in

 Planque N, Bleau A-M & Perbal B

secreted CCN3 protein that is already produced at low, but easily detectable, levels in these cells.

Another fascinating aspect of CCN3 biology has recently emerged from our discovery of biologically active isoforms whose subcellular compartmentalization appears to be directly related to the anarchic proliferation of cancer cells (Perbal, 1999; Planque and Perbal, 2003).

In this model, the secretory form of CCN3 exhibits an antiproliferative activity whereas CCN3 isoforms promote an increased proliferation that might ultimately lead to oncogenic transformation. CCN3 is not the only example of a signaling secretory protein that is suspected to participate in nuclear signaling processes. Evidence for nuclear translocation of full length and truncated plasma membrane growth factors has been previously documented (Wells and Marti, 2002). Although the mechanisms involved in the nuclear translocation and localization of EGFR and CCN3 are still unclear, some similarities can be found between these proteins. For example, both nuclear associated EGFR and CCN3 are not membrane associated (Thomopoulos *et al.*, 2001); EGFR is associated with chromatin while CCN3 physically interacts with a subunit of RNA polymerase II which is involved in cross talk with chromatin remodeling factors and the mediator. In both cases, a sub-domain of the protein is reported to regulate transcription (Wells and Marti; unpublished observations). Furthermore, both nuclear forms of CCN3 and EGFR appear to accumulate in cancer cells, suggesting that a disruption in the balance between secreted and intracellular forms is associated with, or responsible for anarchic proliferation.

Altogether, these observations identify CCN3 as a dual functional protein whose regulatory activities may provide a clue for understanding the complex expression pattern of this protein in normal and pathological conditions.

The new avenues of investigation that these observations open are quite exciting and widen the scope of regulatory functions played by CCN3. Deciphering the mechanisms underlying the nuclear localization of the CCN3 isoforms and identifying the functions of these various isoforms is of prime importance. The next challenge will be to determine whether the duality of functions that we have attributed to CCN3 also applies to other members of the CCN family.

ACKNOWLEDGMENTS

The work performed in Professor Perbal's laboratory was supported in part by the Ministère de L'Education Nationale, de l'Enseignement Supérieur et de la

Recherche, La Ligue Nationale Contre le Cancer (Comités de Paris, du Cher et de l'Indre), and l'Association pour la Recherche contre le Cancer. We are greatful to A. Perbal and M. Le Rigoleur and for their help and support and to Dr. H. Yeger for critical reading of the manuscript.

REFERENCES

Beckwith JB, Kiviat NB and Bonadio JF (1990) Nephrogenic rests, nephroblastomatosis, and the pathogenesis of Wilms' tumor. *Pediatr Pathol* **10**:1–36.

Benini S, Perbal B, Zambelli D, Colombo MP, Manara MC, Serra M, Parenza M, Martinez V, Ficci P and Scotlandi K (2005). In Ewing's sarcoma CCN3(Nov) inhibits proliferation while promoting migration and invasion of the same cell type. *Oncogene*, in press.

Chevalier G, Yeger H, Martinerie C, Laurent M, Alami J, Schofield P and Perbal B (1998) NovH: Differential Expression in Developing Kidney and Wilms' Tumors. *Amer J Pathol* **152**:1563–1575.

Coullin P, Li CL, Ziercher L, Auffray C, Bernheim A, Zoorob R and Perbal B (2002) Assignment of the chicken NOV gene (alias CCN3) to chromosome 2q34 → q36: Conserved and compared synteny between avian, mouse and human. *Cytogenet Genome Res* **97**:140C.

Ellis PD, Chen Q, Barker PJ, Metcalfe JC and Kemp PR (2000) Nov gene encodes adhesion factor for vascular smooth muscle cells and is dynamically regulated in response to vascular injury. *Arterioscler Thromb Vasc Biol* **20**:1912–1919.

Fu CT, Bechberger JF, Ozog MA, Perbal B and Naus CC (2004) CCN3 (NOV) interacts with connexin43 in C6 glioma cells: Possible mechanism of connexin-mediated growth suppression. *J Biol Chem* **279**:36943–36950.

Gellhaus A, Dong X, Propson S, Maass K, Klein–Hitpass L, Kibschull M, Traub O, Willecke K, Perbal B, Lye SJ and Winterhager E (2004) Connexin43 interacts with NOV: A possible mechanism for negative regulation of cell growth in choriocarcinoma cells. *J Biol Chem* **279**:36931–36942.

Glukhova L, Angevin E, Lavialle C, Cadot B, Terrier M, Lacombe J, Perbal B, Bernheim A and Goguel AF (2001) Specific Genomic Alterations Associated With Poor Prognosis In High Grade Renal Cell Carcinomas. *Cancer Genet Cytogenet* **130**:105–110.

Gupta N, Wang H, McLeod TL, Naus CC, Kyurkchiev S, Advani S, Yu J, Perbal B and Weichselbaum RR (2001) Inhibition of glioma cell growth and tumorigenic potential by CCN3 (NOV). *Mol Pathol* **54**:293–99.

Joliot V, Martinerie C, Dambrine G, Plassiart G, Brisac M, Crochet J and Perbal B (1992) Proviral rearrangements and overexpression of a new cellular gene (nov) in myeloblastosis-associated virus type 1-induced nephroblastomas. *Mol Cell Biol* **12**:10–21.

Katsube K, Chuai ML, Liu YC, Kabasawa Y, Takagi M, Perbal B and Sakamoto K (2001) The expression of chicken NOV, a member of the CCN gene family, in early stage development. *Gene Expr Patterns* **1**:61–65.

Kyurkchiev S, Yeger H, Bleau AM and Perbal B (2004) Potential cellular conformations of the CCN3(NOV) protein. *Cell Commun Signal* **2** (9):1.

Kocialkowski S, Yeger H, Kingdom J, Perbal B and Schofield P (2001) Expression of the human nov gene in first trimester fetal tissues. *Anat Embryol* **203**:417–427.

Li CL, Martinez V, He B, Lombet A and Perbal B (2002) Role of the CCN3 (Nov) protein in calcium signaling. *Mol Pathol* **55**:250–261.

Li WX, Martinerie C, Zumkeller W, Westphal M and Perbal B (1996) Differential expression of nov H and CTGF in human glioma cell lines. *J Clin Pathol: Mol Pathol* **49**:91–97.

Lin CG, Leu SJ, Chen N, Tebeau CM, Lin SX, Yeung CY and Lau LF (2003) CCN3 (NOV) is a novel angiogenic regulator of the CCN protein family. *J Biol Chem* **278**:24200–24208.

Lombet A, Planque N, Bleau AM, Li CL and Perbal B (2003) CCN3 and calcium signaling. *Cell Commun Signal* **1**:10.

Mahony D, Kalionis B and Antalis TM (1999) Plasminogen activator inhibitor type-2 (PAI-2) gene transcription requires a novel NF-kappaB-like transcriptional regulatory motif. *Eur J Biochem* **263**:765–772.

Maillard M, Cadot B, Ball RY, Sethia K, Edwards D, Perbal B and Tatoud R (2001) Differential expression of novH proto-oncogene in human prostate cell lines and tissues. *Mol Pathol* **54**:275–280.

Manara MC, Perbal B, Benini S, Strammiello R, Cerisano V, Perdichizzi S, Serra M, Astolfi A, Bertoni F, Alami J, Yeger H, Picci P and Scotlandi K (2002) The expression of ccn3(nov) gene in musculoskeletal tumors. *Am J Pathol* **160**:849–859.

Martinerie C, Chevalier G, Rauscher FJ 3rd and Perbal B (1996) Regulation of nov by WT1: a potential role for nov in nephrogenesis. *Oncogene* **12**:1479–1492.

Perbal B (1994) Contribution of MAV-1-induced nephroblastoma to the study of genes involved in human Wilms' tumour development. *Critical Reviews in Oncogenesis* **5**:589–613.

Perbal B (1995) Pathogenic potential of myeloblastosis-associated viruses. *Infet. Agents Dis.* **4**:212–227.

Perbal B, Martinerie C, Sainson R, Werner M, He B and Roizman B (1999) The C-terminal domain of the regulatory protein NOVH is sufficient to promote interaction with fibulin 1C: A clue for a role of NOVH in cell-adhesion signaling. *Proc Natl Acad Sci USA* **96**:869–874.

Perbal B (1999) Nuclear localization of NOV protein: a potential role for nov in the regulation of gene expression. *Mol Pahol* **52**:84–91.

Perbal B (2001) NOV (nephroblastoma overexpressed) and the CCN family of genes: structural and functional issues. *Mol Pathol* **54**:57–79.

Perbal B (2004) CCN proteins: multifunctionnal signalling regulators. *Lancet* **363**: 62–64.

Planque N and Perbal B (2003) A structural approach to the role of CCN proteins in tumorigenesis. *Cancer Cell International* **3**:15.

Sakamoto K, Yamaguchi S, Ando R, Miyawaki A, Kabasawa Y, Takagi M, Li CL, Perbal B and Katsube K (2002) The nephroblastoma overexpressed gene (NOV/ccn3) protein associates with Notch1 extracellular domain and inhibits myoblast differentiation via Notch signaling pathway. *J Biol Chem* **277**: 29399–29405.

Su BY, Cai WQ, Zhang CG, Martinez V, Lombet A and Perbal B (2001) The expression of ccn3 (nov) RNA and protein in the rat central nervous system is developmentally regulated. *Mol Pathol* **54**:184–191.

Wells A and Marti U (2002) Signalling shortcuts: cell-surface receptors in the nucleus? *Nat Rev Mol Cell Biol* **3**:697–702.

Yu C, Le AT, Yeger H, Perbal B and Alman B (2003) NOV (CCN3) regulation in the growth plate and CCN family member expression in cartilage neoplasia. *J Pathol* **201**:609–615.

CHAPTER 14

CCN PROTEINS AND CONNEXINS: INTERACTIONS AND GROWTH CONTROL

Christine Fu*, Alexandra Gellhaus[†], Elke Winterhager[†]
and Christian C. Naus*[,‡]

*Department of Cellular & Physiological Sciences
The University of British Columbia, Vancouver, BC, Canada
[‡]Christian.naus@ubc.ca

[†]Institute of Anatomy, University Hospital Essen, Germany

1. INTRODUCTION

There has been a longstanding association of connexin expression and its resultant gap junctional communication with cell growth control. Nevertheless, understanding the mechanisms by which such growth is controlled by gap junctions has remained elusive. While correlation of decreased connexin expression and gap junctional coupling in association with increased tumorigenic phenotype can readily be found, several well documented exceptions exist, including: 1) the presence of connexin expression and gap junctions in tumor cells, 2) variability in the tumor suppressing ability of connexins, 3) discrepancy between *in vivo* and *in vitro* growth effects of enhanced connexin expression, and 4) the absence of enhanced gap junctional coupling in spite of increased connexin expression and growth suppression. These discrepancies have motivated the search for other mechanisms by which connexins may alter growth. One avenue we have pursued concerns effects mediated by alteration in expression of other genes which are involved in growth control. Using this approach in two different model systems (i.e. glioma and choriocarcinoma), we have identified enhanced expression of CCN3 following transfection with connexin43. Furthermore, we have found an interaction of connexin43 with CCN3, and propose that such interactions of connexins with other proteins could underlie their role as growth suppressors.

2. CONNEXINS AND GAP JUNCTIONS

Gap junctions are intercellular channels that directly connect the cytoplasm of adjacent cells. They permit the exchange of ions and molecules smaller than 1 kDa in size, such as amino acids, second messengers and metabolites, as well as coordination of cellular activities (Simon and Goodenough, 1998). Gap junctions are formed by the end-to-end apposition of connexons, which are proteinaceous cylinders spanning the plasma membrane with a hydrophilic core in the centre. Each connexon hemichannel is composed of 6 subunit proteins called connexins, which belong to a multigene family consisting of at least 20 members in vertebrates (Sohl and Willecke, 2003). The connexin protein spans the plasma membrane four times, and is oriented so that both its N- and C-termini face the cytoplasm, with extracellular domains involved in connexin-connexin interactions.

3. GAP JUNCTIONS AND TUMOR SUPPRESSION

Gap junctions have long been implicated in cellular growth regulation and the control of tumor progression. Since the first discovery that gap junctional coupling is decreased in tumor cells as compared to their normal counterparts (Loewenstein and Kanno, 1966), downregulation of gap junctions has been repeatedly observed in a variety of tumor tissues and cell lines analyzed (Mesnil, 2002). Altered gap junctional coupling may take the form of aberrant homologous (among tumor cells) and/or heterologous (with surrounding normal cells) communication (Yamasaki and Naus, 1996). In addition, many tumor promoting agents have the ability to down-regulate coupling (Naus, 2002; Ruch and Trosko, 2001). Conversely, antineoplastic agents such as retinoic acid and vitamin D are capable of reversing these effects (Trosko and Chang, 2001; Hossain and Bertram, 1994). Furthermore, gap junctional communication is also modulated by oncogenes (Azarnia *et al.*, 1989; Kalimi *et al.*, 1992). The effect of gap junction proteins on growth has been studied by analyzing both the loss of gap junctions and the forced expression of connexins. The loss of gap junctions in many cases correlates with an increase in growth rate *in vitro* and tumor formation *in vivo* (Yamasaki and Naus, 1996). On the other hand, many transformed cells transfected with connexin cDNAs exhibit suppression of growth and/or tumorigenicity. The connexin family members shown to act as tumor suppressors include Cx43, Cx26, Cx32 and Cx30. Some other connexins fail to exhibit this growth suppressive phenotype (Yamasaki and Naus, 1996; Zhu *et al.*, 1991; Eghbali *et al.*, 1991; Mesnil *et al.*, 1995).

The mechanism by which connexins act as growth inhibitors is unknown. Initial investigation of connexin-induced tumor suppression focused on its channel-forming ability, to allow for either the exchange of normalizing messages, or dispersion of growth-promoting signals (Mesnil, 2002). However, recent evidence contends that functional gap junctions may not be necessary for the growth regulatory effects, but simply the expression of connexin genes may be sufficient. For example, Cx43 inhibits the growth of both human glioblastoma and breast cancer cells *in vivo*, but there is no significant increase in coupling as assessed by dye transfer (Huang *et al.*, 1998; Qin *et al.*, 2003). Moreover, it has been shown that Cx43 may inhibit growth by delaying the G1-S phase transition of the cell cycle (Moorby and Patel, 2001). Along this line, one report indicated that Cx43 promotes the degradation of S phase kinase-associated protein 2 (Skp2) which regulates the ubiquitination of p27, in turn increasing the level of p27 (Zhang *et al.*, 2003). This effect was not influenced by gap junction blockage, suggesting that it is communication-independent.

Previously we reported that transfection of Cx43 led to decreased proliferation of both glioma cells (Zhu *et al.*, 1991) and choriocarcinoma cells (Hellmann *et al.*, 1999). As well, conditioned medium collected from transfected C6 cells (C6-Cx43) alone also possessed anti-proliferative properties (Zhu *et al.*, 1992). Thus, the growth suppressive capacity of Cx43 may be mediated by the secretion of soluble growth inhibitory factors into the surrounding media (Zhu *et al.*, 1992; Goldberg *et al.*, 2000). Using differential gene expression approaches, including differential display for glioma cells (Naus *et al.*, 2000) and gene chip analysis for choriocarcinoma cells (Gellhaus *et al.*, 2004), we have pursued the hypothesis that connexin gene expression might alter the expression of downstream growth regulatory genes. Among the most prominent differentially expressed genes were members of the CYR61/Connective Tissue Growth Factor/Nephroblastoma Overexpressed (CCN) family, which have been shown to play a role in tumorigenesis and growth control. One of these, CCN3 (NOV), is considered a negative growth regulatory protein and stands out as a candidate effector of Cx43-induced growth regulation.

4. CCN3 AND CONNEXIN INTERACTIONS

Evidence that connexins might be active in signaling pathways independent of functional channel formation is accumulating. Elucidating the molecular mechanism of connexin-induced tumor suppression could potentially lead to useful therapeutic intervention applicable to human cancers. In order to investigate how Cx43 elicits growth inhibition, we have explored the possibility that CCN3 is a downstream effector of Cx43 that exerts the suggested tumor suppressive action. The rationale for this hypothesis is based on several lines of evidence. As previously noted, the observation that CCN3 is upregulated in growth-inhibited glioma and choriocarcinoma cells following Cx43 transfection, together with the feature of CCN3 as a secreted protein with known anti-proliferative properties, has led to our proposal that CCN3 serves as one of the soluble factors mediating Cx43-induced tumor suppression. Using immunocytochemistry, we previously demonstrated that CCN3 co-localizes with Cx43 at the plasma membrane in C6-Cx43 cells (McLeod *et al.*, 2001). Our more recent studies (Fu *et al.*, 2004; Gellhaus *et al.*, 2004) have focused on investigating whether a physical interaction exists between Cx43 and CCN3, and how the presence and absence of this interaction correlates with the growth properties of various cell types.

5. CCN3 AND CONNEXIN IN GLIOMA CELLS

The biological functions of CCN3 have been suggested to depend on the size of isoforms expressed, their relative amount and subcellular localization (Planque and Perbal, 2003). Unique localization patterns of CCN3 protein were observed in different cell types. Most cells examined exhibit cytoplasmic staining of CCN3 (Chevalier *et al.*, 1998; Perbal, 1999; Maillard *et al.*, 2001; Kocialkowski *et al.*, 2001). Previous reports using different antibodies have also noted nuclear localization of CCN3 in various tumor cell types (Perbal, 1999; Gupta *et al.*, 2001; Gellhaus *et al.*, 2004). While CCN3 is diffusely present in the cytoplasm of C6 glioma cells, it becomes distinctly localized to areas of Cx43-containing gap junction plaques at the cell membrane in C6-Cx43 cells, which display reduced growth and tumor formation. In addition to C6-Cx43, punctate immunostaining of CCN3 co-localizing with Cx43 plaques was also demonstrated in primary astrocytes. On the other hand, no CCN3 protein was found at the membrane of C6 or human glioma cells, in spite of the presence of Cx43 in some of these cells (Fu *et al.*, 2004). This observation provides a clue into the phenomenon that some tumor cells contain high levels of connexin. The mere expression of connexin protein does not guarantee that the downstream pathways are active.

Following Cx43 transfection into C6 glioma cells, the redistribution of CCN3 from the cytoplasm to the plasma membrane coincides with the appearance of a 48 kDa isoform, which is thought to correspond to the full-length CCN3 protein. Western blot analysis reveals that this isoform is present in C6-Cx43 but not in parental C6 cells. The upregulation of full-length CCN3 could be the consequence of increased level of mRNA transcript, reduced proteolytic degradation into the truncated isoforms, or a combination of both. Considering the concurrence of 48 kDa upregulation and membrane association, it is likely that full-length CCN3 co-localizes with Cx43 in C6-Cx43 cells. However, present evidence does not rule out the potential ability of shorter CCN3 isoforms to interact with Cx43, although the downstream functions of this interaction may not be preserved following the truncation.

In addition to immunolocalization studies, we have also used biochemical approaches, including GST pull-down assay and co-immunoprecipitation, to demonstrate that Cx43 and CCN3 not only co-localize to the same subcellular domains, but that they also physically associate. It is of interest to identify protein domains that mediate the interaction between CCN3 and Cx43. The loss of Cx43 co-localization and co-immunoprecipitation with CCN3 in C6 cells transfected with a C-terminal truncated Cx43 indicates that the C-terminus of Cx43 is necessary for its interaction with CCN3. Despite the presence of phosphorylation sites on the C-terminal tail of Cx43, which are important players in its regulation, treatment of C6-Cx43 lysate with alkaline phosphatase does not abolish the physical association between Cx43 and CCN3, suggesting that binding is phosphorylation-independent. The specific site(s) of interaction between CCN3 and the C-terminus of Cx43 remain to be identified.

The full-length CCN3 has been established as a negative regulator of growth. When expressed in chicken embryo fibroblasts (CEF), it maintains a growth suppressive effect (Joliot *et al.*, 1992). Transfection of CCN3 cDNA into the human glioma cell line G59 significantly reduces their *in vitro* growth rate, as well as tumor size *in vivo* (Gupta *et al.*, 2001). Full-length CCN3 was detected in the conditioned medium of CCN3-expressing cells, such as transfected G59 cells (Gupta *et al.*, 2001) and vascular endothelial cells (Ellis *et al.*, 2000). The presence of 48 kDa CCN3 in C6-Cx43 cells and primary astrocytes in contrast to C6 and human glioma cells agrees with its putative function as a growth suppressor (Fu *et al.*, 2004). Furthermore, several CCN3 isoforms were detected in medium conditioned by C6-Cx43 cells. Conditioned medium of C6-Cx43 causes a reduction in the growth rate of C6 cells in culture (Zhu *et al.*, 1992). Since CCN3 that possesses known anti-proliferative activities is secreted into the extracellular environment, this may account for the growth suppressive effects observed by the conditioned medium of C6-Cx43 cells.

Among the connexins we have expressed in C6 glioma cells, the interaction with CCN3 is specific to Cx43 (Fu *et al.*, 2004). From a structural point of view, this finding comes as no surprise since the C-terminus, which appears essential in the Cx-CCN3 interaction, is also the most variable region among all connexin family members (Sohl and Willecke, 2004). From a functional point of view, it has been discovered that the ability to regulate growth in a particular cell type is specific to limited number of connexins (Mesnil, 2002). Considering that the C-terminus is sufficient to elicit growth suppression and is also the site of association with a variety of molecules, it can be speculated that the connexin specificity of growth control might be mediated by the differential ability of different connexins to interact with downstream effector molecules.

6. CCN3 AND CONNEXIN IN CHORIOCARCINOMA CELLS

Further evidence that CCN3 might be a major player in the Cx43 dependent control of cell growth results from studies with another cell type, the choriocarcinoma cell line Jeg3 (Gellhaus *et al.*, 2004). These malignant trophoblast cells are communication-deficient and therefore suitable for the analysis of the role of different connexins in the control of proliferation and differentiation processes. Using doxycycline (Dox) inducible Cx43, Cx40 and C-terminal truncated Cx43 (trCx43) transfected cell lines, it could be demonstrated that restoration of cell-cell communication via Cx43 protein channels, but not Cx40 and trCx43, is able to reduce cell growth of Jeg3 cells *in vitro* and tumor growth in nude mice. To identify genes responsible for the different cell physiological behavior of these generated connexin cell lines gene array analysis was performed. Among the differently regulated genes obtained from this analysis, the growth regulator gene CCN3 revealed an upregulation only in Cx43 cells which could be validated by RT-PCR. This finding corresponds to the differential display data of the C6-Cx43 glioma cells mentioned above (Naus *et al.*, 2000). Using hierarchical clustering of gene array data another aspect became conspicuous. We found a clear separation of gene expression patterns between the Cx43 cell clones on the one hand and the Cx40 and trCx43 cells on the other hand. This classification strengthens the finding that the C-terminus of Cx43 may act as a mediator of the different signal cascade leading to the observed growth reduction of Cx43 transfected Jeg3 cells.

Parental Jeg3 cells express CCN3 but localization was exclusively found in the nucleus and in perinuclear regions as well as to a lesser extent in the cytoplasm. Upon induction of the different connexins only in the Cx43 cells a switch in CCN3 expression pattern from the nucleus to a distinct localization at areas of cell-cell contact at the cell membrane was observed where CCN3 is co-localized with Cx43 in some gap junction plaques. Because co-localization failed if the Cx43 protein lacks its C-terminus, we hypothesize that this part of the Cx43 molecule might be crucial for the molecular interaction of these proteins. Time course experiments between 0 and 48 hrs after removal of the connexin inducer Dox shed light on the correlation between Cx43 and CCN3 protein expression using co-immunolocalization and RT-PCR. For up to 14 hrs after Dox removal, a co-localization of CCN3 and Cx43 along the lateral cell membranes could still be observed whereas after 24 hrs Cx43 and CCN3 have disappeared from the membranes and only nuclear and cytoplasmic staining of CCN3 was found. These observations are in the same line as the results from RT-PCR analysis of Cx43 and CCN3 mRNA expression. The CCN3 transcripts decreased simultaneously with the reduction of Cx43 expression during this time course. Thus CCN3 upregulation and membrane localization are clearly dependent on the presence of Cx43.

The direct physical interaction of CCN3 with the C-terminus of Cx43 could be confirmed by *in vitro* and *in vivo* co-immunoprecipitation in transfected 293T cells. The loss of co-immunoprecipitation with CCN3 in trCx43 cells along with the identified co-immunoprecipitation between CCN3 and a Cx43 mutant with only the C-terminal tail ($Cx43CT_{257-382}$) defines the region of interaction between 257 and 382 amino acids of the Cx43 C-terminus. This binding is apparently independent from the requirement of any cofactor molecule. However this does not exclude the participation of other signaling molecules and the involvement of phosphorylation in this interaction process.

The binding between CCN3 and Cx43 is associated with a growth reduction in Jeg3 cells, the same could be demonstrated in glioma cells as mentioned above. Preliminary analysis of CCN3 expression in benign placenta reveals that in contrast to a small amount of CCN3 mRNA in the malignant trophoblast cell line Jeg3, CCN3 is strongly expressed in term placenta. Further investigations, however, are needed to evaluate if CCN3 may play a role in the control of placental growth.

It is discussed in previous reports that a N-terminus truncated CCN3 protein, which was detected in the nuclei of tumor cells, leads to an increased cell growth whereas the full length protein, which is secreted or remains at the cell membrane, inhibits proliferation (Perbal, 2001; Planque and Perbal, 2003). However, as already described above, the opposite effect has been shown as well. This suggests that CCN3 may act differently dependent from the cell type and the expressed isoform. Since we found that CCN3 localization in parental and uninduced Jeg3 cells is predominantly restricted to the nucleus we must consider in future experiments whether different CCN3 isoforms may affect growth control in choriocarcinoma cells.

Transfection of human glioma cells with CCN3 results in reduced growth and tumor formation (Gupta *et al.*, 2001). Experiments with parental Jeg3 cells overexpressing full length CCN3 gives us the first evidence that CCN3 can act as a growth regulator in these cells independent from Cx43. In these CCN3 transfected cells, full length CCN3 is located exclusively at the lateral cell membrane and induces growth reduction in Jeg3 cells independent from Cx43 expression, similar to the observation in Cx43 cells after induction of Cx43. This finding clearly demonstrated that CCN3 itself exhibited an antiproliferative effect on Jeg3 cells. CCN3 can act independently from the presence of Cx43, however, the Cx43 protein seems to be able to increase full length CCN3 expression and to change its localization from the nucleus to the membrane by binding CCN3 at the C-terminus. The disappearance of CCN3 from the nucleus may cause CCN3 to no longer be available as a transcriptional regulator. This mechanism could be responsible for the change in cell physiology of the Cx43 expressing Jeg3 cells leading to the reduction in cell proliferation and tumor growth. If other signaling molecules, which may interact with CCN3 and/or Cx43, are involved in this process in Jeg3 cells remains to be clarified. First hints of further interaction partners result from gene array analysis of CCN3 transfected Jeg3 cells compared to the data of Cx43 transfected cells. The data lists contain interesting common upregulated genes like fibronectin-1, S100 calcium binding protein A11 (calgizzarin) and endothelin receptor type B, which may play a role in this signaling pathway (unpublished observation).

In summary, there is accumulating evidence that connexin-mediated growth regulation seems not only dependent on channel properties but on the C-terminus induced signal cascades by interacting with other proteins in an isoform and cell type specific manner. We suggest that Cx43 seems to be responsible to regulate cell growth in Jeg3 cells via the upregulation of full length CCN3 accompanied by a shift of localization of CCN3 from the nucleus to the membrane and a direct binding of CCN3 to the C-terminus of Cx43.

7. SIGNIFICANCE OF INTERACTION BETWEEN CCN3 AND CONNEXIN43

The C-terminus of Cx43 has been shown to be critical for its growth suppressive function. Deletion of the C-terminus in Cx43 abrogates its growth regulation

(Moorby and Patel, 2001). This discovery is consistent with the proposed role of CCN3 as a downstream effector of Cx43-mediated growth inhibition, since CCN3 interacts with the C-terminal tail of Cx43. Additionally, transfection of the C-terminus of Cx43 alone is sufficient in reducing the growth rate of tumor cells (Moorby and Patel, 2001; Zhang *et al.*, 2003). These truncated Cx43 proteins are present diffusely in the cytoplasm of transfected cells, suggesting that membrane localization is not required for growth suppression. This finding does not contradict our proposal that the interaction between Cx43 and CCN3 might be involved in growth control, since C-terminal Cx43 is expected to retain its ability of interacting with CCN3 in the cytoplasm.

Cx43 has been found to bind various other proteins, such as calmodulin (Van Eldik *et al.*, 1985), microtubules (Giepmans *et al.*, 2001b), v-Src (Kanemitsu *et al.*, 1997), c-Src (Giepmans *et al.*, 2001a), β-catenin (Ai *et al.*, 2000) and zonula occludins-1 (ZO-1) (Giepmans and Moolenaar, 1998; Toyofuku *et al.*, 1998). ZO-1 is a peripheral membrane component of tight junctions and adherens junctions. It contains multiple protein interaction domains, including a SH3 domain and three PDZ domains, the second of which interacts with Cx43 (Giepmans and Moolenaar, 1998). We discovered by immunocytochemistry that ZO-1 co-localizes with CCN3 and Cx43 at the plasma membrane in C6-Cx43 cells (unpublished observation). While the significance of this particular complex formation remains to be established, a number of hypotheses can be generated in light of the emerging view of gap junctions as the anchorage of multi-protein complexes called Nexus (Duffy *et al.*, 2002; Herve *et al.*, 2004). It has been proposed that ZO-1 may recruit signaling proteins to gap junction plaques (Giepmans and Moolenaar, 1998). In turn, these proteins could become accessible to CCN3 and other factors that associate with it. Alternatively, CCN3 and ZO-1 may be brought together to interact by both binding to connexins. CCN3 does possess T/SXV (Songyang *et al.*, 1997) and X(V/I)(E/Q)V (Hock *et al.*, 1998) consensus motifs, which have been implicated in the binding to PDZ domains found in ZO-1. However, biochemical evidence needs to be collected to ascertain a physical interaction between CCN3 and ZO-1.

With respect to CCN3, its biological properties have been proposed to depend on combinatorial events resulting from interaction with various partners (Perbal, 2001; Planque and Perbal, 2003). Indeed, the multimodular structure of CCN3 suggests that it may lie at the hub of multiple signaling pathways. By acting as a scaffolding protein, CCN3 could recruit different molecules such as Cx43 into a complex and coordinate crosstalk between different signals, ultimately contributing to growth regulation.

8. MODEL OF CCN3 INTERNALIZATION

The CCN3 protein contains a signal peptide that destines it for secretion, which is confirmed by numerous reports noting its presence in the conditioned medium of CCN3-expressing cells. The implication of the Cx43 C-terminal tail in its interaction with CCN3 suggests that the interaction occurs intracellularly at the plasma membrane. This finding raises the question as to how CCN3 gains access to the cytoplasm. It has been proposed that full-length CCN proteins are secreted into the extracellular matrix, where proteolytic cleavage could take place. The truncated products are subsequently shuttled back into the cell, possibly via the action of an unidentified channel/receptor (Perbal, 2001). This model was suggested to explain the nuclear localization of N-truncated CCN3. Likewise, the same logic can be extended to the cytoplasmic localization of the full-length isoform. While certain truncated isoforms are directed to the nucleus, the full-length protein could be retained in the cytoplasm. One study consistent with this model demonstrated that extracellular CCN2 was internalized in vesicles, translocated into the cytoplasm, and subsequently moved into the nucleus (Wahab *et al.*, 2001). Nuclear targeting of extracellular signaling molecules has been exemplified by several growth factors, including insulin, prolactin and PTHrP (Henderson, 1997). The mechanism by which these molecules are internalized and released into the cytoplasm remains a mystery, although increasing data implicate caveolae in mediating the internalization of regulatory proteins in their bioactive forms (Anderson, 1993). The ability of caveolin-1 to co-immunoprecipitate with CCN3 (unpublished observation) suggests that the possible involvement of caveolae in the internalization of CCN3 is worthy of further exploration. Along the same line, it has been demonstrated that Cx43 localizes to lipid rafts and interacts with caveolin-1 (Lin *et al.*, 2003; Schubert *et al.*, 2002). Therefore, these specialized membrane subdomains might serve as docking points where Cx43 and CCN3 are brought into proximity to interact.

9. CONCLUSION

We have demonstrated that CCN3 is localized to the plasma membrane at Cx43 plaques in growth-suppressed cells. The appearance of CCN3-Cx43 co-localization coincides with the upregulation of a 48 kDa full-length isoform of CCN3, which is also secreted into the conditioned medium of C6-Cx43 cells. CCN3 and Cx43 physically interact, and this interaction requires the C-terminus of Cx43. CCN3 directly interacts with the C-terminus of Cx43.

The potential role of this connexin-specific interaction in growth regulation warrants further investigation.

ACKNOWLEDGMENTS

This work was supported by grants from the Canadian Institutes of Health Research to C. Naus, the Mildred Scheel Foundation (10-1488-WI 2) to E. Winterhager, and the NIH (1R01 HD42558-01) to E. Winterhager and S. J. Lye.

REFERENCES

Ai Z, Fischer A, Spray DC, Brown AM and Fishman GI (2000) Wnt-1 regulation of connexin43 in cardiac myocytes. *J Clin Invest* **105**: 161–171.

Anderson RG (1993) Caveolae: where incoming and outgoing messengers meet. *Proc Natl Acad Sci USA* **90**: 10909–10913.

Azarnia R, Mitcho M, Shalloway D and Loewenstein WR (1989) Junctional intercellular communication is cooperatively inhibited by oncogenes in transformation. *Oncogene* **4**: 1161–1168.

Chevalier G, Yeger H, Martinerie C, Laurent M, Alami J, Schofield PN and Perbal B (1998) novH: differential expression in developing kidney and Wilm's tumors. *Am J Pathol* **152**: 1563–1575.

Duffy HS, Delmar M and Spray DC (2002) Formation of the gap junction nexus: binding partners for connexins. *J Physiol Paris* **96**: 243–249.

Eghbali B, Kessler JA, Reid LM, Roy C and Spray DC (1991) Involvement of gap junctions in tumorigenesis: transfection of tumor cells with connexin 32 cDNA retards growth *in vivo*. *Proc Natl Acad Sci USA* **88**: 10701–10705.

Ellis PD, Chen Q, Barker PJ, Metcalfe JC and Kemp PR (2000) Nov gene encodes adhesion factor for vascular smooth muscle cells and is dynamically regulated in response to vascular injury. *Arterioscler Thromb Vasc Biol* **20**: 1912–1919.

Fu CT, Bechberger JF, Ozog MA, Perbal B and Naus CC (2004) CCN3 (NOV) Interacts with Connexin43 in C6 Glioma Cells: Possible mechanism of connexin-mediated growth suppression. *J Biol Chem* **279**: 36943–36950.

Gellhaus A, Dong X, Propson S, Maass K, Klein-Hitpass L, Kibschull M, Traub O, Willecke K, Perbal B, Lye SJ and Winterhager E (2004) Connexin43 interacts with NOV: A possible mechanism for negative regulation of cell growth in choriocarcinoma cells. *J Biol Chem* **279**: 36931–36942.

Giepmans BN and Moolenaar WH (1998) The gap junction protein connexin43 interacts with the second PDZ domain of the zona occludens-1 protein. *Curr Biol* **8**: 931–934.

Giepmans BN, Hengeveld T, Postma FR and Moolenaar WH (2001a) Interaction of c-Src with gap junction protein connexin-43. Role in the regulation of cell-cell communication. *J Biol Chem* **276**: 8544–8549.

Giepmans BN, Verlaan I, Hengeveld T, Janssen H, Calafat J, Falk MM and Moolenaar WH (2001b) Gap junction protein connexin-43 interacts directly with microtubules. *Curr Biol* **11**: 1364–1368.

Goldberg GS, Bechberger JF, Tajima Y, Merritt M, Omori Y, Gawinowicz MA, Narayanan R, Tan Y, Sanai Y, Yamasaki H, Naus CC, Tsuda H and Nicholson BJ (2000) Connexin43 suppresses MFG-E8 while inducing contact growth inhibition of glioma cells. *Cancer Res* **60**: 6018–6026.

Gupta N, Wang H, Mcleod TL, Naus CCG, Kyurkchiev S, Advani S, Yu J, Perbal B and Weichselbaum RR (2001) Inhibition of glioma cell growth and tumorigenic potential by CCN3 (NOV). *J Clin Patho* **54**: 293–299.

Hellmann P, Grummer R, Schirrmacher K, Rook M, Traub O and Winterhager E (1999) Transfection with different connexin genes alters growth and differentiation of human choriocarcinoma cells. *Exp Cell Res* **246**: 480–490.

Henderson JE (1997) Nuclear targeting of secretory proteins. *Mol Cell Endocrinol* **129**: 1–5.

Herve JC, Bourmeyster N and Sarrouilhe D (2004) Diversity in protein-protein interactions of connexins: emerging roles. *Biochim Biophys Acta* **1662**: 22–41.

Hock B, Bohme B, Karn T, Yamamoto T, Kaibuchi K, Holtrich U, Holland S, Pawson T, Rubsamen-Waigmann H and Strebhardt K (1998) PDZ-domain-mediated interaction of the Eph-related receptor tyrosine kinase EphB3 and the ras-binding protein AF6 depends on the kinase activity of the receptor. *Proc Natl Acad Sci USA* **95**: 9779–9784.

Hossain MZ and Bertram JS (1994) Retinoids suppress proliferation, induce cell spreading, and up-regulate connexin43 expression only in postconfluent 10T1/2 cells: Implications for the role of gap junctional communication. *Cell Growth Differ* **5**: 1253–1261.

Huang RP, Fan Y, Hossain MZ, Peng A, Zeng ZL and Boynton AL (1998) Reversion of the neoplastic phenotype of human glioblastoma cells by connexin 43 (cx43). *Cancer Res* **58**: 5089–5096.

Joliot V, Martinerie C, Dambrine G, Plassiart G, Brisac M, Crochet J and Perbal B (1992) Proviral rearrangements and overexpression of a new cellular gene (nov) in myeloblastosis-associated virus type 1-induced nephroblastomas. *Mol Cell Biol* **12**: 10–21.

Kalimi GH, Hampton LL, Trosko JE, Thorgeirsson SS and Huggett AC (1992) Homologous and heterologous gap-junctional intercellular communication in v-raf-, v-myc-, and v-raf/v-myc-transduced rat liver epithelial cell lines. *Mol Carcinog* **5**: 301–310.

Kanemitsu MY, Loo LWM, Simon S, Lau AF and Eckhart W (1997) Tyrosine phosphorylation of connexin 43 by v-SRC is mediated by SH2 and SH3 domain interactions. *J Biol Chem* **272**: 22824–22831.

Kocialkowski S, Yeger H, Kingdom J, Perbal B and Schofield PN (2001) Expression of the human NOV gene in first trimester fetal tissues. *Anat Embryol (Berl)* **203**: 417–427.

Lin D, Zhou J, Zelenka PS and Takemoto DJ (2003) Protein kinase Cgamma regulation of gap junction activity through caveolin-1-containing lipid rafts. *Invest Ophthalmol Vis Sci* **44**: 5259–5268.

Loewenstein WR and Kanno Y (1966) Intercellular communication and the control of tissue growth: lack of communication between cancer cells. *Nature* **209**: 1248–1249.

Maillard M, Cadot B, Ball RY, Sethia K, Edwards DR, Perbal B and Tatoud R (2001) Differential expression of the ccn3 (nov) proto-oncogene in human prostate cell lines and tissues. *Mol Pathol* **54**: 275–280.

McLeod TL, Bechberger JF and Naus CC (2001) Determination of a potential role of the CCN family of growth regulators in connexin43 transfected C6 glioma cells. *Cell Commun Adhes* **8**: 441–445.

Mesnil M, Krutovskikh V, Piccoli C, Elfgang C, Traub O, Willecke K and Yamasaki H (1995) Negative growth control of HeLa cells by connexin genes: Connexin species specificity. *Cancer Res* **55**: 629–639.

Mesnil M (2002) Connexins and cancer. *Biol Cell* **94**: 493–500.

Moorby C and Patel M (2001) Dual functions for connexins: Cx43 regulates growth independently of gap junction formation. *Exp Cell Res* **271**: 238–248.

Naus CC, Bond SL, Bechberger JF and Rushlow W (2000) Identification of genes differentially expressed in C6 glioma cells transfected with connexin43. *Brain Res Brain Res Rev* **32**: 259–266.

Naus CC (2002) Gap junctions and tumor progression. *Can J Physiol Pharmacol* **80**: 136–141.

Perbal B (1999) Nuclear localisation of NOVH protein: a potential role for NOV in the regulation of gene expression. *Mol Pathol* **52**: 84–91.

Perbal B (2001) NOV (nephroblastoma overexpressed) and the CCN family of genes: structural and functional issues. *Mol Pathol* **54**: 57–79.

Planque N and Perbal B (2003) A structural approach to the role of CCN (CYR61/CTGF/NOV) proteins in tumorigenesis. *Cancer Cell Int* **3**: 15.

Qin H, Shao Q, Thomas T, Kalra J, Alaoui–Jamali MA and Laird DW (2003) Connexin26 regulates the expression of angiogenesis-related genes in human breast tumor cells by both GJIC-dependent and -independent mechanisms. *Cell Commun Adhes* **10**: 387–393.

Ruch RJ and Trosko JE (2001) Gap-junction communication in chemical carcinogenesis. *Drug Metab Rev* **33**: 117–124.

Schubert AL, Schubert W, Spray DC and Lisanti MP (2002) Connexin family members target to lipid raft domains and interact with caveolin-1. *Biochemistry* **41**: 5754–5764.

Simon AM and Goodenough DA (1998) Diverse functions of vertebrate gap junctions. *Trends Cell Biol* **8**: 477–482.

Sohl G and Willecke K (2003) An update on connexin genes and their nomenclature in mouse and man. *Cell Commun Adhes* **10**: 173–180.

Sohl G and Willecke K (2004) Gap junctions and the connexin protein family. *Cardiovasc Res* **62**: 228–232.

Songyang Z, Fanning AS, Fu C, Xu J, Marfatia SM, Chishti AH, Crompton A, Chan AC, Anderson JM and Cantley LC (1997) Recognition of unique carboxyl-terminal motifs by distinct PDZ domains. *Science* **275**: 73–77.

Toyofuku T, Yabuki M, Otsu K, Kuzuya T, Hori M and Tada M (1998) Direct association of the gap junction protein connexin-43 with ZO-1 in cardiac myocytes. *J Biol Chem* **273**: 12725–12731.

Trosko JE and Chang CC (2001) Mechanism of up-regulated gap junctional intercellular communication during chemoprevention and chemotherapy of cancer. *Mutat Res* **480-481**: 219–229.

Van Eldik LJ, Hertzberg EL, Berdan RC and Gilula NB (1985) Interaction of calmodulin and other calcium-modulated proteins with mammalian and arthropod junctional membrane proteins. *Biochem Biophys Res Commun* **126**: 825–832.

Wahab NA, Brinkman H, Mason RM (2001) Uptake and intracellular transport of the connective tissue growth factor: a potential mode of action. *Biochem J* **359**: 89–97.

Yamasaki H and Naus CCG (1996) Role of connexin genes in growth control. *Carcinogenesis* **17**: 1199–1213.

Zhang YW, Kaneda M and Morita I (2003) The gap junction-independent tumor-suppressing effect of connexin 43. *J Biol Chem* **278**: 44852–44856.

Zhu D, Caveney S, Kidder GM and Naus CC (1991) Transfection of C6 glioma cells with connexin 43 cDNA: analysis of expression, intercellular coupling, and cell proliferation. *Proc Natl Acad Sci USA* **88**: 1883–1887.

Zhu D, Kidder GM, Caveney S and Naus CC (1992) Growth retardation in glioma cells cocultured with cells overexpressing a gap junction protein. *Proc Natl Acad Sci USA* **89**: 10218–10221.

CHAPTER 15

THE ROLE OF CCN1 IN TUMORIGENESIS AND CANCER PROGRESSION

James O'Kelly and H. Phillip Koeffler

Division of Hematology/Oncology
Cedars Sinai Medical Center
UCLA School of Medicine
8700 Beverly Boulevard
Los Angeles, CA 90048
Phone: (310) 423-7733; Fax: (310) 423 0225
koeffler@cshs.org
okellyj@cshs.org

1. INTRODUCTION

Cyr61 is a member of the CCN protein family, which also includes CTGF, Nov WISP-1, WISP-2, and WISP-3 (Brigstock, 1999). CCN genes contain four conserved modular domains that share sequence similarity with other protein families: IGF-binding protein (IGFBP), Von Willebrand type C (VWC), Thrombospondin type 1 (TSP1) and a c-terminal domain (CT). The first cloned member of the CCN family is Cyr61 (CCN1), which was identified by differential hybridization screening of a cDNA library from serum-stimulated fibroblasts (Lau and Nathans, 1985). CCN1 is a secreted cysteine-rich heparin binding protein that associates with the extracellular matrix and the cell surface (Yang and Lau, 1991). Integrins, including $\alpha_v\beta_3$, $\alpha_v\beta_5$, $\alpha_6\beta_1$ and $\alpha_{IIb}\beta_3$ have been identified as receptors of CCN1. Through these receptors, CCN1 is thought to exert its effects on cellular activity in a variety of tissue types (Lau and Lam, 1999).

CCN1 plays diverse roles in cellular proliferation, differentiation, adhesion and migration, as well as regulating chemotaxis and angiogenesis (Babic *et al.*, 1998; Kireeva *et al.*, 1996). Dysregulation of these processes contributes to tumorigenesis and cancer progression; therefore, inappropriate expression of CCN1 has the capacity to play a multifactorial role in the development of cancer. CCN1 is an immediate-early response gene that is rapidly induced by a wide range of stimuli encompassing hormones, growth factors and cytokines, many of which have known roles in the regulation of tumor initiation and progression. Coordination of these extracellular signals by CCN1 may determine the fate of cells in a potentially oncogenic environment. CCN1 was originally observed to promote tumor growth, with high expression found in a number of cancer cell lines (Babic *et al.*, 1998). However, subsequent studies by different groups of investigators have shown that CCN1 has divergent functions and expression profiles in cancer from different tissues. For example, high expression has been detected in breast cancer and is associated with increased tumorigenicity, but CCN1 expression is lower in lung cancer where it was found to reduce cellular proliferation (see Table 1 for summary). This chapter will examine the differential expression of CCN1 in cancers, review the current understanding of how CCN1 is regulated in tumor cells, and discuss the consequences of its aberrant expression in relation to cancer pathology.

Table 1.   Differential expression of CCN1 in cancer.

Tissue	Study comparisons	Expression of CCN1 relative to normal[a]
Breast	Cancer *vs.* normal	↑ 36–39%
	Stage II Invasive ductal carcinoma *vs.* matched normal	↑ 70%
Brain	Glioma *vs.* normal	↑ 38%
Uveal melanoma	Fresh tumor biopsies and archival tissue *vs.* cultured normal cells	↑ N.D.[b]
Prostate	Benign prostatic hyperplastic *vs.* normal	↑ 79%
Lung	Non-small cell lung cancer *vs.* matched normal	↓ 80%
Prostate	Cancer *vs.* matched normal	↓ 54%
Endometrial	Cancer *vs.* matched normal	↓ 88%
Uterine leiomyomas	Leiomyoma *vs.* autologous normal myometrium	↓ 100%

[a]Percentage of tumor samples that had expression levels of CCN1 significantly different to normal tissue.
[b]Not described.

2. CCN1 EXPRESSION IN CANCER

2.1. Increased Expression of CCN1 in Cancer

Normal breast tissue expresses low levels of CCN1. Several studies comparing matched samples from breast cancer patients have found elevated expression of CCN1 in tumor compared to normal tissue (Tsai *et al.*, 2000; Sampath *et al.*, 2001a; Xie *et al.*, 2001a). Overexpression of CCN1 was detected in 36% of samples by northern blot analysis (Xie *et al.*, 2001b) and 39% of samples by real-time PCR analysis (Xie *et al.*, 2001b), where overexpression was defined as greater than three-fold that of normal. Univariate statistical analysis showed a significant association between CCN1 overexpression and advanced stage and size of the primary tumor, lymph node involvement at the time of primary tumor removal and greater age at onset of disease. In a separate study, elevated CCN1 protein levels were detected in 70% of tumor tissues from patients with stage II invasive ductal carcinoma tumor compared to matched normal tissue (Sampath *et al.*, 2001a). *In situ* hydridization analysis of tissue from these patients demonstrated that CCN1 is expressed specifically by the tumor epithelial cells rather than by the surrounding stromal cells (Sampath *et al.*, 2001a). These observations suggest that CCN1 is closely associated with the malignant phenotype and may serve as a marker of breast cancer progression.

CCN1 expression was found to be positively correlated with elevated Her2/neu levels and with ER positivity (Sampath *et al.*, 2001a; Xie *et al.*, 2001a), suggesting that it contributes to estrogen and growth factor-driven breast tumorigenesis. However, ER-positive breast tumors are responsive to estrogen blockade, and ER expression in breast cancer is a good prognostic indicator. Therefore, it is paradoxical that CCN1 expression is associated with advanced disease, but these tumors are often ER-positive. In addition, expression of CCN2 and CCN4, but not CCN3, was elevated in breast cancers (Xie *et al.*, 2001b). Significant correlation was found among CCN1, CCN2 and CCN4 expression; however, association of CCN3 with the other three genes was not significant. A classification tree multivariate model predicted that the proportion of patients who have CCN1-positive breast cancer increases with age, tumor stage and ER level, and that ER and Her2 levels and tumor stage were important predictors for having prominent expression of two or three CCN genes.

Differential screening identified CCN1 as a gene up-regulated in the breast cancer cell line MDA-MB-231 as compared with the normal breast line MCF-12A (Xie *et al.*, 2001a; 2002). Prominent expression of CCN1 has been demonstrated in the highly invasive breast cancer cell lines MDA-MB-231,

MDA-MB-436, MDA-MB-157, with lower expression found in less tumorigenic cell lines MCF-7, BT-20 and ZR-57-1, and barely detectable expression in the normal breast cell lines MCF-10A and MCF-12A (Tsai *et al.*, 2000; Xie *et al.*, 2001a; Evtimova *et al.*, 2004). In contrast to that which is observed in primary tumor specimens, expression of CCN1 in breast cell lines correlates inversely with their ER status. However, in highly transformed breast cancer cell lines, other factors may be involved in up-regulation of CCN1, eliminating their requirement for estrogen. For example, some ER-negative cell lines also overexpress heregulin, an activator of Her2 signaling, which has been shown to induce CCN1 expression *in vitro* (Tsai *et al.*, 2000).

CCN1 mRNA was prominently expressed in highly invasive and tumorigenic glioma cell lines U87, U118, U138, U373 and T98G; levels were markedly lower in the less invasive and less tumorigenic cell line, U343, and was barely detectable in the normal brain cell line HBMEC (Xie *et al.*, 2002b; Martinerie *et al.*, 1997). Real-time PCR analysis was used to compare levels of CCN1 between normal brain and glioma samples, and the relationship between its expression and clinical parameters was explored (Xie *et al.*, 2004). CCN1 was significantly overexpressed in nearly half of primary gliomas. Univariate analysis showed a significant association of CCN1 overexpression with tumor grade, and a larger proportion of female patients expressed high CCN1 levels compared to males. Overexpression of CCN1 occurred predominantly in the most malignant samples (glioblastoma multifomae [GBM]). Furthermore, statistical analysis showed that CCN1 expression had a remarkably significant correlation with patient survival. In analysis of either all glioma patients or GBM patients only, high CCN1 expression was a negative prognostic factor for patient survival.

In another study, expression of CCN1 was undetectable in normal uveal melanocytes, but was found at elevated levels in uveal melanoma cell lines and primary uveal melanomas (Walker *et al.*, 2002). Data from RT-PCR experiments was confirmed by immunohistochemical analysis of paraffin-embedded tissue.

In conclusion, these studies suggest that CCN1 may play a role in the progression of cancer and their levels at diagnosis may have prognostic significance.

2.2. Reduced expression of CCN1 in cancer

CCN1 mRNA levels were decreased markedly in four of five non-small cell lung cancer (NSCLC) samples compared with matched normal lung tissue (Tong *et al.*, 2001). Extensive real-time PCR analysis using matched samples from 91 patients has confirmed this initial observation (our unpublished data). CCN1 levels were reduced in 80% of cancer samples compared to normal tissue, and expression was inversely correlated with stage of disease. Expression varied among lung cancer cell lines. It was absent both in the small cell lung cancer cell lines tested, and in two NSCLC cell lines that have a high potential to form tumors in nude mice. In contrast, expression was high in three NSCLC cell lines that have low tumorigenicity *in vivo*. Furthermore, overexpression of CCN1 in NSCLC cell lines caused growth suppression *in vitro* and *in vivo*, suggesting that it may play a role as a tumor-suppressor in NSCLC.

Leiomyomas are thought to arise from dedifferentiated smooth muscle cells in the myometrium. Expression of CCN1 was found to be consistently down-regulated at both the mRNA and protein level in uterine leiomyoma compared to matched normal myometrial tissue (Sampath *et al.*, 2001). These results suggest a role for CCN1 in maintaining a differentiated phenotype in uterine smooth muscle. Interestingly, while estrogen, basic fibroblast growth factor (bFGF) and serum were able rapidly to up-regulate CCN1 expression in normal myometrium explants cultured *ex vivo*, CCN1 mRNA levels in leiomyoma explants could not be up-regulated with these treatments. The lack of CCN1 up-regulation in response to local stimuli may therefore explain its reduced expression in leiomyomas.

Endometrial adenocarcinoma is the most common gynecologic malignancy. Development of these cancers arises from a series of genetic alterations that transform the normal endometrium through the stages of hyperplasia, dysplasia and finally overt carcinoma. Levels of CCN1 were decreased in endometrial tumors compared to matched normal endometrium (Chien *et al.*, submitted 2004). In seven out of eight paired tissue samples, the expression of CCN1 was higher in the normal endometrial tissue compared to the matched tumors. Also, CCN1 expression tended to be higher in the well-differentiated cell lines.

In contrast, CCN1 expression was found to be up-regulated in endometrium of women suffering from endometriosis (Absenger *et al.*, 2004). Endometriosis is a benign disorder characterized by the growth of endometrial cells outside the uterus, which may be caused by inappropriate differentiation of the endometrium. In this study, endometrial CCN1 expression was estrogen-responsive and anti-estrogen sensitive, and was up-regulated in the proliferative phase of the menstrual cycle.

CCN1 expression was found to be down-regulated in the tumor tissue of a small cohort of patients with prostate cancer (Pilarsky *et al.*, 1998). Immuno-histochemical analysis showed that expression was restricted to epithelial cells in the normal prostate, which was reduced in cancer tissue. Additionally, CCN1 was barely detectable in prostate cancer cell lines. In contrast, oligonucleotide array analysis identified CCN1 as being overexpressed in benign prostatic hyperplasia (BPH) compared to normal prostate (Sakamoto *et al.*, 2003). BPH is a non-malignant hyperplasia predominantly of the stromal tissue of the prostate. Lysophosphatidic acid, a known mitogen for prostate cells (Daaka, 2002), up-regulated CCN1 expression, and recombinant CCN1 promoted proliferation and cell adhesion of a prostate epithelial cell line derived from a patient with BPH (Sakamoto *et al.*, 2004). Taken together, these studies conducted thus far do not reveal a consistent role of CCN1 in regulation of cell growth in the prostate, but suggest a cell-specific expression pattern whose significance is yet to be determined.

Microarray analysis of papillary thyroid cancer demonstrated down regulation of CCN1 expression in tumor samples compared to normal thyroid (Wasenius *et al.*, 2003). One study used RT-PCR to measure expression of CCN1 in the livers of patients with hepatocellular carcinoma (Hirasaki *et al.*, 2001). However, expression levels were low, and no difference in expression was detected between the tumor and surrounding normal tissue.

3. REGULATION OF CCN1 EXPRESSION IN CANCER

CCN1 gene expression is regulated by an upstream serum-inducible promoter (Latinkic *et al.*, 1991). The promoter contains a serum response element (SRE) originally identified in the c-fos promoter. The SRE integrates MAPK signaling pathways to confer induction by serum, growth factors, cytokines and other extracellular stimuli (Whitmarsh *et al.* 1995). Consistent with these observations, CCN1 is up-regulated by a variety of growth-related factors, many of which are involved in cancer progression. The following section will discuss the regulation of CCN1 expression as it relates to cancer pathobiology.

The correlation of CCN1 expression with estrogen receptor positivity in breast cancers (Sampath *et al.*, 2001; Xie *et al.*, 2001a) indicates CCN1 mRNA expression may be regulated by estrogen receptor signaling *in vivo*. Also, CCN1 mRNA is induced by 17β-estradiol (E_2) in the uteri of ovariectomized rats (Rivera–Gonzalez *et al.*, 1998). CCN1 mRNA and protein can be induced by E_2 treatment in estrogen-responsive breast cell lines (Sampath *et al.*, 2001; Xie *et al.*, 2001). Induction of CCN1 in MCF-7 breast cancer cells is ER-dependent, as pre-treatment of MCF-7 cells with the estrogen receptor antagonists tamoxifen and ICI182,780 prevents E_2-stimulation of CCN1 expression. Two separate studies have shown that up-regulation of CCN1 by estrogen in

MCF-7 cells occurs in the presence of the protein synthesis inhibitor, cyclohexamide indicating that synthesis of intermediate proteins is not required (Sampath *et al.*, 2001a; Tsai *et al.*, 2002b). Therefore it is apparent that estrogen acts through the ER to induce CCN1 expression without the synthesis of additional proteins through a process that may or may not involve regulation of mRNA stability. CCN1 is an important downstream effector of hormone- and growth factor-dependent cell proliferation. Indeed, enhanced MCF-7 cell growth by either E_2 or EGF was diminished with neutralizing anti-CCN1 polyclonal antibodies, demonstrating that it contributes to E_2 and EGF-stimulated breast cancer proliferation (Sampath *et al.*, 2001). A similar study by the same group showed that the progestin, R5020, transcriptionally up-regulated CCN1 levels in a progesterone receptor- (PR) dependent manner (Sampath *et al.*, 2002). Anti-CCN1 antibodies again suppressed progestin-dependent DNA-synthesis, reinforcing the concept of CCN1 as a mediator of hormone- and growth factor-stimulated proliferation.

Up-regulation of the Her2 pathway is associated with tumor progression (Roskoski, 2004). As discussed above, a strong association exists between CCN1 and Her2 expression in breast tumors, suggesting that Her2 activation may be an important regulator of CCN1 expression in tumors. In support of this hypothesis, EGF and heregulin, ligands that activate the Her-2 kinase pathway, have been shown to induce CCN1 expression (Sampath *et al.*, 2002a; Tsai *et al.*, 2000). MCF-7 cells stably transfected with heregulin also express high levels of CCN1 (Tsai *et al.*, 2000). Furthermore, a neutralizing anti-CCN1 antibody blocked heregulin-dependent chemomigration of MCF-7 cells, demonstrating that CCN1 was a downstream regulator of heregulin action.

Cross talk between growth factor and hormone pathways appears to converge to enhance expression of on CCN1. Co-treatment of MCF-7 cells with E_2 and EGF resulted in prolonged CCN1 mRNA accumulation in comparison to treatment with either agent alone. This synergy correlates well with the overexpression of CCN1 in ER and EGFR positive tumors found in patient samples examined in the same study (Sampath *et al.*, 1998).

The active metabolite of vitamin D, 1,25-dihyroxyvitamin D_3 [1,25(OH)$_2$D$_3$], has been shown to up-regulate expression of CCN1 in breast cancer cell lines and osteoblasts (Scheutze *et al.*, 1998; Tsai *et al.*, 2002a). The 1,25(OH)$_2$D$_3$ has anti-proliferative activity against many cancers, including breast, and therefore would not have been expected to induce CCN1 expression. However, CCN1 is known to play a role in the differentiation of several cell types, and its transient expression may be an important component of 1,25(OH)$_2$D$_3$-induced differentiation. In contrast, retinoic acid, another steroid hormone receptor ligand with antiproliferative activity, was shown to suppress CCN1 expression in retinoid-sensitive MCF-7, cells but not retinoid-resistant MDA-MB-231 cells (Tsai *et al.*, 2002a). Additionally, the PKC activator, TPA, induces CCN1 expression in both MCF-7 and MDA-MB-231 breast cancer cells (Tsai *et al.*, 2002a).

Tumor growth and metastasis depend upon the formation of new blood vessels. Hypoxia within the tumor microenvironment promotes expression of angiogenic factors, including the transcription factor and hypoxia-inducible factor-1α (HIF1α) (Pugh and Ratcliffe, 2003). Microarray analysis identified CCN1 as a hypoxia-inducible gene in malignant melanoma cell lines (Kunz *et al.*, 2003). CCN1 promoter activity was induced by hypoxia in a HIF1α and AP-1 dependent manner. Furthermore, HIF1α and AP-1 physically interacted in response to hypoxia *in vivo*. EMSA analysis revealed that this hypoxia-inducible transcription factor complex formed within four hours of hypoxic stimulation on the CCN1 promoter. Thus, hypoxic conditions found within tumors may up-regulate CCN1 expression, which induces angiogenesis by promoting endothelial cell migration and proliferation (Kireeva *et al.*, 1996; Leu *et al.*, 2002). These results provide a mechanistic explanation for the increased size and vascularity of CCN1 overexpressing tumors grown in nude mice.

An intact AP-1 site is also important in regulation of CCN1 promoter function in response to activation of Rho GTPases by sphingosine-1-phosphate (S1P) in smooth muscle cells (Han *et al.*, 2003). Experiments using specific kinase inhibitors showed that Rho GTPase signals through p38 MAPK phosphorylation to induce CREB and AP-1-mediated CCN1 promoter activity and gene expression. In contrast, activation of JNK, but not p38 MAPK, was important for CCN1 up-regulation during neuronal cell death in response to toxic stimuli (Kim *et al.*, 2003). However, studies in the same cell line demonstrated that CCN1 expression during neuronal cell differentiation was associated with ERK activation (Chung *et al.*, 1998).

Thus, diverse signals of CCN1 are transmitted by different extracellular factors through various kinases and transcription factors in a cell- and context-specific manner. The wide range of biological functions in which CCN1 is involved are reflected by the diversity of environmental stimuli involved in its up-regulation.

4. EFFECT OF CCN1 ON TUMOR CELL LINES

CCN1 plays a significant role in normal development and cell biology (Mo *et al.*, 2002). It is thought to exert its effects through integrin activation.

Table 2.　Summary of effects of forced CCN1 overexpression on cell line behavior.

Tissue	Cell line	Change in phenotype caused by forced expression of CCN1[a]
Breast	MCF-12A	Increased anchorage-independent cell growth
		Stimulated migration *in vitro*
		Allowed tumor formation in nude mice
	MCF-7	Increased anchorage-independent cell growth
		Promoted migration *in vitro*
		Stimulated tumor growth and angiogenesis *in vivo*
		Caused estrogen-independent growth *in vitro* and *in vivo*
		Became resistance to chemotherapeutic drugs
Brain	U343	Increased proliferation *in vitro*
		Increased ability to form tumors in nude mice
		Enhanced angiogenesis *in vivo*
		Enhanced migration *in vitro*
		Activation of integrin signalling pathway
Gastric	RF-1	No change in proliferation *in vitro*
		Increased tumor size and vascularity *in vivo*
Lung	H460	Decreased proliferation *in vitro* associated with G_1
	H520	cell cycle arrest
		Reduced ability to form tumors in nude mice
		Up-regulation of β-catenin-c-myc-p53 pathway
Endometrium	AN3CA	Caused growth retardation *in vitro* and *in vivo*
	Ishikawa	Increased apoptosis

[a] See text for details.

Intracellular signaling pathways activated by CCN1 in normal cells have been identified that may be relevant to its effects on tumor cell growth: CCN1 is able to stimulate Wnt-β-catenin signaling during xenopus development (Latinkic *et al.*, 2003); it promotes a genetic program for wound healing in fibroblasts that includes ERK phosphorylation (Chen *et al.*, 2001). The following section will address the behavioral and molecular changes that occur in cancer cells when they are exposed to high levels of CCN1, by either forced overexpression by cDNA expression vector transfection, or treatment with recombinant protein (Table 2).

4.1.　Breast

Transfection of CCN1 in breast cell lines that do not normally express the gene elicits a more tumorigenic phenotype. MCF-12A (non-transformed) and MCF-7 (transformed) breast cell lines stably transfected with CCN1 proliferate more rapidly *in vitro* and demonstrate enhanced anchorage-independent growth, forming more and larger colonies in soft agar in comparison to control cells (Xie *et al.*, 2001a). Enhanced proliferation caused by CCN1 overexpression was found to occur in an estrogen-independent manner (Tsai *et al.*, 2002b). However, MCF-7/CCN1 cells remained responsive to estrogen, as their proliferation was still stimulated by treatment with E_2, and this stimulation could be reversed by antiestrogen treatment. Consistent with observations made in tumor samples, MCF-7/CCN1 cells retained expression of a functional ER, although levels of the receptor were slightly diminished (Tsai *et al.*, 2002b).

Tumor formation in nude mice was promoted by CCN1 overexpression in breast cells. MCF-7 cells overexpressing CCN1 (MCF-7/CCN1) grew more rapidly *in vivo* in comparison to vector control (MCF-7/V) cells. Significantly, while the normal MCF-12A cell line cannot form tumors in nude mice, transfection of these cells with CCN1 allowed them to form tumors within three weeks of subcutaneous injection, suggesting that CCN1 promotes tumorigenesis as well as tumor progression in breast cancer. Experiments showing that MCF-7 cells are unable to form tumors in ovariectomized, estrogen deficient nude mice, unless they are transfected with CCN1, provides additional evidence for this hypothesis (Tsai *et al.*, 2002b).

CCN1 overexpression also promotes angiogenic activity of breast cancer cells. In the breast cancer cell line MCF-7/CCN1, tumors grown subcutaneously in nude mice demonstrated increased blood vessel formation compared to MCF-7/V tumors as determined histologically and by immunohistochemical detection of CD31 (Xie *et al.*, 2001; Tsai *et al.*, 2002b). Furthermore, expression of the angiogenic growth factor, VEGF, could be detected in tumors derived from MCF-7/CCN1 cells (Tsai *et al.*, 2002b).

In addition, transfection of breast cell lines promotes a metastatic phenotype *in vitro*. MCF-12A and MCF-7 overexpressing CCN1 showed an increased ability to migrate in vitronectin-coated Boyden chambers, and MCF-7/CCN1 cells displayed extensive migration in a Matrigel outgrowth assay (Tsai *et al.*, 2002b; Xie *et al.*, 2001a).

4.2. Brain

Forced overexpression of CCN1 in the U343 brain tumor cells accelerated their proliferation in liquid culture as well as enhanced their anchorage-independent growth in soft agar and significantly increased their ability to form tumors in nude mice (Xie *et al.*, 2004b). Similar to results found in breast cancer cell lines, tumors derived from U343/CCN1 cells demonstrated increased blood vessel density in comparison to those from U343/V controls. In addition, U343/CCN1 cells displayed enhanced migration in vitronectin-coated Boyden chambers, which could be inhibited with an anti-CCN1 antibody. U343/CCN1 cells increased their expression of integrin receptor mRNAs, and expressed higher levels of $\alpha_v\beta_3$ on their cell surface.

Integrins can regulate cell migration and proliferation by signaling through an integrin-linked serine-threonine kinase (ILK) (Dedhar *et al.*, 1999). ILK activity was dramatically increased in U343/CCN1 cells, suggesting that CCN1 overexpression up-regulates levels of its own receptors, leading to activation of the ILK kinase. ILK is an upstream effector of the PI3 kinase-dependent phosphorylation of both AKT and glycogen synthase kinase 3 (GSK-3). Phosphorylation of these kinases results in β-catenin nuclear translocation, where it binds to members of the TCF4/LEF family and serves as a transcriptional activator of a variety of genes (Delcommenne *et al.*, 1998; Behrens, 2000). These genes are often associated with stimulating cell proliferation and include cyclin D1, c-myc and c-jun. This pathway was found to be up-regulated in U343/CCN1 cells, which increased their phosphorylation of both AKT and GSK-3 in relation to control cells and also demonstrated nuclear translocation of β-catenin. Transfection of U343/CCN1 cells with a dominant negative TCF4 caused their severe growth arrest and down-regulation of cyclin D1 levels in comparison to control cells, suggesting that CCN1 stimulates β-catenin to enhance expression of proteins associated with cell proliferation. Taken together, these observations reveal a mechanism whereby secreted CCN1 promotes tumor growth by activating integrin signaling to downstream kinases, which results in nuclear translocation of β-catenin and increased proliferation (Fig. 1).

4.3. Lung

CCN1 overexpression in NSCLC cell lines elicits a decrease in the tumorigenic phenotype, which is consistent with its reduced expression in patient samples (Tong *et al.*, 2001). Both H520 and H460 NSCLC cell lines express very low levels of CCN1. Stable transfection of these cell lines with a CCN1 expression vector resulted in their decreased proliferation relative to controls in both liquid culture and soft agar. These cells exhibited G_1 cell cycle arrest, which was accompanied by up-regulation of the cyclin-dependent kinase inhibitor, p21, and decreased CDK2 kinase activity. Both wild-type p53 and pRB2/p130 were up-regulated in H520/CCN1 and H460/CCN1 cells. More detailed analysis subsequently revealed that the molecular pathway used by CCN1 to suppress growth in these cells involved β-catenin, c-myc and p53

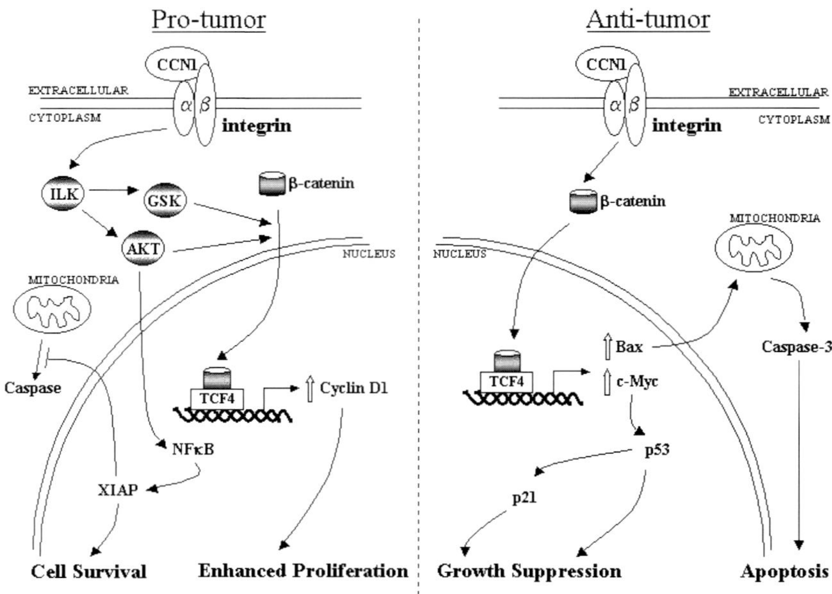

Fig. 1. **Divergent effects of CCN1 in cancer**. CCN1 is able to either promote or repress tumor growth in a tissue-specific manner. Cancerous tissues in which high levels of CCN1 has been detected, and where it increases tumorigenicity include breast and brain. Tissues in which CCN1 expression is decreased in cancer, and where it suppresses proliferation include lung and endometrium. Pro-growth effects of CCN1 in cancer cells are mediated though integrin receptors that activate the integrin linked kinase (ILK), which phosphorylates AKT and glcogen synthase kinase-3 (GSK). This results in β-catenin translocation to the nucleus, where it binds to TCF4 to activate transcription of genes such as cyclin D1 to enhance proliferation. Anti-tumor effects of CCN1 in tumor cells are also associated with β-catenin activity. In these cells, up-regulation of c-myc leads to stimulation of p53 or Bax, which causes growth arrest or apoptosis.

(Tong *et al.*, 2004; Fig. 1). The proto-oncogene, c-myc, is a transcriptional target of β-catenin/TCF4; and it in turn, acts as a transcriptional activator for genes that play important roles in cell proliferation, growth arrest and apoptosis (He *et al.*, 1998). c-myc acts as an oncogene in the absence of functional p53; in the lung cancer cells with wild-type p53, c-myc up-regulates p53 transcription and increases protein stability, leading to p53-mediated growth arrest and/or apoptosis (Hermeking and Eick, 1994; Zindy *et al.*, 1998). Proliferation of NSCLC cell lines expressing a mutant form of p53 was not suppressed by CCN1. In contrast, growth retardation in wild-type p53 expressing H520 and H460 cells transfected with CCN1 could be reversed by inhibiting p53 with either HPV16 E6 or dominant-negative p53. Levels of β-catenin and c-myc were increased in

H520/CCN1 cells and β-catenin was translocated into the nucleus. Transfection of H520/CCN1 cells with a dominant-negative form of TCF4 inhibited expression of c-myc, suggesting that activation of β-catenin signaling by CCN1 resulted directly in c-myc up-regulation. Treatment of H520/CCN1 cells with antisense oligonucleotides against c-myc reduced p53 expression, demonstrating that c-myc was upstream of p53. In addition, forced overexpression of c-myc in wild-type H520 cells caused growth inhibition and up-regulation of p53 and p21, confirming c-myc was a negative growth regulator in these cells. Thus, a molecular model for growth suppression by CCN1 in NSCLC can be proposed. First, secreted CCN1 binds to integrin receptors in an autocrine and paracrine manner and triggers a signal transduction pathway that leads to nuclear accumulation of β-catenin. Nuclear β-catenin associates with TCF4 to promote expression of c-myc, which successively up-regulates p53 and p21, leading to growth suppression. The mechanisms by which CCN1 is down-regulated in NSCLC are currently under investigation in our laboratory.

4.4. Endometrium

Undifferentiated AN3CA endometrial cancer cells transfected with CCN1 proliferated less rapidly and formed smaller tumors in nude mice compared to vector-transfected control cells (Chien *et al.*, submitted 2004). No difference occurred in the cell-cycle profiles between AN3CA/CCN1 and AN3CA/V cells. However, CCN1-mediated growth retardation was associated with increased apoptosis and decreased expression of anti-apoptotic Bcl-2 protein and increased levels of pro-apoptotic Bax. Caspase-3 activity was substantially increased and mitochondrial potential was decreased in well-differentiated Ishikawa endometrial adenocarcinoma cells treated with CCN1 conditioned medium, indicating that CCN1 inhibits growth of endometrial cancer cells by promoting apoptosis. In an analogous fashion to lung cancer cells, CCN1 overexpression in AN3CA cells up-regulated β-catenin activity and c-myc expression. c-myc has been reported to enhance activation of Bax (Juin *et al.*, 2002). We found Bax expression prominently induced in AN3CA/CCN1 cells, suggesting that a β-catenin-mediated tumor suppression pathway was activated by CCN1 in the endometrium (Fig. 1).

4.5. Other

RF-1 gastric adenocarcinoma cells have a low capacity to form tumors in nude mice. Stable transfection with a CCN1 expression vector did not change their growth rate in culture, but did promote their formation of lager, more vascular tumors *in vivo* (Babic *et al.*, 1998). Also, recombinant CCN1 protein promoted the adhesion and proliferation of BRF-55T prostatic epithelial cells (Sakomoto *et al.*, 2004). Furthermore, the serum-stimulated growth of BRF-55T cells was slightly decreased by the addition of an anti-CCN1 polyclonal antibody.

5. CHEMORESISTANCE

Differential expression of CCN1 has been identified by microarray analysis in several studies screening for genes involved in resistance to chemotherapy. In a large study of 55 cell lines, CCN1 was associated with resistance to a panel of chemotherapeutic drugs, suggesting that it may be a common predictive marker of drug resistance (Dan *et al.*, 2002). Further evidence for this hypothesis was provided by the observation that 5-Fluorouracil-resistant HCT116 colon cancer cell lines expressed higher levels of CCN1 (de Angelis *et al.*, 2004). In contrast, CCN1 was down-regulated in three melanoma sublines with acquired resistance to cisplatin, etoposide and fotemustine (Wittig *et al.*, 2002). Overexpression of CCN1 in MCF-7 cells confers resistance to apoptosis induced by paclitaxel, adriamycin and β-lapachone (Lin *et al.*, 2004). Resistance to apoptosis was due to up-regulation of NF-κB activity and consequent increased expression of anti-apoptotic XIAP (Fig. 1). Inhibition of NF-κB or XIAP restored the sensitivity of MCF-7/CCN1 cells to the chemotherapeutic agents. Furthermore, neutralizing antibodies to integrins $\alpha_v\beta_3$ and $\alpha_v\beta_5$ inhibited NF-κB activation and XIAP expression in MCF-7/CCN1 cells, providing a link between CCN1 signaling at the cell surface and apoptosis resistance. Another potential mechanism for drug resistance is demonstrated by the finding that in U343 brain cancer cells transfected with CCN1, increased ILK and AKT activity leads to phosphorylation and suppression of the proapoptotic protein Bad (Xie *et al.*, 2004b).

6. SUMMARY

CCN1 has the potential to regulate coordinately many aspects of cancer activity. It may promote tumor progression in an autocrine or paracrine fashion by stimulating proliferation via ligation and activation of integrin receptor-stimulated intracellular kinase pathways and augmention of growth factor activity, by enhancing the migratory capacity of tumor epithelial cells, by promoting tumor neovascularization though recruitment of endothelial cells, and by causing resistance to chemotherapeutic drugs. Suppression of tumor cell proliferation by CCN1 seems to occur through similar intracellular signaling pathways. Therefore, the ultimate effects of CCN1 expression are both context- and tissue-dependent. These divergent effects may result from a combination of several events: i) the presence or absence of different growth factors in the local tumor environment that interact with CCN1 in a synergistic or antagonistic fashion; ii) the expression of different integrin receptor profiles by cancer cells that results in their activation by CCN1 of signaling pathways specific to either proliferation or growth suppression; and iii) the molecular context in which these signals are received, for example the cells expression of either wild-type or mutant p53.

7. ACKNOWLEDGMENTS

Supported by grants from National Institutes of Health RO1-109295, the NIH Lung Spore Grant and Department of Defense Grant DAMD17-02-1-0337. H.P.K is the recipient of the Mark Goodson Endowed Chair in Oncology Research and is a member of the Jonsson Cancer Center and the Molecular Biology Institute of UCLA.

REFERENCES

Absenger Y, Hess–Stumpp H, Kreft B, Kratzschmar J, Haendler B, Schutze N, Regidor PA and Winterhager E (2004) Cyr61, a deregulated gene in endometriosis. *Mol Hum Reprod* **10**: 399–407.

Babic AM, Kireeva ML, Kolesnikova TV and Lau LF (1998) CYR61, a product of a growth factor-inducible immediate early gene, promotes angiogenesis and tumor growth. *Proc Natl Acad Sci USA* **95**: 6355–6360.

Behrens J (2000) Control of beta-catenin signaling in tumor development. *Ann N Y Acad Sci* **910**: 21–33.

Brigstock DR (1999) The connective tissue growth factor/cysteine-rich 61/nephroblastoma overexpressed (CCN) family. *Endocr Rev* **20**: 189–206.

Chen CC, Mo FE and Lau LF (2001) The angiogenic factor Cyr61 activates a genetic program for wound healing in human skin fibroblasts. *J Biol Chem* **276**: 47329–47337.

Chien WW, Kumagai T, Miller CW, Desmond JC, Frank JM, Said JW and Koeffler HP (2004) Cyr61 suppresses growth of human endometrial cancer cells. Manuscript submitted.

Chung KC and Ahn YS (1998) Expression of immediate early gene Cyr61 during the differentiation of immortalized embryonic hippocampal neuronal cells. *Neurosci Lett* **255**: 155–158.

Daaka Y (2002) Mitogenic action of LPA in prostate. *Biochim Biophys Acta* **1582**: 265–269.

Dedhar S, Williams B and Hannigan G (1999) Integrin-linked kinase (ILK): a regulator of integrin and growth-factor signalling. *Trends Cell Biol* **9**: 319–323.

Delcommenne M, Tan C, Gray V, Rue L, Woodgett J and Dedhar S (1998) Phosphoinositide-3-OH kinase-dependent regulation of glycogen synthase kinase 3 and protein kinase B/AKT by the integrin-linked kinase. *Proc Natl Acad Sci USA* **95**: 11211–11216.

de Angelis PM, Fjell B, Kravik KL, Haug T, Tunheim SH, Reichelt W, Beigi M, Clausen OP, Galteland E and Stokke T (2004) Molecular characterizations of derivatives of HCT116 colorectal cancer cells that are resistant to the chemotherapeutic agent 5-fluorouracil. *Int J Oncol* **24**: 1279–1288.

Evtimova V, Zeillinger R and Weidle UH (2003) Identification of genes associated with the invasive status of human mammary carcinoma cell lines by transcriptional profiling. *Tumour Biol* **24**: 189–198.

Han JS, Macarak E, Rosenbloom J, Chung KC and Chaqour B (2003) Regulation of Cyr61/CCN1 gene expression through RhoA GTPase and p38MAPK signaling pathways. *Eur J Biochem* **270**: 3408–3421.

Hirasaki S, Koide N, Ujike K, Shinji T and Tsuji T (2001) Expression of Nov, CYR61 and CTGF genes in human hepatocellular carcinoma. *Hepatol Res* **19**: 294–305.

He TC, Sparks AB, Rago C, Hermeking H, Zawel L, da Costa LT, Morin PJ, Vogelstein B and Kinzler KW (1998) Identification of c-MYC as a target of the APC pathway. *Science* **281**: 1509–1512.

Hermeking H and Eick D (1994) Mediation of c-Myc-induced apoptosis by p53. *Science* **265**: 2091–2093.

Juin P, Hunt A, Littlewood T, Griffiths B, Swigart LB, Korsmeyer S and Evan G (2002) c-Myc functionally cooperates with Bax to induce apoptosis. *Mol Cell Biol* **22**: 6158–6169.

Kireeva ML, Mo FE, Yang GP and Lau LF (1996) Cyr61, a product of a growth factor-inducible immediate-early gene, promotes cell proliferation, migration, and adhesion. *Mol Cell Biol* **16**: 1326–1334.

Kim KH, Min YK, Baik JH, Lau LF, Chaqour B and Chung KC (2003) Expression of angiogenic factor Cyr61 during neuronal cell death via the activation of c-Jun N-terminal kinase and serum response factor. *J Biol Chem* **278**: 13847–13854.

Kunz M, Moeller S, Koczan D, Lorenz P, Wenger RH, Glocker MO, Thiesen HJ, Gross G and Ibrahim SM (2003) Mechanisms of hypoxic gene regulation of angiogenesis factor Cyr61 in melanoma cells. *J Biol Chem* **278**: 45651–45660.

Latinkic BV, O'Brien TP and Lau LF (1991) Promoter function and structure of the growth factor-inducible immediate early gene Cyr61. *Nucleic Acids Res* **19**: 3261–3267.

Lau LF and Nathans D (1985) Identification of a set of genes expressed during the G0/G1 transition of cultured mouse cells. *EMBO J* **4**: 3145–3151.

Leu SJ, Lam SC and Lau LF (2002) Pro-angiogenic activities of CYR61 (CCN1) mediated through integrins alphavbeta3 and alpha6beta1 in human umbilical vein endothelial cells. *J Biol Chem* **277**: 46248–46255.

Lin MT, Chang CC, Chen ST, Chang HL, Su JL, Chau YP and Kuo ML (2004) Cyr61 expression confers resistance to apoptosis in breast cancer MCF-7 cells by a mechanism of NF-kappaB-dependent XIAP up-regulation. *J Biol Chem* **279**: 24015–24023.

Martinerie C, Viegas–Pequignot E, Nguyen VC and Perbal B (1997) Chromosomal mapping and expression of the human Cyr61 gene in tumour cells from the nervous system. *Mol Pathol* **50**: 310–316.

Menendez JA, Mehmi I, Griggs DW and Lupu R (2003) The angiogenic factor CYR61 in breast cancer: molecular pathology and therapeutic perspectives. *Endocr Relat Cancer* **10**: 141–152.

Mo FE, Muntean AG, Chen CC, Stolz DB, Watkins SC and Lau LF (2002) CYR61 (CCN1) is essential for placental development and vascular integrity. *Mol Cell Biol* **22**: 8709–8720.

Pilarsky CP, Schmidt U, Eissrich C, Stade J, Froschermaier SE, Haase M, Faller G, Kirchner TW and Wirth MP (1998) Expression of the extracellular matrix signaling molecule Cyr61 is down regulated in prostate cancer. *Prostate* **36**: 85–91.

Planque N and Perbal B (2003) A structural approach to the role of CCN (CYR61/CTGF/NOV) proteins in tumourigenesis. *Cancer Cell Int* **3**: 15.

Pugh CW and Ratcliffe PJ (2003) Regulation of angiogenesis by hypoxia: role of the HIF system. *Nat Med* **9**: 677–684.

Rivera–Gonzalez R, Petersen DN, Tkalcevic G, Thompson DD and Brown TA (1998) Estrogen-induced genes in the uterus of ovariectomized rats and their regulation by droloxifene and tamoxifen. *J Steroid Biochem Mol Biol* **64**: 13–24.

Roskoski R Jr. (2004) The ErbB/HER receptor protein-tyrosine kinases and cancer. *Biochem Biophys Res Commun* **319**: 1–11.

Sakamoto S, Yokoyama M, Prakash K, Tsuruha J, Masamoto S, Getzenberg RH and Kakehi Y (2003) Development of quantitative detection assays for CYR61 as a new marker for benign prostatic hyperplasia. *J Biomol Screen* **8**: 701–711.

Sakamoto S, Yokoyama M, Zhang X, Prakash K, Nagao K, Hatanaka T, Getzenberg RH and Kakehi Y (2004) Increased expression of CYR61, an extra-cellular matrix signaling protein, in human benign prostatic hyperplasia and its regulation by lysophosphatidic acid. *Endocrinology* **145**: 2929–2940.

Sampath D, Winneker RC and Zhang Z (2001a) Cyr61, a member of the CCN family, is required for MCF-7 cell proliferation: regulation by 17beta-estradiol and overexpression in human breast cancer. *Endocrinology* **142**: 2540–2548.

Sampath D, Zhu Y, Winneker RC and Zhang Z (2001b) Aberrant expression of Cyr61, a member of the CCN (CTGF/Cyr61/Cef10/NOVH) family, and dysregulation by 17 beta-estradiol and basic fibroblast growth factor in human uterine leiomyomas. *J Clin Endocrinol Metab* **86**: 1707–1715.

Sampath D, Winneker RC and Zhang Z (2002) The angiogenic factor Cyr61 is induced by the progestin R5020 and is necessary for mammary adenocarcinoma cell growth. *Endocrine* **18**: 147–159.

Schutze N, Lechner A, Groll C, Siggelkow H, Hufner M, Kohrle J and Jakob F (1998) The human analog of murine cystein rich protein 61 is a 1alpha,25-dihydroxyvitamin D3 responsive immediate early gene in human fetal osteoblasts: regulation by cytokines, growth factors, and serum. *Endocrinology* **139**: 1761–1770.

Tong X, O'Kelly J, Xie D, Mori A, Lemp N, McKenna R, Miller CW and Koeffler HP (2004) Cyr61 suppresses the growth of non-small-cell lung cancer cells via the beta-catenin-c-myc-p53 pathway. *Oncogene* **23**: 4847–4855.

Tong X, Xie D, O'Kelly J, Miller CW, Muller-Tidow C and Koeffler HP (2001) Cyr61, a member of CCN family, is a tumor suppressor in non-small cell lung cancer. *J Biol Chem* **276**: 47709–47714.

Tsai MS, Bogart DF, Castaneda JM, Li P and Lupu R (2002b) Cyr61 promotes breast tumorigenesis and cancer progression. *Oncogene* **21**: 8178–8185.

Tsai MS, Bogart DF, Li P, Mehmi I and Lupu R (2002a) Expression and regulation of Cyr61 in human breast cancer cell lines. *Oncogene* **21**: 964–973.

Tsai MS, Hornby AE, Lakins J and Lupu R (2000) Expression and function of CYR61, an angiogenic factor, in breast cancer cell lines and tumor biopsies. *Cancer Res* **60**: 5603–5607.

van Ginkel PR, Gee RL, Shearer RL, Subramanian L, Walker TM, Albert DM, Meisner LF, Varnum BC and Polans AS (2004) Expression of the receptor tyrosine kinase Axl promotes ocular melanoma cell survival. *Cancer Res* **64**: 128–134.

Walker TM, Van Ginkel PR, Gee RL, Ahmadi H, Subramanian L, Ksander BR, Meisner LF, Albert DM and Polans AS (2002) Expression of angiogenic factors Cyr61 and tissue factor in uveal melanoma. *Arch Ophthalmol* **120**: 1719–1725.

Wasenius VM, Hemmer S, Kettunen E, Knuutila S, Franssila K and Joensuu H (2003) Hepatocyte growth factor receptor, matrix metalloproteinase-11, tissue inhibitor of metalloproteinase-1, and fibronectin are up-regulated in papillary thyroid carcinoma: a cDNA and tissue microarray study. *Clin Cancer Res* **9**: 68–75.

Whitmarsh AJ, Shore P, Sharrocks AD and Davis RJ (1995) Integration of MAP kinase signal transduction pathways at the serum response element. *Science* **269**: 403–407.

Wittig R, Nessling M, Will RD, Mollenhauer J, Salowsky R, Munstermann E, Schick M, Helmbach H, Gschwendt B, Korn B, Kioschis P, Lichter P, Schadendorf D and Poustka A (2002) Candidate genes for cross-resistance against DNA-damaging drugs. *Cancer Res* **62**: 6698–6705.

Xie D, Jauch A, Miller CW, Bartram CR and Koeffler HP (2002) Discovery of over-expressed genes and genetic alterations in breast cancer cells using a combination of suppression subtractive hybridization, multiplex FISH and comparative genomic hybridization. *Int J Oncol* **21**: 499–507.

Xie D, Miller CW, O'Kelly J, Nakachi K, Sakashita A, Said JW, Gornbein J and Koeffler HP (2001a) Breast cancer. Cyr61 is overexpressed, estrogen-inducible, and associated with more advanced disease. *J Biol Chem* **276**: 14187–14194.

Xie D, Nakachi K, Wang H, Elashoff R and Koeffler HP (2001b) Elevated levels of connective tissue growth factor, WISP-1, and CYR61 in primary breast cancers associated with more advanced features. *Cancer Res* **61**: 8917–8923.

Xie D, Yin D, Wang HJ, Liu GT, Elashoff R, Black K and Koeffler HP (2004a) Levels of expression of CYR61 and CTGF are prognostic for tumor progression and survival of individuals with gliomas. *Clin Cancer Res* **10**: 2072–2081.

Xie D, Yin D, Tong X, O'Kelly J, Mori A, Miller C, Black K, Gui D, Said JW and Koeffler HP (2004b) Cyr61 is overexpressed in gliomas and involved in integrin-linked kinase-mediated Akt and beta-catenin-TCF/Lef signaling pathways. *Cancer Res* **64**: 1987–1996.

Zindy F, Eischen CM, Randle DH, Kamijo T, Cleveland JL, Sherr CJ and Roussel MF (1998) Myc signaling via the ARF tumor suppressor regulates p53-dependent apoptosis and immortalization. *Genes Dev* **12**: 2424–2433.

CHAPTER 16

CCN4 AND CCN6 VARIANTS IN WNT-INDUCIBLE SIGNALING PATHWAY

Shinji Tanaka

Department of Hepato-Biliary-Pancreatic Surgery
Tokyo Medical and Dental University
1-5-45 Yushima, Bunkyo-Ku, Tokyo 113-8519, Japan

Extracellular stimulation activates the particular receptors to transduce intracellular signals into the nucleus. It is well recognized that abnormal signal transduction occurs in cancer cells. Recent investigations revealed one of the major pathways implicated in gastrointestinal malignancies has been found to involve the Wnt-Wingless system. Indeed, we isolated a Wnt receptor gene designated Frizzled that represents a key signaling molecule for beta-catenin regulation in cancers. Among Wnt-inducible genes downstream of beta-catenin, it is noteworthy that three homologues of CCN family members has been identified as Wnt-inducible signaling pathway 1 (WISP1)/CCN4, WISP2/CCN5, and WISP3/CCN6. Recently, we isolated novel variants of WISP1/CCN4 and WISP3/CCN6 in human cancer tissues. This review focuses on Wnt in oncogenic signal transduction pathways and especially WISP/CCN variants, and emphasizes their regulation as an important concept that may influence future cancer treatments.

1. WNT/BETA-CATENIN SIGNALING CASCADE

The proto-oncogene Wnt-1 (mammalian Wingless homologue) is a key regulator of cell and segment polarity during the development of *Drosophila*. Activation of mouse Wnt-1 (proto-oncogene) via integration of the mouse mammary tumor virus (MMTV) caused murine breast tumors leading to further investigation on its role in human tumorogenesis (Nusse and

Tel: 81-3-5803-5928, Fax: 81-3-5803-0263
Email: shinji.msrg@tmd.ac.jp

Varmus, 1982). There are numerous reports citing Wnt overexpression, and underexpression in human tumors but mRNA expression levels are only correlative at best. Thus far, the Wnt ligands, of which there are at least 16 members in vertebrates, are secreted glycoproteins that can be loosely categorized according to their ability to promote neoplastic transformation (Wong *et al.*, 1994). For example, Wnt-1, Wnt-3, or Wnt-10b activation, following retroviral insertion into murine mammary epithelial cell DNA, promoted tumor formation (Wong *et al.*, 1994). Because Wnt overexpression resulted in morphologic transformation, the oncogenic potential was assessed in cultured fibroblasts and in mammalian cells; cellular transformation occurred following Wnt-1, Wnt-2, Wnt-3a transfection, but was negative upon use of Wnt-4, Wnt-5a, and Wnt-6 cDNAs. Interestingly, transforming Wnt genes promoted nuclear translocation of beta-catenin in some cultured mammalian cells by mechanisms described in Fig. 1 (Peifer and Polakis, 2000).

Signaling is initiated by the secreted Wnt proteins, which bind to a class of seven transmembrane receptors encoded by the Frizzled genes (Bhanot *et al.*, 1996). Previous studies have found a human Frizzled homologue, FzE3, to be overexpressed in esophageal tumor cell lines, but not in adjacent normal mucosal tissue (Tanaka *et al.*, 1998). It was characterized as the specific LDL-receptor-related protein (LRP) ligand necessary for optimal Wnt/Frizzled signaling (Tamai *et al.*, 2000). Activation of the Frizzled receptor leads to Dishevelled protein interaction, which prevents glycogen synthase

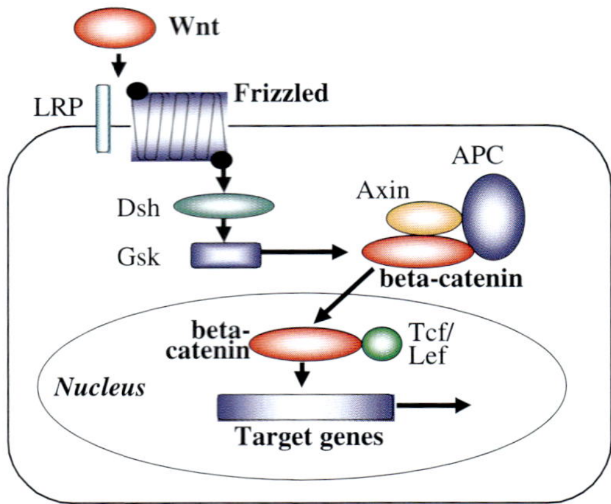

Fig. 1.   Wnt signaling pathway. LRP; LDL-receptor-related protein. Dsh; dishevelled. Gsk; glycogen synthase kinase. APC; adenomatous polyposis coli.

kinase 3 (GSK3) from phosphorylating critical downstream substrates (Itoh *et al.*, 1998). The GSK3 substrates include the negative regulators Axin and APC (adenomatous polyposis coli tumor suppressor), as well as beta-catenin (Rubinfeld *et al.*, 1997). This serine/threonine kinase GSK3 binds to and phosphorylates several intracellular proteins in the Wnt pathway and appears instrumental to the down-regulation of beta-catenin (*Drosphila* Armadillo). Beta-catenin phosphorylation by GSK3 allows recognition by beta-TRCP, an essential component of an E3 ubiquitin ligase system with subsequent degradation of beta-catenin within the cell. The frequency of beta-catenin mutations at the GSK3 recognition or binding site in various cancer cells (Morin *et al.*, 1987).

Based on activation of the Wnt signal transduction cascade and/or substrate mutations, unphosphorylated beta-catenin escapes recognition by beta-TRCP, and translocates to the nucleus where it engages TCF and LEF transcription factors (Behrens *et al.*, 1996). There have been reported numerous genes that are downstream targets of beta-catenin transcriptional activity. The diverse roles of Wnts in both development and tumorigenesis have fostered the search for Wnt responsive genes in these processes. Wg signaling in *Drosophila* is known to transcriptionally activate the expression of *engrailed* and *Ultrabithorax* through the Armadillo/dTCF complex. Besides these *Drosophila* homeobox genes, Wnt signaling through beta-catenin in *Xenopus* results in the transcriptional induction of homeobox genes. Other responsive genes of Wnt/Wg that are transcriptionally activated by beta-catenin/Lef-TCF include the oncogenes such as cyclin D1, c-myc, c-jun, fra-1, and matrix metalloproteases, as well as the p53 tumor suppressor gene (Moon *et al.*, 2002). Recently, Pennica *et al.* isolated novel Wnt-inducible genes downstream of beta-catenin; Wnt-inducible signaling pathway 1 (WISP1), WISP2, and WISP3 (Pennica *et al.*, 1998). It is noteworthy that these WISPs are homologous to CCN family proteins, and classified as CCN4, CCN5, and CCN6, respectively.

2. CCN FAMILY AND WISPs

CCN is a novel secreted cysteine-rich peptide family (Perbal, 2004), which stands for Cysteine-rich 61 (Cyr61)/CCN1, connective tissue growth factor (CTGF)/CCN2, and nephroblastoma overexpressed (Nov)/CCN3 molecule, and appears to be involved in human fibrotic disorders associated with stromal tissues. CTGF mRNA is overexpressed in numerous fibrotic disorders both in skin and in internal organs, such as in scleroderma. Similarly, high protein levels of Cyr61 are detected throughout the reparative phase of the callus, particularly in fibrous tissue and periosteum, and in proliferating chondrocytes during repair of bone fracture. Some CCN family members belong to immediate-early genes expressed after induction by certain oncogenes and growth factors such as EGF and TGF-beta.

A characteristic CCN protein is composed of signal sequences (SS) responsible for protein secretion, and four distinct modules that exhibit homology to conserved regions in a variety of extracellular proteins (Bork, 1993). Module I is an amino-terminal domain that has homology to the insulin-like growth factor binding proteins (IGFBP). Module II is a domain with homology to Von Willebrand factor type C (VWC) repeats and may participate in protein complex formation. Module III is a thrombospondin (THBS) type I domain and may be involved in the binding of CCNs to sulfated glycosaminoglycans either on the cell surface or in the extracellular matrix. Module IV is a cysteine knot (CK) domain, which has been identified in several other signaling peptides (such as TGF-beta, platelet derived growth factor and nerve growth factor), and may participate in dimerization and receptor binding.

Although several CCN family proteins such as Cyr61 and CTGF appear to enhance invasive cell properties (Kireeva *et al.*, 1996; Babic *et al.*, 1999), the WISP1 gene was first isolated as a suppressor of metastasis (Hashimoto *et al.*, 1998). WISP2 lacks the fourth module of CK2 that positively regulates osteoblast function (Kumar *et al.*, 1999) and breast carcinoma cell growth (Zoubine *et al.*, 2001). Finally, WISP3 has been independently reported as a gene (*LIBC*) that was lost in inflammatory breast cancers (van Golen *et al.*, 1999) suggesting a potential tumor suppressor function for wild-type WISP3.

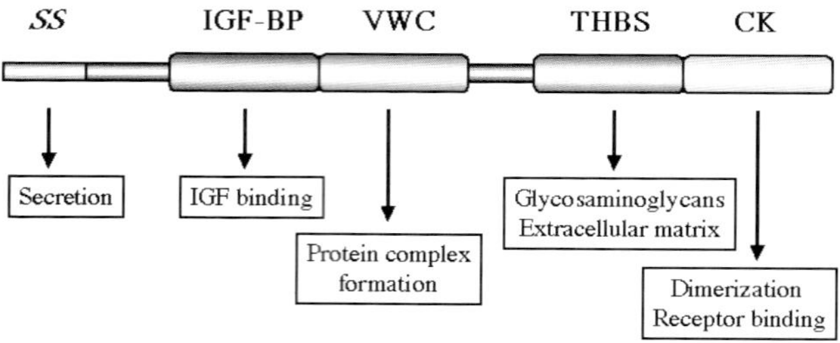

Fig. 2. Schematic structure of the typical CCN protein and potential functions. The CCN protein is composed of signal sequences (SS), a domain with homology to insulin-like growth factor binding proteins (IGFBP), a domain homologous to von Willebrand factor type C (VWC) repeats, a thrombospondin (THBS) type I domain, and a cysteine knot (CK) domain.

While the WISP subfamily might be closely associated with the function of either Wnt signals or matrix proteins, it remained to be identified how to contribute to carcinogenesis and cancer progression.

3. WISP1 VARIANT (WISP1v)

Scirrhous carcinoma of the stomach is characterized by rapid growth with a vast fibrous stroma, high invasiveness, and substantially a poor prognosis. Little is known of the molecular pathogenesis of this disease. Using targeted differential displays, we identified a novel variant of WISP1, named WISP1v, as overexpressed in scirrhous gastric carcinomas (Tanaka *et al.*, 2001). Predicted protein of the WISP1v completely lacks the second module of VWC that is thought to participate in protein complex formation (Fig. 3). According to the human genomic analysis, the WISP1 coding region was found to consist of five exons. Interestingly, each module of WISP1 was encoded by each exon, as noted for other CCN members (Perbal, 2004). Alternative splicing of the entire third exon revealed production of WISP1v mRNA. No genomic mutation in the intron:exon junction was detected in all of the examined samples expressing the WISP1v, hence another mechanism for alternative splicing of the third exon is involved.

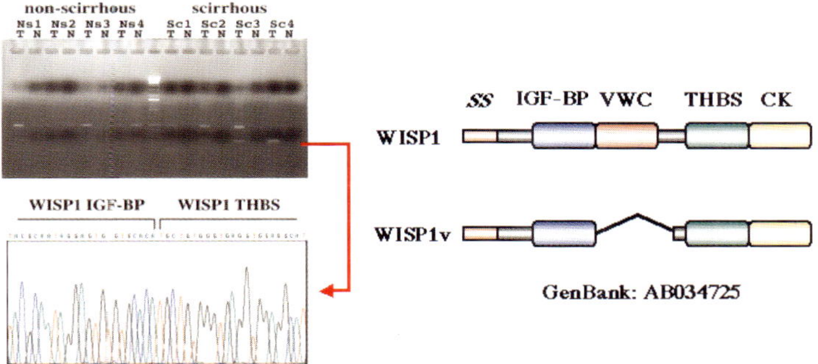

Fig. 3. Identification of cancer-specific genes of the CCN family in non-scirrhous or scirrhous gastric carcinoma tissues. *Left upper:* Expression pattern of CCN family by RT-PCR in tissue samples of gastric carcinoma (T) compared to normal adjacent mucosa (N) derived from non-scirrhous (Ns) or scirrhous types (Sc). The number indicates the clinical sample. A small product amplified from the scirrhous type of gastric carcinoma, but not from the non-scirrhous carcinoma. *Left lower:* Direct sequence analysis on the small PCR product from the scirrhous gastric carcinoma, revealed a sole fragment lacking the second module VWC of WISP1. *Right:* Schematic structure of WISP1v in comparison with the wild-type protein of WISP1.

Ectopic expression revealed WISP1v to be a secreted oncoprotein inducing a striking cellular transformation and rapid piling-up growth. It is noteworthy that WISP1v transfectants enhanced the invasive phenotype of co-cultured gastric carcinoma cells, while wild-type WISP1 had no such potential. These findings suggest that CCN protein WISP1v is involved in the aggressive progression of scirrhous gastric adenocarcinoma. The WISP1v stimulated the invasive phenotype of cancer cells with activation of both p38 and p42/p44 mitogen-activated protein kinases (MAPKs) (Fig. 4). Furthermore, WISP1v-induced cancer invasion was significantly suppressed by the p38 MAPK inhibitor SB203580 but not by the p42/p44 MAPK kinase (MEK) inhibitor PD98059. Our findings suggest that WISP1v mediated signaling is involved in the generation of invasive cellular properties and leads to progression of human carcinoma.

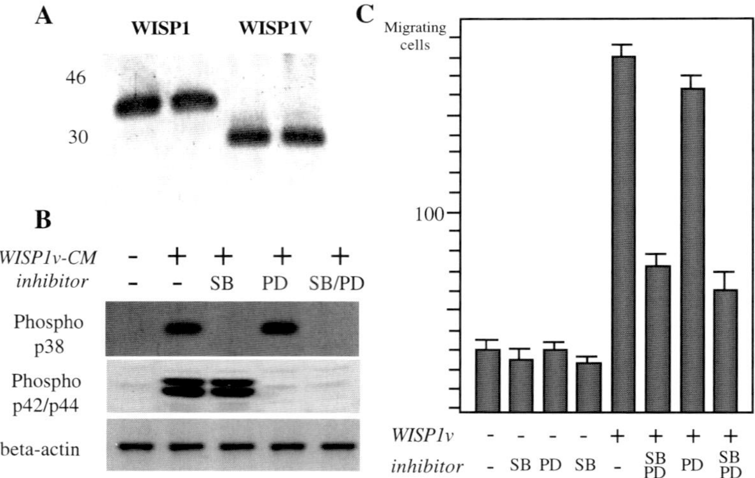

Fig. 4.   Biological effects of WISP1v. *A:* Expression analysis of the WISP1s protein of the transfected cell lysate under non-reducing conditions. The molecular weights shown in the left lane. *B:* Immunoblot detection of threonine phosphorylated MAPKs in WISP1v-stimulated HuCCT1 cholangiocarcinoma cells (upper: p38 MAPK, lower: p42/p44 MAPK). SB and PD indicate the administration of kinase inhibitors specific for p38 (1×10^{-2} mM of SB203580) and for MEK1/2 (5×10^{-2} mM of PD98059), respectively. Equal expression of beta-actin certified the quality of protein in each sample. *C:* The migration of cholangiocarcinoma cells stimulated by WISP1v-CM (+) or mock-CM (−), with or without kinase inhibitors. Cell movement was evaluated by the number of cells migrating into Matrigel. Results are derived from three individual experiments and error bars indicate the standard deviation from the mean. There is a statistical difference in migrated cell numbers between WISP1v-positives and WISP1v-negatives ($p < 0.05$).

4. WISP3 VARIANT (WISP3v)

Microsatellites are repetitive DNA sequences distributed throughout the genome and known to be unstable in a subset of human tumors referred to as being microsatellite instable (Ionov *et al.*, 1993). The microsatellite instability phenotype is caused by inactivating alterations of mismatch repair genes. Stretches of short repetitive sequences are prone for mutation in a microsatellite instable cell. These sequences are most common in noncoding parts of the genome. However, some genes contain repeats within their coding region, and mutations in these genes may provide a selective advantage for tumor cell growth. Several genes have been found to be targeted by this mechanism and to be mutated in microsatellite instable carcinomas of the gastrointestine.

WISP3 contains an (A)9 repeat within its coding region between the second and third modules, and a frameshift mutation in this repeat leads to a premature stop codon at 69-bp downstream. The truncated protein WISP3(A)8 lacks both of the third module THBS and the fourth module CK (Fig. 5). *WISP3(A)8* gene was recognized in 31% of microsatellite instable colorectal carcinomas (Thorstensen *et al.*, 2001), and 21% of microsatellite instable gastric carcinomas in our experiments (Tanaka *et al.*, 2002a), while none of the microsatellite instable esophageal carcinomas exhibited such mutation. To evaluate the effects

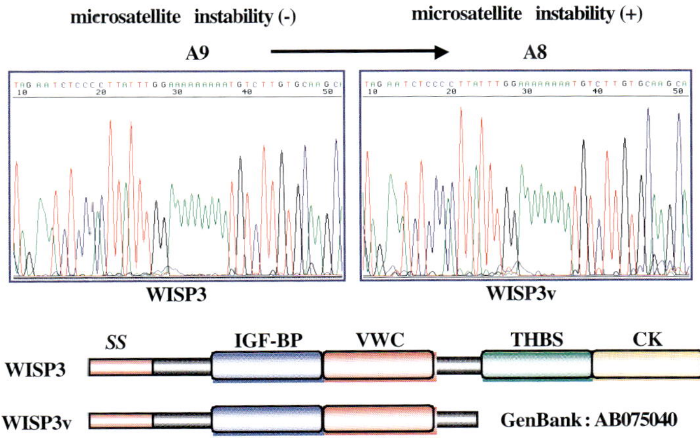

Fig. 5.   Identification of a variant WISP3 gene associated with microsatellite instability. *Upper:* Direct sequence analysis on the A9 repeat of the WISP3 gene, revealed a A8 frameshift mutation in microsatellite instable carcinoma of the stomach. *Lower:* Schematic structure of WISP3v in comparison with the wild-type protein of WISP3.

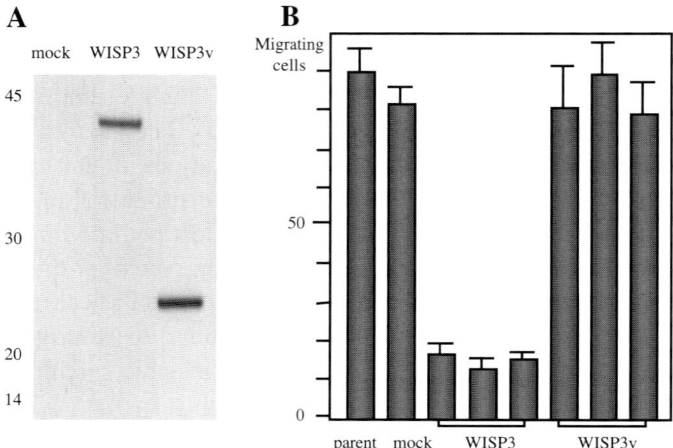

Fig. 6. *A:* Ectopic expression of WISP3 in Kato-III human gastric carcinoma cells. Stable transfectants of WISP3(A)9 cDNA or WISP3(A)8 cDNA were prepared with a bicistonic expression
vector pIRESneo (Clontech, Palo Alto, CA), that contains the internal ribosome entry site (IRES)
of the encephalomyocarditis virus. The expression cassette contains a single promoter which, in
combination with the IRES, permits the translation of both the gene of interest and the neomycin-
resistant gene from the same mRNA. Specific expression of WISP3 was detected in cell lysates of
the transfectants using a polyclonal antibody against the WISP3 IGFBP domain (sc-8871; Santa
Cruz Biochemistry, Santa Cruz, CA). The mock-vector transfectants as a negative control. The
molecular weights shown in the left lane. *B:* Matrigel invasion assays on the transfectants or the
parental gastric carcinoma Kato-III cells. The number of cells invading into the Matrigel matrix
were determined for each clone, following 24 hr stimulation of chemotaxis by serum. To avoid
clonal deviation, three independent clones of WISP3(A)9 or WISP3(A)8 transfectants were used,
and the results were derived from three individual experiments. Error bars indicate the standard
deviation from the mean.

of the mutation in *WISP3*, stable transfection of the cDNA was performed on
human gastric carcinoma cell line Kato-III that expresses no WISP3 mRNA.
Western blot analysis of cell lysates of the transfectants with WISP3(A)9 cDNA
and WISP3(A)8 cDNA identified specific proteins of approximately 40 kDa
and 25 kDa, respectively (Fig. 6A), using anti-N-terminus WISP3 antibody.
Since CCN family proteins have been reported to regulate cellular invasiveness (Kleer *et al.*, 2002), invasion analysis of the transfected gastric carcinoma
cells was assessed using Boyden Chamber methods (Tanaka *et al.*, 2003). Compared to the parental or mock-transfected Kato-III cells, invasion potentials
of WISP3(A)9-transfectants were remarkably reduced, as shown in Fig. 6B.
On the other hand, no inhibitory effect was not recognized in the transfectants
with WISP3(A)8 missing both the third and fourth domain.

WISP3 has been independently reported as a gene *LIBC* that was lost in inflammatory breast cancers (van Golen *et al.*, 1999; Kleer *et al.*, 2002), indicating a tumor suppressor function of wild-type WISP3. In addition, frequent mutations in the *WISP3* gene were reported to be associated with autosomal recessive skeletal disorder progressive pseudorheumatoid dysplasia, characterized by unrestrained invasive phenotype of synovial fibroblasts into cartilage (Hurvitz *et al.*, 1999). A better understanding of the WISP3 function will likely increase awareness of its importance in the pathological cell invasion. Identification of such targets of microsatellite instability as well as the functional alteration should be warranted.

5. CONCLUSION AND PERSPECTIVES

It has been suggested that the different modules of CCNs can function independently in different cell types (Perbal, 2004). The oncogenic form of Nov, isolated from myeloblastomatosis-associated virus-induced nephroblastoma, had a truncated structure lacking the first module of IGFBP, and capable of inducing cell transformation in fibroblasts (Fig. 7). In contrast, wild-type Nov that is negatively regulated by growth factors (Scholz *et al.*, 1996), has suppressor activity for cell growth (Joliot *et al.*, 1992). Xu *et al.* reported that WISP1 induced morphological transformation, accelerated cell growth, and enhanced saturation density in normal rat kidney fibroblasts (Xu *et al.*, 2000), while the suppression of melanoma growth by a mouse orthologue of WISP1 was reported (Hashimoto *et al.*, 1998). The fourth module of CK is lost in WISP2 (Pennica *et al.*, 1998) that functions as a secreted protein which positively

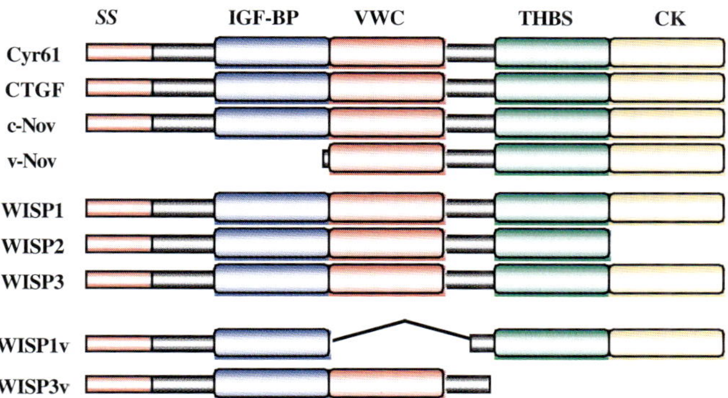

Fig. 7.　Schematic structures of CCN family proteins; Cyr61, CTGF, c-Nov (wild-type), v-Nov (oncogenic), WISP1, WISP2, WISP3, WISP1v and WISP3v.

regulates osteoblast functions (Kumar *et al.*, 1999). Unlike all other known CCN proteins, 4 of 10 conserved cysteine residues are not recognized in the VWC module of WISP3 that is frequently mutated in patients with progressive pseudorheumatoid dysplasia (Hurvitz *et al.*, 1999).

In our experiments, WISP1v lacking the second module of VWC induced a striking cellular transformation and a rapid piling-up forms of growth. WISP1v transfectants enhanced the invasive phenotype of co-cultured carcinoma cells, while wild-type WISP1 had no such potential. The difference in biological effects between the WISP1 and WISP1v expressing cells may reflect the ability of the cells to better process the variant form. Additionally, we identified a WISP3 mutant lacking both the third THBS and the fourth CK domains in microsatellite instable cancers. The secreted CCN protein may stimulate various receptors on the cell surface such as integrin (Lau and Lam, 1999), LRP (Segarini *et al.*, 2001) and proteoglycans (Desnoyers *et al.*, 2001). Interaction with matrix proteins of the cancer stroma have major effects on cellular function in the context of the malignant phenotype.

Targeting antitumor therapy to stromal elements required for tumor growth may prove to be an attractive strategy for retarding the growth of solid neoplasms, since normal tissue is unlikely to show genetic plasticity that accompanies malignant transformation, an event that allows transformed cells to rapidly acquire resistance to chemotherapeutic agents. Together with the increased importance of Wnt signaling in human carcinogenesis (Tanaka *et al.*, 2002b), a better understanding of the CCN contributions will likely increase our knowledge of the importance of molecules of WISP1v and WISP3v, and permit identification of therapeutic agents that focus on the CCN signaling as a target.

ACKNOWLEDGMENTS

I thank Prof. Jack R. Wands (Brown University) for helpful discussions. This work was supported by Grant-in-Aid from the Ministry of Education, Science, Technology, Sports and Culture of Japan. S.T. is a recipient of the Japan Cancer Society Incitement Award.

REFERENCES

Babic AM, Chen CC and Lau LF (1999) Fisp12/mouse connective tissue growth factor mediates endothelial cell adhesion and migration through integrin alphavbeta3, promotes endothelial cell survival, and induces angiogenesis *in vivo*. *Mol Cell Biol* **19**: 2958–2966.

Behrens J, *et al.* (1996) Functional interaction of beta-catenin with the transcription factor LEF-1. *Nature* **382**: 638–642.

Bhanot P, *et al.* (1996) A new member of the frizzled family from Drosophila functions as a wingless receptor. *Nature* **282**: 225–230.

Bork P (1993) The modular architecture of a new family of growth regulators related to connective tissue growth factor. *FEBS Lett* **327**: 125–130.

Desnoyers L, Arnott D and Pennica D (2001) WISP-1 binds to decorin and biglycan. *J Biol Chem* **276**: 47599–47607.

Hashimoto Y, *et al.* (1998) Expression of the Elm1 gene, a novel gene of the CCN family, suppresses *in vivo* tumor growth and metastasis of K-1735 murine melanoma cells. *J Exp Med* **187**: 289–296.

Hurvitz JR, *et al.* (1999) Mutations in the CCN gene family member WISP3 cause progressive pseudorheumatoid dysplasia. *Nat Genet* **23**: 94–98.

Ionov Y, Peinado MA, Malkhosyan S, Shibata D and Perucho M (1993) Ubiquitous somatic mutations in simple repeated sequences reveal a new mechanism for colonic carcinogenesis. *Nature* **363**: 558–561.

Itoh K, Krupnik VE and Sokol SY (1998) Axis determination in Xenopus involves biochemical interactions of axin, glycogen synthase kinase 3 and beta-catenin. *Curr Biol* **8**: 591–594.

Joliot V, *et al.* (1992) Proviral rearrangements and over expression of a new cellular gene (nov) in myeloblastosis-associated virus type 1-induced nephroblastomas. *Mol Cell Biol* **12**: 10–21.

Kireeva ML, Mo FE, Yang GP and Lau LF (1996) Cyr61, a product of a growth factor-inducible immediate-early gene, promotes cell proliferation, migration, and adhesion. *Mol Cell Biol* **16**: 1326–1334.

Kleer CG, *et al.* (2002) WISP3 is a novel tumor suppressor gene of inflammatory breast cancer. *Oncogene* **21**: 3172–3180.

Kumar S, *et al.* (1999) Identification and cloning of a connective tissue growth factor-like cDNA from human osteoblasts encoding a novel regulator of osteoblast functions. *J Biol Chem* **174**: 17123–17131.

Lau LF and Lam SC (1999) The CCN family of angiogenic regulators: the integrin connection. *Exp Cell Res* **248**: 44–57.

Moon RT, Bowerman B, Boutros M and Perrimon N (2002) The promise and perils of Wnt signaling through beta-catenin. *Science* **296**: 1644–1646.

Morin P, *et al.* (1997) Activation of beta-catenin-Tcf signaling in colon cancer by mutations in beta-catenin or APC. *Science* **275**: 1787–1790.

Nusse R and Varmus HE (1982) Many tumors induced by the mouse mammary tumor virus contain a provirus integrated in the same region of the host genome. *Cell* **31**: 99–109.

Peifer M and Polakis P (2000) Wnt signaling in oncogenesis and embryogenesis-a look outside the nucleus. *Science* **287**: 1606–1616.

Perbal B (2004) CCN proteins: multifunctional signalling regulators. *Lancet* **363**: 62–64.

Pennica D, *et al.* (1998) WISP genes are members of the connective tissue growth factor family that are upregulated in Wnt-1-transformed cells and aberrantly expressed in human colon tumors. *Proc Natl Acad Sci USA* **95**: 14717–14722.

Rubinfeld B, Robbins P, El-Gamil M, Albert I, Porfiri E and Polakis P (1997) Stabilization of beta-catenin by genetic defects in melanoma cell lines. *Science* **275**: 1790–1792.

Scholz G, Martinerie C, Perbal B and Hanafusa H (1996) Transcriptional downregulation of the nov proto-oncogene in fibroblasts transformed by p60v-src. *Mol Cell Biol* **16**: 481–486.

Segarini PR, Nesbitt JE, Li D, Hays LG, Yates JR and Carmichael DF (2001) The low density lipoprotein receptor-related protein/alpha2-macroglobulin receptor is a receptor for connective tissue growth factor. *J Biol Chem* **276**: 40659–40667.

Tamai K, *et al.* (2000) LDL-receptor-related proteins in Wnt signal transduction. *Nature* **407**: 530–535.

Tanaka S, Akiyoshi T, Mori M, Wands JR and Sugimachi K (1998) A novel frizzled gene identified in human esophageal carcinoma mediates APC/beta-catenin signals. *Proc Natl Acad Sci USA* **95**: 10164–10169.

Tanaka S, *et al.* (2001) A novel variant of WISP1 lacking a Von Willebrand type C module overexpressed in scirrhous gastric carcinoma. *Oncogene* **20**: 5525–5532.

Tanaka S, Sugimachi K, Shimada M, Maehara Y and Sugimachi K (2002a) Variant WISPs as targets for gastrointestinal carcinomas. *Gastroenterology* **123**: 392–393.

Tanaka S, *et al.* (2002b) Oncogenic signal transduction and therapeutic strategy for hepatocellular carcinoma. *Surgery* **131**: S142–S147.

Tanaka S, Sugimachi K, Maehara Si, Shimada M and Maehara Y (2003) A loss of function mutation in *WISP3* derived from microsatellite unstable gastric carcinoma. *Gastroenterology* **125**: 1563–1564.

Thorstensen L, *et al.* (2001) WNT1 inducible signaling pathway protein 3, WISP-3, a novel target gene in colorectal carcinomas with microsatellite instability. *Gastroenterology* **121**: 1275–1280.

van Golen KL, *et al.* (1999) A novel putative low-affinity insulin-like growth factor-binding protein, LIBC (lost in inflammatory breast cancer), and RhoC GTPase correlate with the inflammatory breast cancer phenotype. *Clin Cancer Res* **5**: 2511–2519.

Wong GT, Gavin BJ and McMahon AP (1994) Differential transformation of mammary epithelial cells by Wnt genes. *Mol Cell Biol* **14**: 6278–6286.

Xu L, Corcoran RB, Welsh JW, Pennica D and Levine AJ (2000) WISP-1 is a Wnt-1- and beta-catenin-responsive oncogene. *Genes Dev* **14**: 585–595.

Zoubine MN, Banerjee S, Saxena NK, Campbell DR and Bannerjee SK (2001) WISP-2: a serum-inducible gene differentially expressed in human normal breast epithelial cells and in MCF-7 breast tumor cells. *Biochem Biophys Res Commun* **282**: 421–425.

INDEX